Groups, Representations and Physics

Second Edition

Groups, Representations and Physics

Second Edition

H F Jones

Department of Physics, Imperial College of Science, Technology and Medicine, London

Published in 1998 by
Taylor & Francis Group
270 Madison Avenue
New York, NY 10016

Published in Great Britain by
Taylor & Francis Group
2 Park Square
Milton Park, Abingdon
Oxon OX14 4RN

Printed in the United States of America on acid-free paper
10 9 8 7 6 5 4 3

International Standard Book Number-10: 0-7503-0504-5 (Softcover)
International Standard Book Number-13: 978-0-7503-0504-4 (Softcover)

Library of Congress Cataloging-in-Publication Data

Catalog record is available from the Library of Congress

informa

Taylor & Francis Group
is the Academic Division of Informa plc.

Visit the Taylor & Francis Web site at
http://www.taylorandfrancis.com

Contents

*'Starred' sections are somewhat specialized, and may be omitted at a first reading.

Preface to the Second Edition

In this second edition I have taken the opportunity to expand the scope of the book somewhat by including a new chapter (Chapter 9) on the Cartan–Weyl–Dynkin approach to Lie algebras. This is a generalization to more complicated Lie algebras of the method of raising and lowering operators used to obtain the irreps of SU(2). It is a systematic and unified approach, which allows one to classify all possible simple Lie algebras and, in principle, to find all their irreducible representations. I hope that inclusion of this topic, albeit in a necessarily rather condensed form, will extend the usefulness of the book.

I am grateful to those readers who took the trouble to contact me pointing out various errors in the previous edition, and I have corrected those remaining errors of which I am aware.

H F Jones

London, February 1998

Preface to the First Edition

It was with some trepidation that I decided to write yet another book on group theory and physics. There are, after all, quite a large number already. However, in the first place some of the best of these were, inexplicably, out of print, and secondly I thought I perceived a niche for a book which, while by no means skimping the physical applications, exposes the student to the power and elegance of abstract mathematics.

There is, indeed, a fascinating interplay between mathematics and physics. In some cases the need to understand and formulate a physical problem provides a stimulus for the development of the relevant mathematics, as for example in Newton's development of the calculus as a tool for calculating planetary orbits. In others a mathematical formalism already developed turns out to be tailor-made for physics. Thus the theory of vector spaces is exactly what one needs for the general formulation of quantum mechanics, and the representation theory of groups is precisely the mathematical framework needed when one considers the action of symmetry transformations on quantum systems.

With this in mind, I have developed the mathematical theory in a more formal fashion than is strictly necessary for physical applications. Thus, for example, I have dealt with representations in the coordinate-free language of linear transformations acting on vector spaces, rather than simply as matrices. It is my experience that students, far from being put off by this, are fascinated by what may well be their first exposure to the rigorous axiomatic approach, and ultimately obtain a deeper understanding of the subject. However, since the book is aimed primarily at physicists, I have been careful to include many examples and illustrations of the various mathematical structures and theorems. I have also included for the reader's convenience a short glossary of the mathematical symbols used.

The first four chapters give a systematic development of the theory of finite groups and their representations. Applications to various physical systems are given in Chapter 5 and the problems which follow. Although finite groups are of direct importance in solid state and molecular physics, the bulk of the applications to physics are concerned with continuous groups. In dealing with these groups I have adopted an intermediate level of rigour, not worrying overduly about proofs of convergence which are the proper concern of mathematicians. The theory of the rotation groups SO(2) and SO(3) is developed in Chapter 6, following which I am able to extend the range of the physical examples.

My own interests as a field theorist have influenced the subject matter and scope of the last three chapters. Chapter 8 is concerned with the SU(N) groups, Chapter 9 with the Lorentz and Poincaré groups, with particular emphasis on the concept of helicity, which leads to remarkable simplifications even in a non-relativistic context. In view of the widespread interest in gauge theories, due to their remarkable successes in recent years, I have included a final chapter on gauge groups. This is necessarily somewhat impressionistic, particularly on the field theory aspect, but I hope it conveys the essential points.

The book, based loosely on lectures given to physics students at Imperial College, is aimed primarily at third-year physics undergraduates with a good mathematical background, but it should also be useful to first-year postgraduates in solid state, atomic or elementary particle physics. The first four chapters might also be helpful to mathematics students.

In view of this intended readership, a reasonable acquaintance with quantum mechanics is assumed. Initially only the wave mechanical formulation is used, but later on this becomes increasingly restrictive and I go over to Dirac notation. For readers not familiar with this formalism, a brief account is given in Appendix A. For similar reasons Appendix B gives a review of the standard quantum mechanical treatment of angular momentum, in the Dirac formalism. Appendix C is a derivation of the invariant integration measure for SO(3). This is rather technical, and can certainly be skipped at a first reading, but does make use of a rather elegant construction in spherical geometry. Appendix D fills in the necessary background for the last section of Chapter 9 on relativistic scattering, and Appendix E is a crash course in Lagrangian mechanics for those interested in gauge field theories.

A short bibliography is provided so that the interested reader is able to pursue any particular topic in greater detail, having, I hope, obtained a firm grounding in the basics. He or she can test that grounding against the problems at the end of each chapter†. Sketch answers are given at the end of the book.

H F Jones
London, January 1989

†'Starred' problems are rather difficult, and for the dedicated reader only.

Acknowledgments

I am indebted to the University of London for permission to include some problems which originally appeared in the degree examinations at Imperial College. My thanks are due to Chris Isham for giving the final push as I hesitated on the brink, and to my wife Peggy for her unfailing encouragement and support.

1

Introduction

1.1 Symmetry in Physics; Groups and Representations

There are many physical systems whose underlying dynamics has some symmetry. A good example is provided by the water molecule, treated in greater detail in §5.2. There is clearly a symmetry between the two hydrogen ions, which may be interchanged without affecting the energy of the system. Again, there is a translation symmetry: the interaction between any two ions situated at $\boldsymbol{r}_1$ and $\boldsymbol{r}_2$ depends only on their relative separation $\boldsymbol{r}_1 - \boldsymbol{r}_2$ and not on their absolute positions. That is, the potential energy $V(\boldsymbol{r}_1, \boldsymbol{r}_2, \boldsymbol{r}_3)$ is actually a function of $\boldsymbol{r}_1 - \boldsymbol{r}_2$ and $\boldsymbol{r}_1 - \boldsymbol{r}_3$ only, and the same is true of the kinetic energy, once the centre-of-mass energy has been subtracted off. Furthermore, the system has a rotational invariance whereby its energy is independent of its absolute orientation: $V = V(|\boldsymbol{r}_1 - \boldsymbol{r}_2|, |\boldsymbol{r}_1 - \boldsymbol{r}_3|)$. Thus the underlying Hamiltonian, the classical expression for the energy, which becomes an operator in quantum mechanics, is invariant under a set of transformations of the coordinates, which includes reflections, translations and rotations.

These different types of transformation all have the common property that they form a group. The formal definition of a group will be given in the next section, but the essential property is that the successive application of two such transformations gives another one, which we call the product. There is an identity transformation, which is simply to do nothing, and each transformation has an inverse, which 'undoes' the operation, i.e. the product of a transformation and its inverse is the identity.

The underlying symmetry of the dynamics may or may not be

explicitly manifest in a given physical realization. In the realm of classical mechanics it is more often than not the case that a particular dynamical system does not exhibit the full symmetry. A good example of this is the orbit of a planet under gravity. Any given particular planetary orbit is an ellipse. It is not even circular, which worried the ancients greatly, but perhaps they should have been more worried that even a circle does not respect the perfect spherical symmetry of the underlying Hamiltonian. The explanation is that the symmetry has somehow been broken by the initial conditions of the motion, which picked out first of all a plane in which the motion would take place, and then in that plane a direction, say that of the semi-major axis. Paradoxically, the facts that the motion remains in a plane and that the direction of the major axis remains fixed are due to conservation laws (of angular momentum and of the Runge–Lentz vector discussed in §7.2) which are consequences of the underlying symmetries!

It turns out in fact that the most important applications of group theory in physics are found not in classical mechanics but rather in quantum mechanics. There the ground state of a system usually *does* exhibit the full symmetry of the Hamiltonian, though a very important and interesting exception to this occurs in the phenomenon of spontaneous symmetry breaking, where, again because of some uncontrollable perturbation of the initial conditions, one asymmetric solution is picked out of an infinite set of possible ones. Thus the fundamental interactions of the spins in a ferromagnet are rotationally symmetric, but when one is formed they align themselves in some particular direction. But as an example of the more usual scenario, consider the ground state of the hydrogen atom, the quantum mechanical equivalent of the planetary orbit problem. There, as students of elementary quantum mechanics will know, the ground state wavefunction gives a spherically symmetric probability distribution which indeed respects the spherical symmetry of the $1/r$ potential.

As far as the excited states are concerned, the rotational symmetry of the problem means that they can be classified by the total angular momentum quantum number l and the magnetic quantum number m, which refers to the eigenvalue of its z component. Moreover, the energy does not depend on m (nor, for that matter, on l, but this is a special feature of the $1/r$ potential). This makes perfect sense physically, since there is no preferred direction: the choice of the z axis was completely arbitrary. As far as mathematics is concerned, it means that we have a degenerate space of eigenfunctions u_{nlm}, with m ranging from $+l$ to $-l$, which all have the same energy and can be transformed into each other by rotations. We can take arbitrary linear combinations of the $2l + 1$ eigenfunctions which are still eigenfunctions with the same energy and total angular momentum. They therefore form what is technically known as a *vector space*. The group of rotations in ordinary 3-dimensional space induces transformations within this vector space, giving what is known as a *representation*, which can be realized by matrices, in this case of dimension $(2l + 1) \times (2l + 1)$.

Thus consideration of symmetry transformations leads us naturally to the study of groups, many properties of which can be established generally, without reference to the particular nature of the group elements; and when the transformations act on a quantum mechanical system, the appropriate mathematical framework is that of representation theory. This theory, when developed, will enable us to classify energy levels, their degeneracy, and how this is changed when the symmetry is reduced, to obtain selection rules which tell us when certain matrix elements are zero, and in general to obtain relations between matrix elements, thus reducing to a minimum the number of quantities which have to be calculated independently. In short, it enables us to exploit to the full the invariance of the underlying dynamics.

1.2 Definition of a Group; Some Simple Examples

Since groups are to be the main focus of our attention, it is now time to define exactly what they are. Following the very fruitful practice of modern mathematics, the definition will be an abstract one, in terms of certain objects, the elements of the group, and a certain composition law, commonly referred to as 'multiplication', whereby two members of the group combine to give a third. In order for the whole structure to form a group, the law of composition must satisfy certain axioms. There may be many concrete and quite different sets of objects which combine in this particular way, but what they have in common is the group structure. From the axioms alone one can derive general theorems about groups which are applicable to each of the particular examples: we do not have to prove them over and over again for each particular case. This is the power of the axiomatic approach, but its abstract nature makes it sometimes seem rather remote. In order to counteract this we shall constantly refer to specific examples to illustrate the general constructs and theorems.

Definition

A *group* G is a set of elements $\{a, b, \ldots\}$ with a law of composition (multiplication) which assigns to each ordered pair $a, b \in G$ another element, written ab, of G. The law of composition satisfies the following conditions:

G1 (associative law)
For all $a, b, c \in G$

$$a(bc) = (ab)c \tag{1.1}$$

G2 (unit element)
G contains an element, the identity element, denoted by e, such that for all $a \in G$

$$ae = ea = a \tag{1.2}$$

G3 (existence of inverse)
For all $a \in G$ there is an element, denoted by a^{-1}, such that

$$aa^{-1} = a^{-1}a = e \tag{1.3}$$

Comments

(i) Note that in general $ab \neq ba$, i.e. the operation of 'multiplication' is not commutative. That is why the order of the elements of the pair a,b is important. If $ab = ba$ for all $a,b \in G$ the group is called *Abelian*.
(ii) The definition includes the fact that the product ab is also a member of the set G. This is sometimes stated as a separate axiom G0 (closure).

Examples

	Group G	Composition law	Order	Remarks
1.	C_2: $\{e, a\}$ with $a^2 = e$ (e.g. rotation through π	Unspecified Successive operation)	2	Simplest non-trivial example Abelian
2.	S_n: permutations of n objects	Successive operation	$n!$	Non-Abelian in general
3.	Z_n: integers modulo n	Addition (mod n)	n	Abelian $Z_2 \cong C_2$
4.	Positive rational numbers	Multiplication	∞	Abelian

Here the order of the group is the number of elements in the set G. You should check the axioms in each case, identifying the unit element e and the inverse a^{-1}. Note that in only one of the above examples does the composition law we have called 'multiplication' correspond to multiplication in the usual sense.

The integers modulo n are the integers 0, 1, 2, . . ., $n - 1$. Addition modulo n is defined by adding the two integers together in the usual way and then subtracting off n to bring the result into the above range. Thus, for example, $1 + 1 \pmod 2 = 2 - 2 = 0$. The groups Z_2 and C_2 are *isomorphic* (denoted by $\cong$). We shall define this term more precisely later on, but in essence it means that the two groups have exactly the same structure, although the elements and composition law can be quite different.

Counter-examples (sets plus composition laws which are not groups)

Set S	Composition law	Order	Remarks
1. Z: integers	Multiplication	∞	$1/n \notin Z$
2. R: real numbers	Multiplication	∞	0 has no inverse

For the most part we will be concerned with symmetry transformations of geometrical objects and dynamical systems. In this case the law of composition is always successive operation:

$c = ab$ is the result of first applying b

then applying a

and the associative law is automatically satisfied (check!).

1.3 Some Simple Point Groups

The groups C_n and D_n introduced below are interesting in their own right and will be used to illustrate many of the general properties of groups. They are of direct physical relevance in solid state physics since they are symmetry groups of regular solids, but only for the special values $n = 2, 3, 4$ and 6.

(i) ***The Cyclic Group C_n*** (the symmetry group of rotations of a regular polygon with n directed sides)

By 'directed' we mean that each side has a direction associated with it, as shown in figure 1.1, and for a rotation to be counted as a symmetry it must transform the polygon into itself, respecting the directionality.

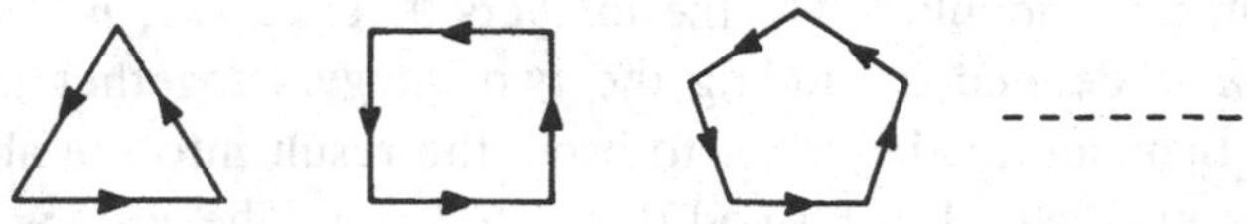

Figure 1.1 Regular polygons with directed sides which are transformed into themselves by the rotations of the group C_n.

The elements of the group are rotations through an angle $2\pi r/n$ ($r = 0, 1, \ldots, n-1$) about an axis through the centre and normal to the plane of the figure. Calling these elements C_n^r, it is clear that we can achieve the same result as C_n^r, a rotation through $2\pi r/n$, by repeating the smallest non-trivial rotation C_n^1 r times, i.e.

$$C_n^r = (c)^r \tag{1.4}$$

where $c \equiv C_n^1$ is a rotation through $2\pi/n$. Note that c satisfies

$$c^n = e \tag{1.5}$$

since a rotation through 2π is equivalent to the identity operation, i.e. no rotation at all.

Mathematically this group is known as the *cyclic group* of order n, *generated* by the element c, and consisting of the elements $\{e, c, c^2, \ldots, c^{n-1}\}$. By 'generated' we mean that each element of the group can be obtained by multiplication of c by itself the appropriate number of times, and this is expressed by the notation $C_n = \text{gp}\{c\}$.

It is an Abelian group since

$$c^r c^s = c^{r+s} = c^{s+r} = c^s c^r \tag{1.6}$$

and in fact is *isomorphic* (i.e. there is a 1:1 correspondence) to Z_n, the group of integers modulo n introduced above. The correspondence is precisely

$$c^r \in C_n \leftrightarrow r \in Z_n \tag{1.7}$$

That is, multiplication of elements of C_n is entirely equivalent to adding the exponents of c, but modulo n, since an exponent of n is not to be distinguished from an exponent of 0.

Another way of specifying a finite group, which in this particular case, and indeed in most cases, is much more cumbersome, is by writing down the multiplication table. For example, the table for C_3 is:

Table 1.1

$g_1 \backslash g_2$	e	c	$b\ (= c^2)$
e	e	c	b
c	c	b	e
b	b	e	c

Here the convention is that the element shown as the product is $g_1 g_2$ in that order, although in the case of C_n the order is in fact irrelevant. The $n \times n$ box of products forms what is known as a Latin square, with the property that each element appears once only in each row and in each column. [Problem: show from the group axioms that this must be the case.] This property is necessary, but not sufficient, for such a square to represent a valid multiplication table for a group. The extra property that needs to be checked is the associative law.

(ii) The Dihedral Group D_n (the symmetry group of rotations of a regular polygon with n undirected sides)

Group D_3

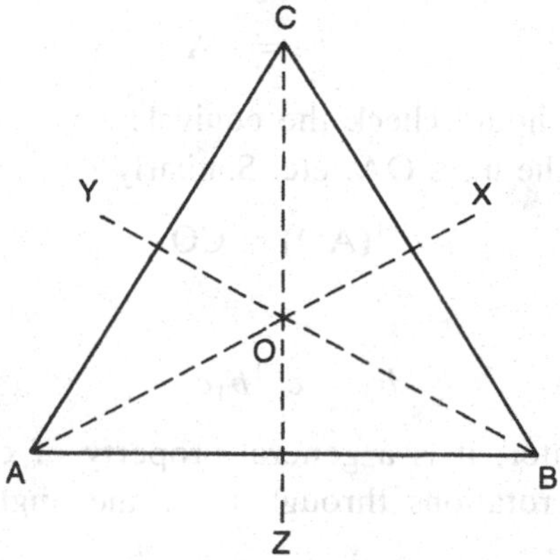

Figure 1.2 Equilateral triangle with undirected sides which is transformed into itself by the rotations of the group D_3.

This group certainly contains the elements e, c, c^2 of C_3 which transform the triangle into itself under the more restricted condition of preserving directionality. However, having now dropped this condition we can identify three additional elements, namely rotations through π about the axes OX, OY and OZ. Let us call these b_1, b_2 and b_3 respectively. They are '2-fold' rotations, satisfying $b_i^2 = e$, and the corresponding axes are termed 2-fold axes.

Note that c, an anticlockwise rotation in the plane of the figure through $2\pi/3$ about the centre O, rotates these axes into each other. Thus, for example,

$$c(\mathrm{AO}) = \mathrm{BO}$$

This implies that b_1 and b_2 are related to each other by *conjugation* with c:

$$b_2 = cb_1c^{-1} \tag{1.8}$$

That is, the effect of b_2 can be reproduced by first performing the inverse of c, namely a clockwise rotation through $2\pi/3$, then a π rotation about the axis OY (considered as fixed in space) and finally 'undoing' the first rotation, a sequence of operations of a type familiar to practitioners of the Rubik cube! In order to verify this assertion we need to check the operation of the RHS on (all) the sides of the triangle to see whether it indeed matches up with that of b_2.

Consider, for example, AC. Under the action of b_2 it becomes $b_2(\mathrm{AC}) = \mathrm{CA}$. On the other hand

$$\begin{aligned} cb_1c^{-1}(\mathrm{AC}) &= cb_1(\mathrm{CB}) \\ &= c(\mathrm{BC}) \\ &= \mathrm{CA} \end{aligned}$$

As an exercise you should check the equivalence for the other sides of the triangle and for the lines OA, etc. Similarly

$$c^{-1}(\mathrm{AO}) = \mathrm{CO}$$

which implies that

$$b_3 = c^{-1}b_1c \tag{1.9}$$

As we shall show later, it is a general property of conjugate rotations that they represent rotations through the same angle about a rotated axis.

In this particular case there is another relation between b_2 and b_1, namely

$$b_2 = b_1 c \tag{1.10}$$

as we can check by considering the action of the RHS on AC, for example. Thus

$$\begin{aligned} b_1 c(\mathrm{AC}) &= b_1(\mathrm{BA}) \\ &= \mathrm{CA} \end{aligned}$$

as required. In fact, since these rotations satisfy the axiom of closure (G0), the product $b_1 c$ must be *some* member of the group: it just happens to be b_2.

From now on let us write b_1 simply as b. The above relationship means that the group is *generated* in the sense explained before by the elements b and c. That is, every element can be represented as the product of b's and c's.

We have already shown that

$$b_2 = bc \tag{1.10'}$$

Similarly it is easy to see that

$$b_3 = bc^{-1} = bc^2 \tag{1.11}$$

Thus we can write

$$\mathrm{D}_3 = \mathrm{gp}\{c, b\} \qquad c^3 = b^2 = e$$

This, however, does not completely specify the group. Now that we have two generators rather than just one, we need to know how to commute them. That is, we can express, as we have done above, each element of the group as $b^m c^n$. But we can very well encounter products in the opposite order, for example cb, and we need to be able to write these in the canonical form.

The answer lies in the two equivalent expressions we have derived for b_2, namely

$$b_2 = cbc^{-1} = bc \tag{1.12}$$

Multiplying this relation on the right by c gives

$$cb = bc^2 \tag{1.13}$$

An alternative way of codifying this information is to note that

$$(bc)^2 = b(cb)c = b(bc^2)c = b^2c^3 = e \tag{1.14}$$

which is in fact nothing more than the statement that b_2 $(= bc)$ is a 2-fold rotation satisfying $b_2^2 = e$.

The complete specification of the group is therefore given by

$$D_3 = \text{gp}\{c, b\} \qquad c^3 = b^2 = (bc)^2 = e \tag{1.15}$$

and the six independent elements can be taken as $\{e, c, c^2, b, bc, bc^2\}$.

Group D_4

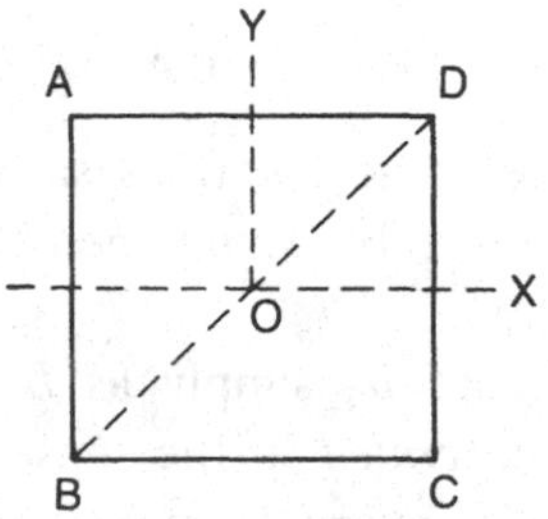

Figure 1.3 Square with undirected sides which is transformed into itself by the rotations of the group D_4.

If c is an anticlockwise rotation through $\pi/2$, satisfying $c^4 = e$, about an axis through the centre O perpendicular to the plane of the figure, and $b \equiv b_1$ is the 2-fold rotation about the axis OX, it is easy to see, by examination of their action on the sides of the square, that the other 2-fold rotations, b_2 about AC, b_3 about OY and b_4 about BD, can be identified as

$$\begin{aligned} b_2 &= bc \\ b_3 &= bc^2 \\ b_4 &= bc^3 \end{aligned} \tag{1.16}$$

so that D_4 can be specified as

$$D_4 = \text{gp}\{c, b\} \qquad c^4 = b^2 = (bc)^2 = e \tag{1.17}$$

The group is of order 8, and the elements can be written in standard form as $\{e, c, c^2, c^3, b, bc, bc^2, bc^3\}$. Any other product can be reduced to the canonical form by judicious use of the defining relations. In particular, cb can be re-expressed as bc^3. For we know that $bcbc = (bc)^2 = e$. Multiplying by b on the left and c^3 on the right gives $cb = bc^3$. [Problem: show that $c^2b = bc^2$ and $c^3b = bc$.]

The generalization to larger values of n should now be reasonably clear: the group D_n, the group of symmetry rotations of a regular polygon with n sides, is generated by an n-fold rotation about an axis through the centre and perpendicular to the plane of the figure and any one of the n 2-fold rotations about axes in the plane:

$$D_n = \mathrm{gp}\{c, b\} \qquad c^n = b^2 = (bc)^2 = e \tag{1.18}$$

The group is of order $2n$, and the $2n$ elements can be written in canonical form as $\{e, c, c^2, \ldots, c^{n-1}, b, bc, bc^2, \ldots, bc^{n-1}\}$. It is illustrated geometrically in figure 1.4.

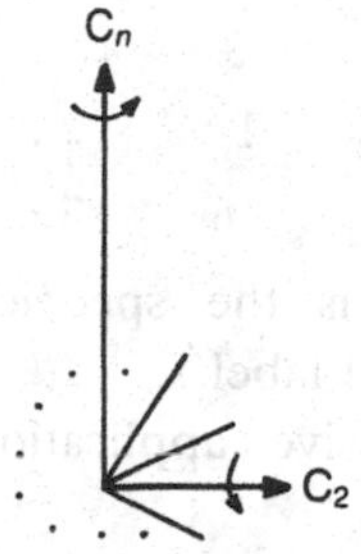

Figure 1.4 The group C_n.

Note that the rotations c^r comprising C_n form a *subgroup* of D_n. This is a term we shall define more precisely later, but what it essentially means is that the subset of elements $\{e, c, c^2, \ldots, c^{n-1}\}$ is closed under the group operation. This is clear algebraically, but also geometrically, since any product of rotations about the perpendicular axis will be another rotation about that axis. Each of the 'b' rotations generates a C_2 subgroup consisting of the rotation itself and its square, which is equivalent to the identity.

The algebraic definition of D_n can also be extended to smaller n to define the group D_2, which from the geometrical point of view is a degenerate case. The group, of order 4, is also known as the 4-group (*Vierergruppe*) V.

1.4 The Permutation Group S_n

This group, besides serving to illustrate some of the general properties of groups which we will treat more formally later, is of direct physical relevance to physical systems involving several identical particles, e.g. multi-electron atoms, and has a special status in the theory of finite groups by virtue of Cayley's theorem (q.v.). It consists of the permutations of n objects, or the labels of those objects, and is of order n! Multiplication of two permutations is defined as successive application.

A permutation P under which the object with label i ($i = 1, \ldots, n$) is changed into that with label p_i can be written as

$$P = \begin{pmatrix} 1 & 2 & \cdots & n \\ p_1 & p_2 & \cdots & p_n \end{pmatrix} \tag{1.19}$$

The order of the columns is irrelevant: P could equally well be written as

$$P = \begin{pmatrix} 2 & 1 & \cdots & n \\ p_2 & p_1 & \cdots & p_n \end{pmatrix} \tag{1.20}$$

since all that is important is the specification of the new label p_i corresponding to each original label i.

Multiplication, i.e. successive application of permutations, is not commutative:

$$PQ = \begin{pmatrix} 1 & 2 & \cdots & n \\ p_1 & p_2 & \cdots & p_n \end{pmatrix} \begin{pmatrix} 1 & 2 & \cdots & n \\ q_1 & q_2 & \cdots & q_n \end{pmatrix}$$
$$\neq QP \text{ in general} \tag{1.21}$$

For example,

$$\begin{pmatrix} 1 & 2 & 3 \\ 1 & 3 & 2 \end{pmatrix} \begin{pmatrix} 1 & 2 & 3 \\ 3 & 1 & 2 \end{pmatrix} = \begin{pmatrix} 1 & 2 & 3 \\ 2 & 1 & 3 \end{pmatrix} \tag{1.22}$$

whereas

$$\begin{pmatrix} 1 & 2 & 3 \\ 3 & 1 & 2 \end{pmatrix} \begin{pmatrix} 1 & 2 & 3 \\ 1 & 3 & 2 \end{pmatrix} = \begin{pmatrix} 1 & 2 & 3 \\ 3 & 2 & 1 \end{pmatrix} \tag{1.23}$$

(The way to evaluate these products is to follow each label through the two permutations in turn. Thus in the last example we have $1 \rightarrow 1$ under P and then $1 \rightarrow 3$ under Q. Similarly $2 \rightarrow 3 \rightarrow 2$ and $3 \rightarrow 2 \rightarrow 1$.)

The identity element is obviously

$$e = \begin{pmatrix} 1 & 2 & \cdots & n \\ 1 & 2 & \cdots & n \end{pmatrix} \tag{1.24}$$

and the inverse of P is

$$P^{-1} = \begin{pmatrix} p_1 & p_2 & \cdots & p_n \\ 1 & 2 & \cdots & n \end{pmatrix} \tag{1.25}$$

An alternative and very useful way of writing permutations is the *cycle notation*. In this notation we follow all the permutations of the label 1 until we get back to 1, giving one *cycle*. If the permutations of all the labels have been thereby included we stop: the entire permutation is one cycle. If, however, there are some labels not accounted for, we take one of these and follow all its permutations, forming another cycle. If this now accounts for all the labels we stop: the original permutation has been decomposed into two cycles. If not we take one of the labels still not included, follow all its permutations, etc. The process is continued until all the labels are accounted for and the original permutation has been decomposed into a certain number of *disjoint* cycles. 'Disjoint' means that the cycles have no elements in common, which is a consequence of the method of procedure.

For example, the permutation

$$\begin{pmatrix} 1 & 2 & 3 & 4 \\ 2 & 4 & 3 & 1 \end{pmatrix}$$

of S_4 decomposes into the disjoint cycles $1 \to 2 \to 4 \to 1$ and $3 \to 3$. Reordering the columns we can write it as

$$\begin{pmatrix} 1 & 2 & 3 & 4 \\ 2 & 4 & 3 & 1 \end{pmatrix} = \left(\begin{array}{ccc|c} 1 & 2 & 4 & 3 \\ 2 & 4 & 1 & 3 \end{array}\right) = \begin{pmatrix} 1 & 2 & 4 \\ 2 & 4 & 1 \end{pmatrix}\begin{pmatrix} 3 \\ 3 \end{pmatrix}$$

In a cycle the bottom row is really superfluous: all the information, that $1 \to 2 \to 4 \to 1$, etc., is encoded in the order of the labels in the top row. In cycle notation we simply omit the bottom row, and the above example would be written as

$$\begin{pmatrix} 1 & 2 & 3 & 4 \\ 2 & 4 & 3 & 1 \end{pmatrix} = (1\;2\;4)(3) \qquad (1.26)$$

As a further abbreviation of the notation we omit 1-cycles, such as the (3) above, it being understood that any labels not appearing explicitly just transform into themselves.

To give an illustration of the use of this notation let us rewrite the example given above for $PQ \neq QP$. In cycle notation it reads

$$(2\;3)(1\;3\;2) = (1\;2) \qquad (1.22')$$

versus

$$(1\;3\;2)(2\;3) = (1\;3) \qquad (1.23')$$

A general permutation can always be written as the product of disjoint cycles. Since they operate on different indices these cycles *commute*, and so the order in which they appear is immaterial. In enumerating the individual permutations of S_n it is convenient to group them by cycle structure, i.e. by the number and length of cycles. In fact, this grouping has a deeper significance, as we shall see shortly.

Group S_2

This group is very simple: it consists of the identity permutation, which we can write as ( ), and the *transposition* $1 \leftrightarrow 2$, or in cycle notation (1 2). As a group it is identical to C_2.

Group S_3

S_3 is of order 6. It consists of the identity, three 2-cycles and two 3-cycles:

$$S_3 = \{(\,);\,(1\,2),\,(1\,3),\,(2\,3);\,(1\,2\,3),\,(1\,3\,2)\} \tag{1.27}$$

It turns out to be isomorphic to D_3. In fact, referring to figure 1.2, all the rotations of D_3 can be considered as permutations of the three vertices A, B, C. Relabelling these as 1, 2, 3 respectively we can identify c with the permutation (1 2 3) and b with the permutation (2 3). Then

$$bc = (2\,3)(1\,2\,3) = (1\,3) = b_2$$

a result we obtained previously from geometrical considerations. Continuing in this way we can establish the complete correspondence

$$\{(\,);\,(2\,3),\,(3\,1),\,(1\,2);\,(1\,2\,3),\,(1\,3\,2)\} \leftrightarrow \{e;\,b,\,bc,\,bc^2;\,c,\,c^2\} \tag{1.28}$$

It is often convenient to consider the point groups C_n, D_n, etc. as subgroups of the permutation group S_n of the n vertices. Relationships such as $b_2 = bc$ are easily established this way. The *order* of an element, i.e. the smallest power to which it must be raised in order to give the identity, can be read off immediately from the cycle structure, since a cycle of length r has order r (check!).

Group S_4

With $n = 4$ the number of elements increases to 24 and the number of different cycle structures to five. These are ( ), (. .), (. .) (. .), (. . .) and (. . . .). Here we have indicated the labels in the cycles by dots, which are to be filled in in all possible ways, giving the complete list of elements as

$$\begin{aligned} S_4 = \{&(\,);\,(1\,2),\,(1\,3),\,(1\,4),\,(2\,3),\,(2\,4),\,(3\,4);\\ &(1\,2)(3\,4),\,(1\,3)(2\,4),\,(1\,4)(2\,3);\\ &(1\,2\,3),\,(1\,3\,2),\,(1\,2\,4),\,(1\,4\,2),\,(1\,3\,4),\,(1\,4\,3),\,(2\,3\,4),\,(2\,4\,3);\\ &(1\,2\,3\,4),\,(1\,2\,4\,3),\,(1\,3\,2\,4),\,(1\,3\,4\,2),\,(1\,4\,2\,3),\,(1\,4\,3\,2)\}\end{aligned} \tag{1.29}$$

Cayley's Theorem

We have discussed above how any of the groups C_m, D_m can be considered as subgroups of the group S_m of permutations of the m vertices. The permutation group has a more fundamental importance in the theory of finite groups by virtue of Cayley's theorem, which states that *every* finite group of order n can be considered as (is isomorphic to) a subgroup of S_n. In general this is quite a different embedding from the above, since the order of D_m is $2m$, which is equal to $m!$ only for the case $m = 3$.

The theorem arises from the observation that for a finite group G, multiplication of all the elements $\{a_i\}$ by a given element g simply permutes them, i.e.

$$\{ga_1, ga_2, \ldots, ga_n\} = \{a_{\pi_1}, a_{\pi_2}, \ldots, a_{\pi_n}\} \tag{1.30}$$

so that to each element g of the group there corresponds a permutation $\Pi(g)$ of S_n:

$$g \to \Pi(g) = \begin{pmatrix} 1 & 2 & \cdots & n \\ \pi_1 & \pi_2 & \cdots & \pi_n \end{pmatrix} \tag{1.31}$$

This is indeed a true permutation, i.e. each index appears once and only once on the bottom row. For there are n elements ga_i and they must all be distinct, since $ga_j = ga_k \Rightarrow a_j = a_k$ on multiplication from the left by g^{-1}. However, this is impossible since the elements $\{a_i\}$ are themselves distinct. Moreover, the correspondence is invertible, i.e. a given permutation $\Pi(g)$ cannot arise from any other group element g'. For $ga_i = g'a_i \Rightarrow g = g'$ on multiplication from the right by a_i^{-1}.

Thus we have a 1:1 correspondence $g \leftrightarrow \Pi(g)$. Furthermore, this correspondence *respects the group structure* of the two groups G and S_n. Thus if $g_1 \leftrightarrow \Pi(g_1)$ and $g_2 \leftrightarrow \Pi(g_2)$ then it is clear that $g_1g_2 \leftrightarrow \Pi(g_1)\Pi(g_2)$, the permutation resulting from successive left multiplications by g_2 and g_1. There are n distinct permutations $\Pi(g_i)$, and in general they will comprise only a small subset of the $n!$ permutations of S_n. It is a subset closed under the composition law of S_n, thus forming a subgroup thereof. What we have established is that this subgroup is, from the group theoretical point of view, identical to G. There is a 1:1 correspondence between the elements compatible with the composition law of the two groups. This is what is meant by saying that two groups are isomorphic.

The permutation $\Pi(g)$ is not something entirely new. In fact it can be read off from the multiplication table of G. Quite simply, the ith row of the multiplication table gives the permutation of the elements $g_1, g_2, \ldots, g_n$ induced by multiplication on the left by g_i. Thus in the case of

C_3, $\Pi(e) = (\,)$, as it must be, $\Pi(c) = (1\,2\,3)$ and $\Pi(b) = (1\,3\,2)$. These form a subgroup of S_3 called the 'alternating group' A_3.

The Alternating Group A_n

We have introduced above the concept of a 'transposition'. This is the simplest kind of permutation, a 2-cycle (ij), whereby just two labels are interchanged. *Any* permutation can be built up from products of such elementary exchanges. Consider, in particular, an r-cycle $(n_1 n_2 n_3 \ldots n_r)$. This can be written as the product of $r - 1$ overlapping 2-cycles:

$$(n_1 n_2 n_3 \ldots n_r) = (n_1 n_2)(n_2 n_3) \ldots (n_{r-1} n_r) \qquad (1.32)$$

For most purposes this is not a very useful decomposition, but it allows us to associate with a general permutation the property of 'evenness' or 'oddness'. A single transposition is counted as odd. We can just take this as a definition, but it can be motivated by setting up (in the case of S_n) a correspondence between the index r and a column vector of n entries, all of which are 0 except for a 1 in the rth row. A permutation of these column vectors can then be effected by an $n \times n$ matrix.

Thus in S_2 we have the following correspondence:

Index	Column vector
1	$\begin{pmatrix}1\\0\end{pmatrix}$
2	$\begin{pmatrix}0\\1\end{pmatrix}$

Permutation	Matrix
(1 2)	$\begin{pmatrix}0 & 1\\1 & 0\end{pmatrix}$

If the transposition (1 2) were in fact an element of S_n, the corresponding $n \times n$ matrix would be filled out with 1's along the leading diagonal and 0's elsewhere.

Such a matrix is termed even or odd according to whether its determinant is $+1$ or -1. This agrees with our previous definition, since it is clear that the matrix for a transposition such as (1 2) has determinant -1. Since the determinant of a product of matrices is the product of the individual determinants, we can easily determine whether a general permutation is even or odd. As we have shown in (1.32), a cycle of length r can be written as the product of $r-1$ 2-cycles, each of which is odd. Hence a cycle of length r is even/odd according to whether r is odd/even.

The alternating group A_n consists of the *even* permutations of S_n. These indeed form a group, since the product of two even permutations is again an even permutation. The order of A_n is precisely half that of S_n (see problem 2.4).

Problems for Chapter 1

1.1 Work out the multiplication table for the dihedral group D_3, defined as gp$\{b, c\}$ with $b^2 = c^3 = (bc)^2 = e$.

1.2 Write the permutations

$$\begin{pmatrix} 1 & 2 & 3 & 4 & 5 & 6 & 7 & 8 \\ 6 & 1 & 4 & 8 & 5 & 7 & 2 & 3 \end{pmatrix}, \begin{pmatrix} 1 & 2 & 3 & 4 & 5 & 6 & 7 & 8 & 9 \\ 3 & 5 & 4 & 1 & 8 & 9 & 6 & 7 & 2 \end{pmatrix}$$

in cycle notation.

1.3 The *order* of an element of a group is defined as the smallest integer n such that $g^n = e$. What is the order of the two permutations in problem 1.2? What is the order of the permutation $(a_1a_2 \ldots a_{2r})(b_1b_2 \ldots b_r)$?

1.4 The tetrahedral group T is the group of symmetry rotations of a regular tetrahedron, which can be inscribed in a cube as shown below, where the tetrahedron is ABCD:

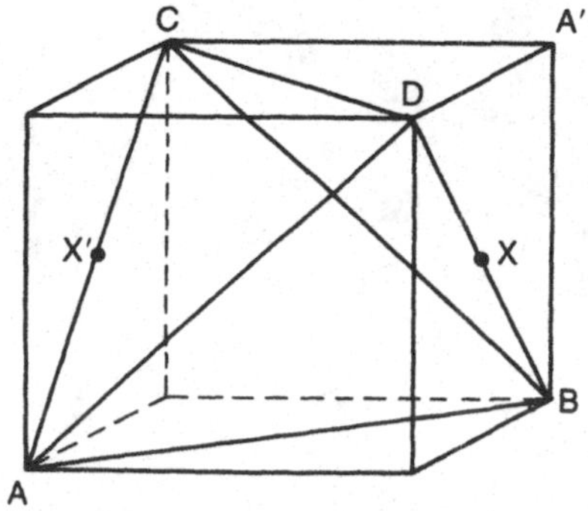

Show that $T = gp\{b, c\}$ with $b^2 = c^3 = (bc)^3 = e$, where b is a 2-fold rotation about the x axis X′X and c is a 3-fold rotation about the cube diagonal AA′. (It may be useful to consider b and c as permutations of the four vertices.) Show that T is isomorphic to A_4.

1.5 State which of the following groups are isomorphic to each other, giving the explicit correspondence where an isomorphism exists:

(i) the complex numbers (1, i, −1, −i) with respect to multiplication;
(ii) the integers (2, 4, 6, 8) with respect to multiplication modulo 10;
(iii) the permutations ( ), (1 2), (3 4), (1 2)(3 4);
(iv) the permutations ( ), (1 2 3 4), (1 4 3 2), (1 3)(2 4);
(v) the four matrices

$$\begin{pmatrix} \pm 1 & 0 \\ 0 & \pm 1 \end{pmatrix}$$

with respect to multiplication.

1.6 The *centre* Z of a group G is defined as the set of elements z which commute with all elements of the group, i.e. $Z = \{z \in G | zg = gz \text{ for all } g \in G\}$. Show that Z is an Abelian subgroup of G.

1.7 Using the multiplication table for D_3 worked out in problem 1.1, write down the isomorphism of Cayley's theorem between the elements b, c and the permutations $\Pi(b)$, $\Pi(c) \in S_6$ induced by left multiplication. Check that $(\Pi(c))^3 = (\Pi(b))^2 = (\Pi(bc))^2 = e$.

2

General Properties of Groups and Mappings

2.1 Conjugacy and Conjugacy Classes

Conjugacy

We have already met this concept in connection with the dihedral group D_3. The formal definition is as follows.

Definition

Two elements a and b of a group G are *conjugate* if there exists an element $g \in G$ such that $a = gbg^{-1}$. The element g is called the *conjugating element*.

The conjugating element will, in general, depend on a and b, but need not be unique. Referring back to D_3, c and c^2 are conjugate, with $c = bc^2b^{-1}$. However, $b_2(= bc)$ or b_3 $(= bc^2)$ are equally good conjugating elements.

Conjugacy is in fact only a particular example of an *equivalence relation*, one of the most fundamental and powerful concepts in mathematics. Such a relation is a set of rules which define whether two elements a and b of some set S are equivalent, written $a \sim b$. The relation so defined must be

(i) reflexive (every element is equivalent to itself)

$$a \sim a \tag{2.1}$$

(ii) symmetric (if a is equivalent to b then b is equivalent to a)

$$a \sim b \Rightarrow b \sim a \tag{2.2}$$

(iii) transitive (if a is equivalent to b and b is equivalent to c then a is equivalent to c)

$$a \sim b \text{ and } b \sim c \Rightarrow a \sim c \tag{2.3}$$

One can think of many examples of equivalence relations in everyday life. We can define two people to be equivalent if they have the same parents, i.e. are brothers or sisters, or have the same nationality (check the axioms!). A counter-example which does not work would be to define two people to be equivalent if they belonged to the same golf club. This fails to satisfy (iii), because b might happen to belong to *two* golf clubs!

Having claimed conjugacy to be an equivalence relation, we had better check that it indeed satisfies the conditions (i)–(iii).

(i) $a \sim a$, since by virtue of G2 we can write $a = eae^{-1}$.

(ii) $a \sim b$ means that $a = gbg^{-1}$ for some $g \in G$. Using G3 we can rewrite this as $b = g^{-1}ag$. That is, $b \sim a$, with g^{-1} as the conjugating element.

(iii) $a \sim b \Rightarrow a = gbg^{-1}$ and $b \sim c \Rightarrow b = hch^{-1}$ for some $h \in G$. So $a = g(hch^{-1})g^{-1} = (gh)c(h^{-1}g^{-1}) = (gh)c(gh)^{-1}$. Thus $a \sim c$, with conjugating element gh, which by G0 is a member of G.

Notice how all the group axioms (we have used associativity implicitly) are needed in the proof.

So conjugate elements are somehow similar or have some property in common. In the case of D_3 we saw that the elements b_1, b_2 and b_3, representing rotations through the same angle about different axes, were conjugate. This is in fact a general result. For let

$$b\boldsymbol{v} = \boldsymbol{v}' \tag{2.4}$$

be the result of rotating a general vector $\boldsymbol{v}$ by b. Then

$$(gbg^{-1})(g\boldsymbol{v}) = gb\boldsymbol{v} = g\boldsymbol{v}' \tag{2.5}$$

Comparing equations (2.4) and (2.5) we see that whereas b takes $\boldsymbol{v}$ to $\boldsymbol{v}'$, the conjugate rotation gbg^{-1} takes $g\boldsymbol{v}$ to $g\boldsymbol{v}'$. So they actually represent the same rotation, through the same angle θ, say; it is just that in the second case the entire space of vectors has first been bodily rotated by g (see figure 2.1). In particular, the axis $\boldsymbol{n}$ of b will also have been rotated by g, so that the axis of the conjugate rotation gbg^{-1} is $g\boldsymbol{n}$. As additional confirmation of this we may note that the axis of a rotation is characterized by being unchanged by the rotation. Thus $b\boldsymbol{n} = \boldsymbol{n}$. But then $(gbg^{-1})(g\boldsymbol{n}) = (g\boldsymbol{n})$, as it should be.

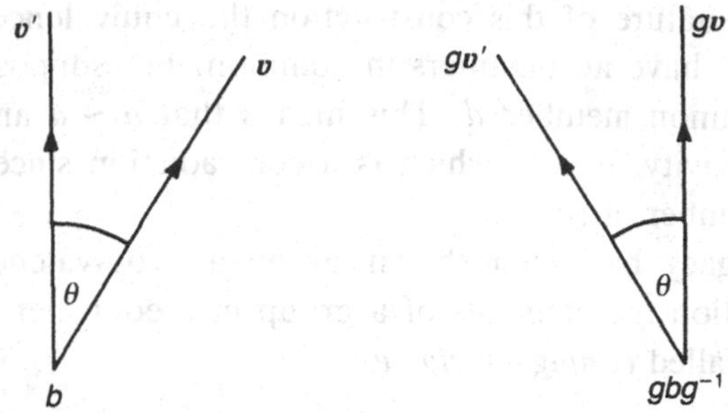

Figure 2.1 The effect of the conjugate rotations b and gbg^{-1}. If b takes $\boldsymbol{v}$ into $\boldsymbol{v}'$, gbg^{-1} takes $g\boldsymbol{v}$ into $g\boldsymbol{v}'$.

However, a word of caution is in order here. If we are talking about the group of all (proper) rotations of 3-dimensional space, then any two rotations through the same angle are conjugate. But if we are considering a *subgroup* of these rotations, such as the group D_4 for example, it is not necessarily true that two such rotations are conjugate. The reason is that the required conjugating rotation, which takes the one axis into the other, may lie outside the subgroup under consideration. Thus in the case of D_4, c^2 and b are not conjugate, even though they are both rotations through π, because there is no element of D_4 which takes the axis of c into that of b (see figure 1.3). For a similar reason b_1 and b_2 are not conjugate either. [Problem: find out which elements *are* conjugate to each other!]

Conjugacy Classes

Any equivalence relation defined on a set gives rise to a *partition* of that set into disjoint (i.e. non-overlapping) *equivalence classes*. The equivalence class of an element a, written (a), is simply the set of elements equivalent to a:

$$(a) = \{b | b \sim a\} \tag{2.6}$$

The procedure for partitioning the set is similar to that used in §1.4 to break up a general permutation into disjoint cycles. That is, we start with any element a and construct its equivalence class (a). If this does not exhaust the set we pick another element b which is not a member of (a) and construct the equivalence class (b). If there are still some elements of the set not accounted for we take one of those elements, c say, and construct *its* equivalence class (c), etc.

By the very nature of this construction the equivalence classes must be disjoint, i.e. have no members in common. For suppose (a) and (b) do have a common member d. This means that $d \sim a$ and $d \sim b$. But then, by transitivity, $a \sim b$, which is a contradiction since b was taken not to be a member of (a).

Since conjugacy has been shown to be an equivalence relation, it serves to partition the elements of a group into equivalence classes, not unreasonably called *conjugacy classes*:

$$(a) = \{b \mid b = gag^{-1} \text{ for some } g \in G\} \tag{2.7}$$

Examples

(1) C_n
Each individual element constitutes a conjugacy class by itself. This is a result common to all Abelian groups. For then† $ga = ag$ $\forall a$ and $g \in G$, and hence

$$gag^{-1} = agg^{-1} = a$$

(2) D_3 $(= \text{gp}\{c, b\}$ with $c^3 = b^2 = (bc)^2 = e)$
The conjugacy classes are

(e)	This is always true since $geg^{-1} = e$
(c, c^2)	For example, $bcb^{-1} = b(cb) = b(bc^2) = c^2$
(b, bc, bc^2)	$(= (b_1, b_2, b_3))$

Note that these correspond to rotations through different angles, namely 0, $2\pi/3$, π respectively. c^2 is a $4\pi/3$ rotation about the axis of c, but this is equivalent to a $2\pi/3$ rotation about the reversed axis.

(3) S_n
It turns out that for S_n the conjugacy classes are in 1:1 correspondence with the cycle structure.

Any general permutation P can be written as

$$P = (a_1 a_2 \ldots a_{l_1})(b_1 b_2 \ldots b_{l_2})(c_1 c_2 \ldots c_{l_3}) \ldots \tag{2.8}$$

ordering the length of the cycles such that $l_1 \geqslant l_2 \geqslant l_3 \ldots \geqslant l_r$ to give what is called a *partition* of n, a set of ordered integers l_i such that $l_1 + l_2 + l_3 + \ldots + l_r = n$. In this decomposition we must explicitly include 1-cycles, rather than omitting them as we have sometimes done above.

†From now on we will employ the useful symbol $\forall$, meaning 'for all'.

Then consider a general conjugation QPQ^{-1}:

$$\begin{pmatrix} a_1a_2 \ldots & a_{l_1}b_1b_2 \ldots & b_{l_2} \ldots \\ A_1A_2 \ldots & A_{l_1}B_1B_2 \ldots & B_{l_2} \ldots \end{pmatrix} \times (a_1a_2 \ldots a_{l_1})(b_1b_2 \ldots b_{l_2}) \ldots$$
$$\times \begin{pmatrix} A_1A_2 \ldots A_{l_1}B_1B_2 \ldots B_{l_2} \ldots \\ a_1a_2 \ldots \quad a_{l_1}b_1b_2 \ldots \quad b_{l_2} \ldots \end{pmatrix} \quad (2.9)$$

(Note that the last permutation is indeed the inverse of the first.) Multiplying this out we find that

$$QPQ^{-1} = (A_1A_2 \ldots A_{l_1})(B_1B_2 \ldots B_{l_2})(C_1C_2 \ldots C_{l_3}) \ldots \quad (2.10)$$

That is, the conjugate permutation has exactly the same cycle structure: it is merely that the entries a_i, b_i, etc. have been replaced by A_i, B_i, etc.

The proof is reversible. Given two permutations with identical cycle structure but different entries, A_i versus a_i, etc., we can construct Q as in (2.9) and show them to be conjugate.

Thus the conjugacy classes of S_n are given by enumerating the different partitions of n. If all the cycle lengths l_i are different, this is written as $l_1 \, l_2 \ldots l_r$. In general, however, there may be several cycles of the same length. In this case we list the distinct cycle lengths l_i and the number of times p_i they occur, and write the partition as $(l_1)^{p_1} \, (l_2)^{p_2} \ldots (l_r)^{p_r}$. For example, in the case of S_3 we have:

Partition	Conjugacy class	Correspondence with D_3
1^3	( )	e
1.2	(2 3), (1 3), (1 2)	b_1, b_2, b_3
3	(1 2 3), (1 3 2)	c, c^2

2.2 Subgroups

This is again a concept we have already met informally. The formal definition is as follows.

Definition
A *subgroup* H of a group G is a subset of G which itself forms a group under the composition law of G.

In the case of a finite group an equivalent definition is that H should be *closed* (H0) under the composition law (multiplication) of G, i.e.

$$h_1 h_2 \in H \qquad \forall h_1, h_2 \in H \subset G \tag{2.11}$$

The associative law (H1) is inherited from the larger group G, and the existence of the identity in H (H2) arises from the closure and finite order of G. For every h, considered as an element of G, has a finite order r such that $h^r = e$. So, by the closure of H, $e \in H$. Similarly $h^{-1} = h^{r-1} \in H$, proving H3.

According to the above definition the single element e forms a subgroup, as does the whole group G. A subgroup which is not one of these two rather trivial cases is called a *proper* subgroup, written $H < G$. For example, $C_2 = \{e, b\}$ and $C_3 = \{e, c, c^2\}$ are both proper subgroups of D_3.

Cosets

Given a subgroup $H = \{h_1, h_2, \ldots, h_r\}$ of a group G, the (left) *coset* of an element $g \in G$, written gH, is defined as the set of elements obtained by multiplying all the elements of H on the left by g:

$$gH := \{gh_1, gh_2, \ldots, gh_r\} \tag{2.12}$$

As we shall now show, these cosets either completely overlap or are completely disjoint, and they provide another, quite different, partition of the group G.

The proof proceeds by setting up another equivalence relationship:

$$a \sim b \qquad \text{if } b \in aH \tag{2.13}$$

or, in words, two elements are equivalent if one lies in the coset of the other. We must now proceed to check that (2.12) indeed defines an equivalence relation satisfying the conditions (2.1)–(2.3).

(i) (reflexivity) Is $a \in aH$?
Yes, for $a = ae$ and $e \in H$! (by H2)

(ii) (symmetry) If $b \in aH$, is $a \in bH$?
Well, $b \in aH \Rightarrow b = ah$ for some $h \in H$.
$\therefore a = bh^{-1}$, with $h^{-1} \in H$. (by H3)

(iii) (transitivity) If $b \in aH$ and $c \in bH$, is $c \in aH$?
Well, $b \in aH$ and $c \in bH \Rightarrow b = ah$ and $c = bh'$ for some $h, h' \in H$.
But then $c = ahh'$, with $hh' \in H$. (by H0)

Because this is indeed an equivalence relation it partitions the group into disjoint equivalence classes, which are just the distinct cosets. For a finite group we can enumerate these as $g_1H, g_2H, \ldots, g_sH$, say. For reasons that will become apparent, the set of cosets, considered as objects in their own right, is denoted by G/H.

Lagrange's Theorem

Again, in a finite group each coset contains exactly the same number, r say, of distinct elements. This is the number of distinct elements of H itself, i.e. the order of H, also written as $[h]$. For consider the coset gH, containing the elements $gh_1, gh_2, \ldots, gh_r$. The number of distinct elements here could be less than r if some of them were the same, for example if $gh_1 = gh_2$. But, multiplying by g^{-1} on the left, we could then deduce that $h_1 = h_2$, contrary to hypothesis.

Thus all the elements of the group G, $[g]$ in number, can be grouped into s cosets each containing r $(= [h])$ elements. Hence

$$[g] = s[h] \tag{2.14}$$

Thus the order $[h]$ of any subgroup of G must be a divisor of $[g]$, the order of G. This is the content of Lagrange's theorem.

As an immediate corollary we can deduce that groups of prime order have no proper subgroups (e.g. C_p for p prime).

Example
D_3 has three C_2 subgroups $\{e, b_i\}$. Take $H = \{e, b_1\} = \{e, b\}$. The cosets of H are then

$$\begin{aligned} eH &= H &&= \{e, b\} \\ cH &= \{c, cb\} &&= \{c, bc^2\} \\ c^2H &= \{c^2, c^2b\} &&= \{c^2, bc\} \end{aligned} \tag{2.15}$$

Note that we could have used different 'representatives' to label these cosets. Thus it would be equally valid to write eH as bH, cH as bc^2H, c^2H as bcH. In general any member of a given coset can be used to label it.

2.3 Normal Subgroups

In general, subgroups and conjugacy classes have very little to do with one another. However, the two concepts are brought together for a very special type of subgroup, variously called a *normal*, an *invariant* or a *self-conjugate* subgroup. Although special, such subgroups do arise quite naturally in the context of mappings of groups, as we shall see in the next section. A normal subgroup H of G is one which satisfies

$$gHg^{-1} = H \quad \forall g \in G \tag{2.16}$$

There are various equivalent formulations of this definition. Multiplying (2.16) by g on the right gives $gH = Hg$. Here Hg is a right coset, consisting of the elements $\{h_1g, h_2g, \ldots, h_rg\}$. So an alternative definition is that a normal subgroup is a subgroup whose right and left cosets coincide.

From the first definition one can see why H should be called self-conjugate, by analogy with a self-conjugate element, satisfying $ghg^{-1} = h$. However, this is not what (2.16) means. It means rather that all the conjugates of h must also be members of H, i.e.

$$h \in H \Rightarrow ghg^{-1} \in H \quad \forall g \in G$$

Another definition is therefore that a normal subgroup is one which is made up of *complete* conjugacy classes.

Example: D_3
$C_2 = \{e, b\}$ is not a normal subgroup of D_3, since the conjugacy class of b is $(b) = \{b, bc, bc^2\}$. However, $C_3 = \{e, c, c^2\}$ *is* normal, since the conjugacy class of c is $(c) = \{c, c^2\}$. Thus C_3 is made up of the complete conjugacy classes (e), (c).

Quotient Group

A remarkable property of a normal subgroup H is that the set of cosets G/H can be endowed with a *group structure* by a suitable definition of the product of two cosets.

Definition
The product of two cosets g_1H and g_2H is defined to be the coset g_1g_2H:

$$(g_1H)(g_2H) := g_1g_2H \tag{2.17}$$

Given this definition, we must check to see whether it indeed satisfies the group axioms G0–G3.

G0 (closure) Is $g_1g_2H \in G/H$?
Yes, by closure of G itself: $g_1g_2 \in G$.

G1 (associativity) This is assured by the associativity of G:

$$
\begin{aligned}
(g_1H)[(g_2H)(g_3H)] &= (g_1H)(g_2g_3H) \\
&= g_1(g_2g_3)H \\
&= (g_1g_2)g_3H \\
&= \dots \\
&= [(g_1H)(g_2H)](g_1H)
\end{aligned}
$$

G2 (identity) The coset $E := eH = H$ acts as the identity, for following the definition,

$$(eH)(gH) = (eg)H = gH$$

and similarly for $(gH)(eH)$.

G3 (inverse) The inverse of the coset gH is $g^{-1}H$, since

$$(gH)(g^{-1}H) = (gg^{-1})H = eH = E$$

The essential feature of the definition (2.17) is that the group properties of G/H are inherited from G. However, there is still one crucial point to check, namely that it is a *consistent* definition. The danger of inconsistency arises from the fact that the representatives g_1, g_2 which we have used to label the cosets g_1H_1, g_2H are by no means unique. We could equally well have labelled them by other members. g_1', g_2' say. If coset multiplication is to be well-defined we must get the same result for the product regardless of which representatives we choose. That is, the coset $g_1'g_2'H$ must be the same coset as g_1g_2H.

Well, if g_1' is a member of the coset g_1H, then $g_1' = g_1h_1$ for some $h_1 \in H$. Similarly $g_2' = g_2h_2$ for some $h_2 \in H$. Therefore

$$g_1'g_2'H = g_1h_1g_2h_2H = g_1h_1g_2H$$

since $h_2H = H$ (as in Cayley's theorem). But because H is a normal subgroup, $g_2H = Hg_2$. Thus

$$g_1'g_2'H = g_1h_1Hg_2 = g_1Hg_2 = g_1g_2H$$

Another, and perhaps somewhat more transparent, proof of consistency comes from the remark that the composition law defined by (2.17) actually corresponds to multiplication of the two sets element by element, but only when H is a normal coset. For then

$$g_1Hg_2H = g_1HHg_2 = g_1Hg_2 = g_1g_2H$$

as required. Here all multiplications are those defined by G. We have used the closure of H in the form $HH = H$ and self-conjugacy in the form $g_2H = Hg_2$.

Example: $H = \{e, c, c^2\}$ in D_3
The two cosets are

$$eH = H = \{e, c, c^2\} = E$$

$$bH \quad = \{b, bc, bc^2\} = B, \text{ say}$$

Under the coset multiplication defined by (2.17) the two cosets E and B form a C_2 group isomorphic to $\{e, b\}$. Thus

$$E^2 = (eH)(eH) = e^2H = H = E$$

$$EB = (eH)(bH) = ebH = bH = B$$

Similarly

$$BE = \quad \ldots \quad = B$$

and

$$B^2 = (bH)(bH) = b^2H = eH = E$$

It is instructive to check this by multiplication element by element. Thus

$$E^2 = \{e, c, c^2\}\{e, c, c^2\} = \{e, c, c^2; c, c^2, e; c^2, e, c\} = E$$

$$BE = \{b, bc, bc^2\}\{e, c, c^2\} = \{b, bc, bc^2; bc, bc^2, b; bc^2, b, bc\} = B$$

[Problem: check EB, B^2 by the same method.]

Counter-example: $H = \{e, b\}$ in D_3
As we have already seen, $\{e, b\}$ is not a normal subgroup of D_3, so we expect something to go wrong. In fact, if we multiply the cosets element by element the result is not always another coset. For example,

$$cHcH = \{c, cb\}\{c, cb\} = \{c^2, c^2b; b, e\} = c^2H \cup eH$$

Direct Product

In the above example we have shown that $D_3/C_3 = C_2$, up to isomorphism. We might be tempted by the notation to 'cross-multiply' and say that $D_3 = C_3 \times C_2$, with some meaning to be assigned to the $\times$ operation. There are in fact cases where it is meaningful to write a group G as a 'direct product' $G = A \times B$ of two subgroups A and B, but D_3 is not one of them.

Definition

A group G is the *direct product* of its subgroups A and B, written $G = A \times B$, if

(i) all the elements of A commute with all those of B,
(ii) every element of G can be written in a unique way as $g = ab$ with $a \in A$ and $b \in B$.

From (i) the product of any two elements $g_1 = a_1b_1$ and $g_2 = a_2b_2$ is

$$g_1g_2 = (a_1b_1)(a_2b_2) = (a_1a_2)(b_1b_2) \tag{2.18}$$

i.e. it is given by multiplying the 'a' components and 'b' components separately.

A further consequence of the definition is that both A and B are *normal* subgroups of G. Thus

$$ga_ig^{-1} = aba_ib^{-1}a^{-1} = aa_ia^{-1} \in A \qquad \forall\ a_i \in A$$

and similarly for B. Moreover, the quotient groups G/A and G/B are in 1:1 correspondence with B and A respectively.

That is, if $B = \{b_1, b_2, \ldots, b_r\}$, the cosets b_jA are all distinct. For if not, suppose, for example, that $b_1A = b_2A$. This would mean that $b_1a_1 = b_2a_3$, say. But then, from (ii), $b_1 = b_2$ ⊗. Then the multiplication of the cosets b_jA exactly follows that of the elements b_j, so that $G/A \cong B$. Similarly $G/B \cong A$. Thus in this case we *can* cross-multiply in both relations to obtain $G \cong A \times B$.

Example: $D_2 = \text{gp}\{a, b\}$ with $a^2 = b^2 = (ab)^2 = e$

D_2 has two normal subgroups isomorphic to C_2, namely $A = \text{gp}\{a\}$ and $B = \text{gp}\{b\}$. They completely commute with each other, since the last condition gives $ab = ba$, and every element of D_2 can be written uniquely as a_ib_j. Thus $C_2 = D_2/C_2$, which can be cross-multiplied to give $D_2 = C_2 \times C_2$.

Counter-example: D_3

As we have seen above, $D_3/C_3 = C_2$. However, $D_3 \neq C_2 \times C_3$. This is easy to verify: $C_2 \times C_3$ is an Abelian group, which D_3 is not.

2.4 Homomorphisms

A homomorphism in general is a mapping f from one set A to another set B (written $f : A \to B$) *preserving some structure* or other.

Each element a of A is mapped into a unique image, a single element $b = f(a)$ of B. However, as illustrated in figure 2.2, the reverse is not in general true, since a given element $b \in B$ may

(i) be the image of several elements of A,
(ii) not be the image of any element of A.

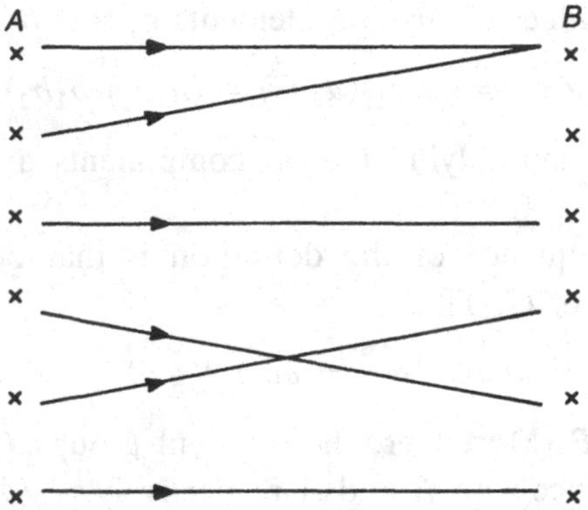

Figure 2.2 General mapping f from one set A to another set B.

The second situation can be rather trivially avoided by restricting our attention to the *image* of f, written $\operatorname{Im} f$ or $f(A)$, consisting of those elements of B which *are* maps of elements of A:

$$\operatorname{Im} f := \{b \in B \mid b = f(a) \text{ for some } a \in A\} \tag{2.19}$$

In that case the mapping is said to be 'onto' or 'surjective', as opposed to merely 'into' or 'injective'.

For example, in the case of the integers Z, the mapping f: $Z \to Z$ defined by

$$f(n) = 1 \qquad n \text{ odd}$$
$$= 0 \qquad n \text{ even}$$

is injective as it stands. However, it can easily be redefined as f: $Z \to Z_2$ to make it surjective.

We have said above that in order to qualify as a homomorphism a mapping must preserve some mathematical structure or other. In the context of groups the structure which must be respected is the group multiplication: the map f: $A \to B$ is a *group homorphism* if, for all a_1, a_2 in A,

$$f(a_1 a_2) = f(a_1) f(a_2) \tag{2.20}$$

In words, the image of a product must be the same as the product of the images. Note that these two 'products' are the composition laws of the respective groups: on the LHS of (2.20) a_1a_2 is formed by the composition law of A, while on the RHS $f(a_1)$ and $f(a_2)$, being elements of B, are combined according to the composition law of B, whatever that is.

In the case of a group homomorphism $f: A \to B$, the image of the mapping, $f(A)$, forms a subgroup of B, as the reader should check. Another very important subgroup, this time of A, is formed by the *kernel* of the homomorphism, written $\mathrm{Ker} f$ or $f^{-1}(e_B)$, consisting of those elements of A which are mapped onto the identity element in B:

$$\mathrm{Ker} f := \{a \in A | f(a) = e_B \in B\} \tag{2.21}$$

In the case of finite groups it is sufficient to prove closure, which follows almost immediately. For $k_1, k_2 \in \mathrm{Ker} f \Rightarrow f(k_1) = f(k_2) = e_B$, and therefore $f(k_1k_2) = f(k_1)f(k_2) = e_Be_B = e_B$, showing that k_1k_2 is also a member of $\mathrm{Ker} f$. To show that $e \in \mathrm{Ker} f$ consider $f(a) = f(ae) = f(a)f(e)$. Multiplying on the left by $(f(a))^{-1}$ gives $f(e) = e_B$, as required. Finally we note that for a general element g of A the image of the inverse is the inverse of the image. Thus $e_B = f(gg^{-1}) = f(g)f(g^{-1})$, so that $f(g^{-1}) = (f(g))^{-1}$. Specializing to an element k of $\mathrm{Ker} f$, $f(k^{-1}) = (f(k))^{-1} = e_B$, proving that $k^{-1} \in \mathrm{Ker} f$.

The fact that the kernel must be a subgroup places an important restriction on the form of any possible homomorphism. For a finite group, at least, there are a limited number of such candidates for the kernel. The choice becomes even more restricted when one realizes that the kernel is in fact a *normal* subgroup, which follows from the observation that $f(gkg^{-1}) = f(g)f(k)f(g^{-1}) = f(g)e_B(f(g))^{-1} = e_B$.

The choice not only of the kernel, but also of the *image*, is dictated by the structure of the group A. Given a normal subgroup K we have already seen that the coset space A/K can be endowed with a group structure in a rather natural way. The content of the (first) isomorphism theorem, which follows, is that the image of any homomorphism f with kernel K is isomorphic to that quotient group!

Isomorphism Theorem

If $f: G \to G'$ is a homomorphism of G into G', with kernel K, then

$$\mathrm{Im}(f) \cong G/K \tag{2.22}$$

We show this by setting up a 1:1 correspondence between the two which respects the two group structures. This correspondence is

$$f(g) \leftrightarrow gK \tag{2.23}$$

whereby the image of any element g is associated with the coset gK. To establish the properties claimed, we need to show the following:

(i) The map† $f(g) \mapsto gK$ is well-defined. There is room for doubt here, since an element of Im f could easily be the image of two different elements, g and g' say, of G. In that case (2.23) would variously associate the element with gK and $g'K$. However, if $f(g') = f(g)$, then $f(g'g^{-1}) = e$. Therefore $g'g^{-1} \in K$, and so the cosets $g'K$ and gK are in fact the same.

(ii) The map $gK \mapsto f(g)$ is well-defined. Again there is room for doubt, since the mapping appears to depend on the representative g. But the above argument works in reverse: if g' is another representative of gK, then $g'g^{-1} \in K$ and hence $f(g') = f(g)$.

(iii) The group structure is preserved. This means that we must have $f(g)f(g') \leftrightarrow (gK)(g'K)$. Well, since f is a homomorphism, $f(g)f(g') = f(gg')$, which maps onto $gg'K$. But by the composition law of cosets this is precisely $(gK)(g'K)$, as required.

Thus a homomorphic mapping from G onto $f(G)$ is a very regular affair. The normal subgroup K is mapped onto the identity $e \in G'$, and entire cosets g_iK are mapped into the single element $f(g_i)$, as shown schematically in figure 2.3.

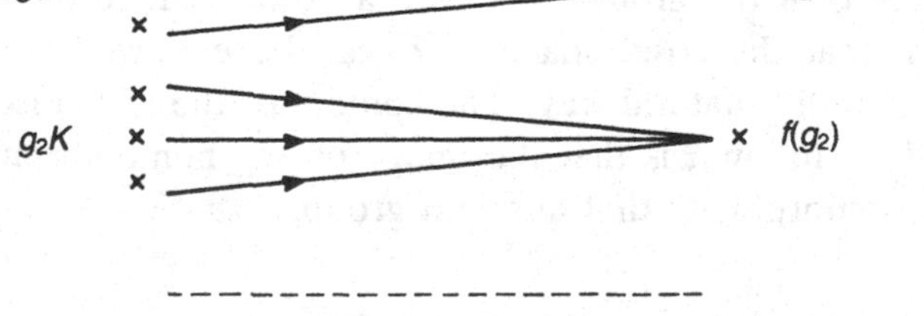

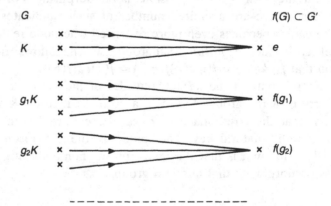

Figure 2.3 The very ordered structure of a homomorphism f from one group G into another G'. Entire cosets g_iK of G map onto the element $f(g_i) \in G'$.

†The notation $\mapsto$ is used when it is the mapping of individual elements which is being defined.

For *finite* groups we have the immediate corollaries:

(i) The possible orders of homomorphic images of the group G are $[g]/r$, where r is the order of a normal subgroup of G.
(ii) The mapping is then $r \to 1$, since cosets are mapped *en bloc* and each coset g_iK has exactly r elements.

Example: D_3 (yet again!)
As we have seen above, the only proper normal subgroup of D_3 is $C_3 = \{e, c, c^2\}$. A homomorphism of which this was the kernel would have an image consisting of two elements, E and B say, with a structure isomorphic to $D_3/C_3 = C_2$. The mapping would be $3 \to 1$, with $\{e, c, c^2\} \to E$ and $\{b, bc, bc^2\} \to B$.

There are two special cases worthy of note: when the kernel consists of the entire group or, at the other extreme, of the identity element alone:

(i) $K = G$, corresponding to the trivial mapping $f(g) = e \quad \forall g \in G$.
(ii) $K = e$. The mapping is in this case an *isomorphism*, otherwise known as a 'bijective' map, a 1:1 correspondence preserving multiplication.

The target group G' does not necessarily have to be a different group: it can be the original group G itself. For this case there is a special nomenclature. A homomorphism of a group into itself is called an *endomorphism*, while an isomorphism becomes an *automorphism*. For example, the mapping defined by $c \to e$ in C_3 is the trivial endomorphism, $c \to c$ is the identity automorphism, while $c \to c^2$ defines a non-trivial automorphism.

Problems for Chapter 2

2.1 Let A and B be two subgroups of a group G. Show that the relation

$$g \sim g' \text{ if } g' = agb \text{ for some } a \in A,\ b \in B$$

is an equivalence relation, partitioning G into *double cosets* Ag_iB. Exhibit this partition for the case $G = D_4 = \text{gp}\{b, c\}$ with $c^4 = b^2 = (bc)^2 = e$, $A = B = C_2 = \text{gp}\{b\}$.

2.2 Find the conjugacy classes of D_4:

(i) by considering the elements as permutations $\in S_4$ of the vertices of the square;
(ii) using the generators and defining relations;
(iii) by geometrical considerations.

2.3* Ditto for the tetrahedral group T.

2.4 Show that A_n is a normal subgroup of S_n. Construct the coset space S_n/A_n and hence deduce that A_n is of order $(\frac{1}{2}n!)$. What is the structure of the group S_n/A_n?

2.5 If H is a subgroup of G with order $[h] = \frac{1}{2}[g]$, prove that H is necessarily a self-conjugate subgroup of G. (Hint: the conjugating element must be in either H or g_1H. Consider the two cases separately.)

2.6 Show that there are only two proper self-conjugate subgroups of S_4. (Remember that conjugacy classes of S_n are given by the different partitions of n.) What is their structure?

2.7 Show that the centre Z of a group G (see problem 1.6) is in fact a normal subgroup. Find the centre of D_4 and construct the group D_4/Z. Is $D_4 \cong D_4/Z \times Z$?

2.8 Show that $C_6 \cong C_3 \times C_2$.

2.9 Show that the map f: $G \to G$ defined by $f(g) = g^{-1}$ is an automorphism if G is an Abelian group.

2.10 Is it possible to map D_4 homomorphically into D_3 in a non-trivial way?

2.11 Show that C_n is a normal subgroup of D_n and that $D_n/C_n \cong C_2$.

2.12 An *inner automorphism* of a group G is the mapping f_a given by conjugation by any fixed element a of G: $g \to aga^{-1}$. Verify that the set of all inner automorphisms of G, denoted by $I(G)$, forms a group, and show that $I(G) \cong G/Z$, where Z is the centre of G.

3

Group Representations

3.1 A Simple Example; Formal Definition

In the physical context in which they arise, the elements of the point groups C_n, D_n, T, etc. are not just abstract objects satisfying certain laws of combination, but are actually spatial rotations, through a restricted set of angles. Under such a rotation the coordinates (x, y, z) of any given point P in ordinary [3] space are transformed into a new set (x', y', z') linearly related to the first. This transformation can be effected by an appropriate 3×3 matrix determined by the rotation in question. So to each element of the group there corresponds such a matrix, and moreover the matrix corresponding to the product of two rotations will just be the product of the individual matrices, since in both cases the law of composition is given by successive application. Though we shall give the formal definition a little later, this is essentially what is meant by a group representation.

Example: C_3
Figure 3.1 shows a general vector OP, with its components along the three axes O1, O2 and O3 and projection OQ onto the x–y plane. Recall that c is a rotation about the z axis O3 through an angle $2\pi/3$. Any rotation about this axis will leave the z component of a vector unchanged, so under c:

$$z \to z' = z \tag{3.1}$$

The transformation of x and y can be deduced by a study of figure 3.2, which shows the position of Q before and after a rotation through a general angle θ about the z axis. The transformation is very easily expressed in polar coordinates: the radial coordinate r is unchanged, while the polar angle φ becomes $\varphi + \theta$.

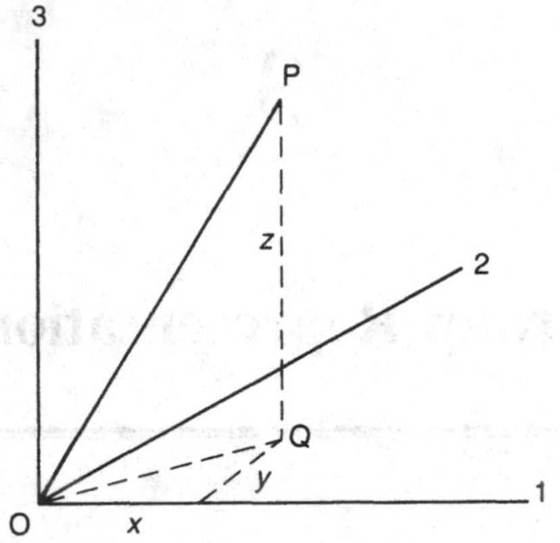

Figure 3.1 A general vector OP and its projection OQ onto the x–y plane.

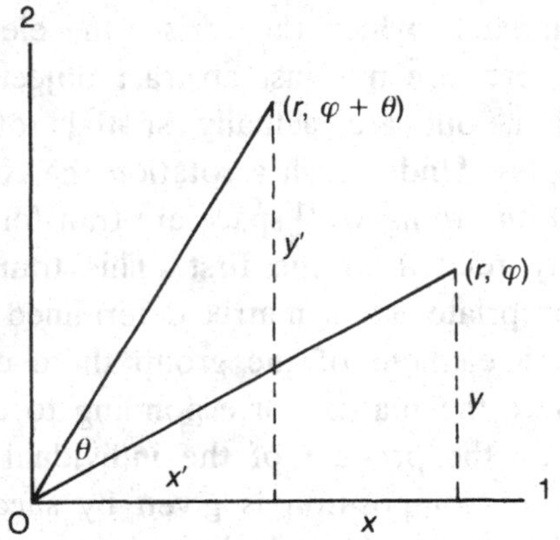

Figure 3.2 Rotation of OQ through φ.

Translating this into Cartesian components:

$$\begin{aligned} x' &= r\cos(\varphi + \theta) \\ &= r(\cos\varphi\cos\theta - \sin\varphi\sin\theta) \\ &= x\cos\theta - y\sin\theta \end{aligned} \tag{3.2}$$

$$\begin{aligned} y' &= r\sin(\varphi + \theta) \\ &= r(\sin\varphi\cos\theta + \cos\varphi\sin\theta) \\ &= y\cos\theta + x\sin\theta \end{aligned} \tag{3.3}$$

i.e.

$$\begin{pmatrix} x' \\ y' \\ z' \end{pmatrix} = \begin{pmatrix} x\cos\theta - y\sin\theta \\ x\sin\theta + y\cos\theta \\ z \end{pmatrix}$$

$$= \begin{pmatrix} \cos\theta & -\sin\theta & 0 \\ \sin\theta & \cos\theta & 0 \\ 0 & 0 & 1 \end{pmatrix} \begin{pmatrix} x \\ y \\ z \end{pmatrix} \quad (3.4)$$

Thus a physical rotation through an angle θ about the z axis is *represented* by the matrix

$$R(\theta) = \begin{pmatrix} \cos\theta & -\sin\theta & 0 \\ \sin\theta & \cos\theta & 0 \\ 0 & 0 & 1 \end{pmatrix} \quad (3.5)$$

For the particular angles 0, $2\pi/3$ and $4\pi/3$ we obtain the matrices $D(e)$, $D(c)$ and $D(c^2)$ which represent those elements of C_3:

$$D(e) = R(0) = \begin{pmatrix} 1 & 0 & 0 \\ 0 & 1 & 0 \\ 0 & 0 & 1 \end{pmatrix}$$

$$D(c) = R(2\pi/3) = \begin{pmatrix} -1/2 & -\sqrt{3}/2 & 0 \\ \sqrt{3}/2 & -1/2 & 0 \\ 0 & 0 & 1 \end{pmatrix} \quad (3.6)$$

$$D(c^2) = R(4\pi/3) = \begin{pmatrix} -1/2 & \sqrt{3}/2 & 0 \\ -\sqrt{3}/2 & -1/2 & 0 \\ 0 & 0 & 1 \end{pmatrix}$$

The crucial property of these matrices which makes them a representation is that $D(c^2) = (D(c))^2$ and $(D(c))^3 = D(e)$, the unit matrix, as you should check! They therefore provide a concrete realization in matrix form of the abstract structure $C_3 = \text{gp}\{c\}$, $c^3 = e$.

Following this specific example we now give the formal definition, which is couched in the language of mappings developed in the previous chapter.

Definition

A *representation* of dimension n of the abstract group G is defined as a homomorphism D: $G \to \text{GL}(n, \mathbb{C})$, the group of non-singular $n \times n$ matrices with complex entries.

In other words it is a mapping under which $g \mapsto D(g)$, preserving the group structure:

$$D(g_1 g_2) = D(g_1)\, D(g_2) \tag{3.7}$$

The mapping is necessarily into the set of non-singular matrices, since each matrix must be invertible: $D(g^{-1}) = (D(g))^{-1}$.

The rotation matrices of (3.5) actually belong to a more restricted subgroup of GL(n, $\mathbb{C}$). In the first place they are *real*, since the coordinates x, y, z are themselves real. But they are also *orthogonal* (check!), since a rotation leaves invariant the squared length $\boldsymbol{x}^2 = x^2 + y^2 + z^2$ or, more generally, the scalar product of any two vectors $\boldsymbol{a}\cdot\boldsymbol{b}$. (Recall that the matrix R is orthogonal if $R^{\mathrm{T}}R = 1$. The scalar product $\boldsymbol{x}^2$ can be written in matrix notation as $\boldsymbol{x}^2 = x^{\mathrm{T}}x$. So $(\boldsymbol{x}')^2 = (x^{\mathrm{T}}R^{\mathrm{T}})(Rx) = \boldsymbol{x}^2$, as required.) Such matrices have $(\det R)^2 = 1$ and hence $\det R = \pm 1$. In fact the matrices of (3.5) all have determinant $+1$, since they represent 'proper' rotations continuously connected to the identity, as opposed to 'improper' rotations, which include reflections such as $x \to -x$, $y \to y$, $z \to z$. The technical name for the group of such matrices is SO(3), the *special orthogonal* group in three dimensions.

A representation is termed 'faithful' if the homomorphism is in fact an isomorphism. From §2.4 we know that this is the case if the kernel of the homomorphism consists solely of the identity: $\mathrm{Ker}\, D = e \in G$, i.e. if the only element represented by the unit matrix is the identity element of the group. This is in fact the case for the example of C_3 discussed above, and indeed is even true of the 2×2 submatrices forming the top left-hand blocks of the matrices (3.6). In contrast, the 1×1 matrices forming the bottom right-hand corners provide an example of an 'unfaithful' representation, which in this case is the trivial representation, with each member of the group being mapped onto the number 1.

3.2 Induced Transformation of the Quantum Mechanical Wavefunction

The representation considered in the previous section was nothing more than an explicit realization, at the classical level, of how the elements of C_3 were originally defined, namely as rotations about a certain axis in ordinary 3-dimensional space. If that were all there were to it, one would hardly need the whole powerful apparatus of representation theory. The latter really comes into its own when we go from the classical to the *quantum* level and consider how the state vector or wavefunction of a physical system is transformed when we perform such a rotation.

For definiteness let us consider the wavefunction $\psi(\boldsymbol{x})$ of the electron in a hydrogen atom centred at the origin. For example, the electron could be in a 2p state with angular dependence proportional to $\cos\theta$ in spherical polar coordinates. The probability density $|\psi(\boldsymbol{x})|^2$ of such a state would then be concentrated near the z axis, as illustrated in the polar plot of figure 3.3(a), where the length of the radius vector is proportional to $\cos^2\theta$.

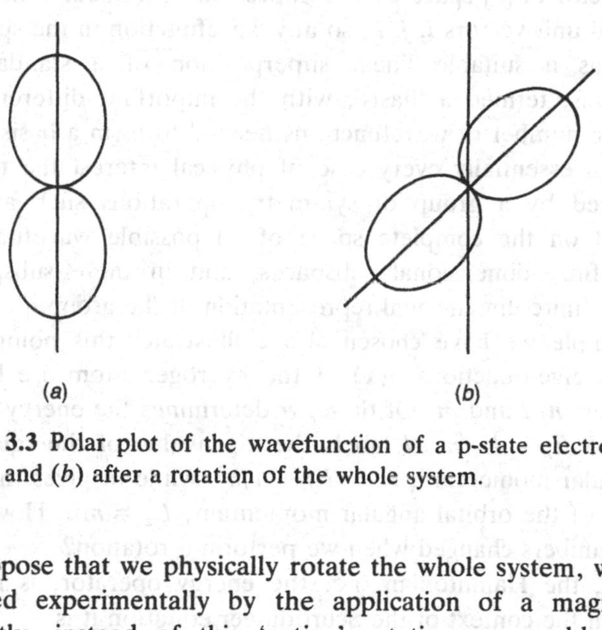

Figure 3.3 Polar plot of the wavefunction of a p-state electron (a) before and (b) after a rotation of the whole system.

Now suppose that we physically rotate the whole system, which could be achieved experimentally by the application of a magnetic field. (Equivalently, instead of this 'active' rotation we could consider a 'passive' rotation, whereby the system is left alone but the axes are rotated in the opposite direction.) After this rotation the system will have a new wavefunction $\psi'(\boldsymbol{x})$, with $|\psi'(\boldsymbol{x})|^2$ concentrated around a displaced axis, as illustrated in figure 3.3(b).

We can determine the relationship between ψ' and ψ by the observation that *the value of the new wavefunction at a rotated point is the same as that of the old wavefunction at the original point*, i.e.

$$\psi'(\boldsymbol{x}') = \psi(\boldsymbol{x}) \tag{3.8}$$

where $\boldsymbol{x}' = R\boldsymbol{x}$. Therefore

$$\psi'(x') = \psi(R^{-1}x') \tag{3.9}$$

But in this equation x' is a dummy label: it is just the position of a general point which we could equally well label as x. So we can remove the primes to give

$$\boxed{\psi'(x) = \psi(R^{-1}x)} \tag{3.10}$$

What has happened is that the rotation R in physical space has *induced* a transformation in the space of quantum mechanical wavefunctions. This space is formally a vector space, a concept we shall discuss in more detail later on, but which essentially means that we can form linear combinations subject to certain rules. Moreover, just as an ordinary vector in [3] space can be expressed as a linear combination of the standard unit vectors $\boldsymbol{i}, \boldsymbol{j}, \boldsymbol{k}$, so any wavefunction in the space can be expressed as a suitable linear superposition of a standard set of wavefunctions, termed a 'basis', with the important difference that in this case the number of wavefunctions needed to form a basis is infinite. However, in essentially every case of physical interest the transformations induced by a group of symmetry operations such as rotations operate not on the complete space of all possible wavefunctions but rather on finite-dimensional subspaces, and in these subspaces they constitute a finite-dimensional representation of the group.

The example we have chosen above illustrates this point precisely. The energy eigenfunctions $u(x)$ of the hydrogen atom are labelled by three integers n, l and m. Of these, n determines the energy eigenvalue according to $E_n = E_1/n^2$, l labels the eigenvalue of the square of the orbital angular momentum, $L^2 = l(l+1)\hbar^2$, while m gives that of the z component of the orbital angular momentum, $L_z = m\hbar$. How are these quantum numbers changed when we perform a rotation?

(*a*) Well, the Hamiltonian, i.e. the energy operator, is rotationally invariant. In the context of the Schrödinger equation it is

$$\hat{H} = \hat{p}^2/2m + V(r) \tag{3.11}$$

with $\hat{\boldsymbol{p}} = -i\hbar\boldsymbol{\nabla}$. The kinetic term involves the square of the vector operator $\hat{\boldsymbol{p}}$, and the potential is a 'central' potential, depending only on $r = |\boldsymbol{r}|$. Thus the principal quantum number n is unchanged by a rotation.

(*b*) Again $\hat{\boldsymbol{L}}^2$ is the square of the vector operator $\hat{\boldsymbol{L}} = \boldsymbol{r} \times \hat{\boldsymbol{p}}$. So l is also unchanged by a rotation.

(*c*) In fact the only quantum number which is changed is m. This makes perfect sense, since it was only m which made reference to a particular direction in space, the z axis, and this direction will be changed under a rotation.

Altogether, then, the transformed wavefunction $u'_{nlm}(\boldsymbol{x})$ can be expressed as a linear superposition of old wavefunctions with the same values of n and l but differing values of m:

$$u'_{nlm}(\boldsymbol{x}) = \sum_{m'} D^l_{m'm}(R) u_{nlm'}(\boldsymbol{x}) \tag{3.12}$$

Here the indices m and m' are restricted to lie between $-l$ and l. So the matrix $D^l_{m'm}(R)$ is a square matrix of dimension $(2l+1) \times (2l+1)$. It does not depend on n because of the factorization of the wavefunction into radial and angular parts: $u_{nlm}(\boldsymbol{x}) = R_n(r) Y_{lm}(\theta, \varphi)$. Such matrices are termed *rotation matrices*, for obvious reasons. By considering two successive rotations it is easy to see that they satisfy the important property

$$D(R_1 R_2) = D(R_1) D(R_2) \tag{3.13}$$

and therefore constitute a *representation*, of dimension $2l + 1$, of the group of 3-dimensional rotations.

Examples

(i) The case we considered above, a 2p electron with an angular wavefunction proportional to $\cos\theta$, has quantum numbers $n = 2$, $l = 1$, $m = 0$. So the transformed wavefunction will be in general a superposition of terms with $m' = -1, 0, 1$. The correctly normalized angular wavefunctions are in fact

$$\begin{aligned} Y_{11} &= -(3/8\pi)^{1/2} \sin\theta \, \mathrm{e}^{\mathrm{i}\varphi} \\ Y_{10} &= (3/4\pi)^{1/2} \cos\theta \\ Y_{1,-1} &= (3/8\pi)^{1/2} \sin\theta \, \mathrm{e}^{-\mathrm{i}\varphi} \end{aligned} \tag{3.14}$$

Consider a rotation $R_2(\beta)$ through the angle β about the y axis. Under such a rotation y is unchanged, while z and x are transformed into

$$z' = z\cos\beta - x\sin\beta$$

$$x' = x\cos\beta + z\sin\beta$$

in direct analogy to equations (3.2) and (3.3). The point $R_2^{-1}\boldsymbol{x}$ required in equation (3.10) therefore has z coordinate

$$\begin{aligned} (R_2^{-1}\boldsymbol{x})_z &= z\cos\beta + x\sin\beta \\ &= r(\cos\beta\cos\theta + \sin\beta\sin\theta\cos\varphi) \end{aligned} \tag{3.15}$$

when expressed in spherical polar coordinates. In the new wavefunction $\cos\theta$ is therefore replaced by the combination $\cos\beta\cos\theta + \sin\beta\sin\theta\cos\varphi$, which can be rewritten in terms of the spherical harmonics $Y_{1m'}$ by noting that $Y_{1,-1} + Y_{11} = 2(3/8\pi)^{1/2}\sin\theta\cos\varphi$. Thus equation (3.12) reads explicitly

$$u'_{210}(\boldsymbol{x}) = \cos\beta\, u_{210} + (1/\sqrt{2})\sin\beta\,(u_{21,-1} - u_{211}) \tag{3.16}$$

In other words we have identified the matrix elements $D^1_{m'0}(R_2(\beta))$ as

$$D^1_{00} = \cos\beta \qquad D^1_{-1,0} = -D^1_{10} = (1/\sqrt{2})\sin\beta \tag{3.17}$$

This representation is actually nothing more than the defining representation in disguise: it acts on the spherical harmonics Y_{1m}, which are just simple linear combinations of x/r, y/r and z/r. However, if we consider any value of l other than 1, we will encounter genuinely different representations.

(ii) One such example, rather a simple one, is to consider the transformation of the ground-state wavefunction u_{100}. In fact this wavefunction is spherically symmetric ($Y_{00} = (1/4\pi)^{1/2}$) and so does not transform at all. Here we have an example of the trivial representation, in which all rotations R are mapped onto the number 1.

(iii) Going up in dimensions, the wavefunctions for the $n = 2$ states can have $l = 0$, 1 or 2. In the last case the transformation for a general rotation is rather complicated, but it is easy to write down if we restrict our attention to a rotation about the z axis, through an angle α, say. The point is that, just as in (3.14), the angular wavefunctions $Y_{2m}(\theta, \varphi)$ have the simple φ dependence $e^{im\varphi}$. So we have immediately, taking account of the inverse in equation (3.10),

$$u'_{22m}(\boldsymbol{x}) = e^{-im\alpha}u_{22m} \tag{3.18}$$

There is in fact nothing special about $l = 2$. For a general value of l we can identify the matrix elements of $D^l_{m'm}(R_3(\alpha))$ as

$$D^l_{m'm}(R_3(\alpha)) = \delta_{m'm}e^{-im\alpha} \tag{3.19}$$

In the above we have considered for definiteness just one particular quantum system, the hydrogen atom, and just one particular kind of transformations, those induced by 3-dimensional rotations. It should be clear that this is just a single example among many. For a general symmetry group of transformations acting on a general quantum mechanical system we will need to know how to:

(1) classify and enumerate the possible representations,
(2) combine representations (the generalization of 'addition of angular momentum'),

(3) relate the representations of a subgroup to those of the whole group (important for the breaking of degeneracy).

To achieve these ends we now go on to develop the theory of representations in a rather formal way. Once this has been accomplished we will apply the theory to a variety of physical systems in Chapter 5.

3.3 Equivalence of Representations; Characters; Reducibility

Definition

Two $[n]$ representations $D^{(1)}$ and $D^{(2)}$ of a group G are *equivalent* if all the matrices $D^{(1)}(g)$ and $D^{(2)}(g)$ are related by the same similarity transformation:

$$D^{(1)}(g) = SD^{(2)}(g)S^{-1} \tag{3.20}$$

$\forall g \in G, \qquad S$ independent of g.

Note that this is consistent with the group property of the two representations: $D^{(i)}(gg') = D^{(i)}(g)D^{(i)}(g')$. Thus

$$\begin{aligned} SD^{(2)}(gg')S^{-1} &= SD^{(2)}(g)D^{(2)}(g')S^{-1} \\ &= (SD^{(2)}(g)S^{-1})(SD^{(2)}(g')S^{-1}) \\ &= D^{(1)}(g)D^{(1)}(g') \\ &= D^{(1)}(gg'), \text{ as required} \end{aligned}$$

In the classification of representations, equivalent representations are regarded as being essentially the same, and we only count the distinct equivalence classes. (Equation (3.20) indeed defines an equivalence relation satisfying the axioms (2.1)–(2.3).) The point is, as we shall see later, that they are the matrices of the same linear transformations of the $[n]$ vector space, but referred to different bases. (Recall the interpretation of conjugacy in the context of rotations (§2.1).)

Let us note immediately that if we want some way of distinguishing between representations which treats equivalent representations as the same, we are led naturally to the idea of the *character*.

Definition

The *character* of a representation D of a group G is the set $\chi = \{\chi(g) | g \in G\}$, where $\chi(g)$ is the trace of the representation matrix $D(g)$:

$$\chi(g) = \mathrm{Tr}(D(g)) \tag{3.21}$$

Recall that the trace of a matrix is the sum of its diagonal elements:

$$\mathrm{Tr}(A) = \sum_i A_{ii}$$

The fact that χ does not distinguish between equivalent representations, i.e. is a function of equivalence classes, follows from the cyclic property of the trace. That is, for any matrices A, B, C,

$$\begin{aligned}\mathrm{Tr}(ABC) &= \sum_{ijk} A_{ij}B_{jk}C_{ki} \\ &= \sum_{ijk} B_{jk}C_{ki}A_{ij} \\ &= \mathrm{Tr}(BCA)\end{aligned} \tag{3.22}$$

So in particular

$$\mathrm{Tr}(SD(g)S^{-1}) = \mathrm{Tr}(D(g)S^{-1}S) = \mathrm{Tr}(D(g))$$

Hence if $D^{(1)}$ and $D^{(2)}$ are equivalent representations, $\{\chi^{(1)}(g)\} = \{\chi^{(2)}(g)\}$. One of the principal results of representation theory is in fact that the argument works in reverse; if two representations have the same character they are necessarily equivalent.

Reducibility

Example: C_3
In the case of C_3 we had matrices of the form

$$R(c) = \begin{pmatrix} \cdot & \cdot & 0 \\ \cdot & \cdot & 0 \\ 0 & 0 & 1 \end{pmatrix} \quad \text{etc.}$$

This 'block diagonal' form comes about because z is invariant under the transformations of C_3; only x and y are changed. That is, the vector space $\boldsymbol{x} = x\boldsymbol{i} + y\boldsymbol{j} + z\boldsymbol{k}$ splits up into two decoupled spaces according to

$$\boldsymbol{x} = \boldsymbol{u} + \boldsymbol{v}$$

where $\boldsymbol{u} = x\boldsymbol{i} + y\boldsymbol{j}$ and $\boldsymbol{v} = z\boldsymbol{k}$.

Thus the above representation effectively 'decomposes' into two separate representations, a 2-dimensional representation $D^{(2)}$ acting on $\boldsymbol{u}$ and the trivial representation $D^{(1)}(g) = 1$ acting on $\boldsymbol{v}$, which are completely independent:

$$RR' = \begin{pmatrix} A & O \\ O & 1 \end{pmatrix}\begin{pmatrix} A' & O \\ O & 1 \end{pmatrix} = \begin{pmatrix} AA' & O \\ O & 1 \end{pmatrix}$$

The special symbol $\oplus$ is used to denote such a block diagonal decomposition: we write

$$R(c) = D^{(1)}(c) \oplus D^{(2)}(c)$$

Let us now go back to the general case and give a precise definition of the terms used above.

Definition

A representation of dimension $n + m$ is said to be *reducible* if $D(g)$ takes the form

$$D(g) = \begin{pmatrix} A(g) & C(g) \\ O & B(g) \end{pmatrix} \tag{3.23}$$

$\forall g \in G$, where A, C and B are submatrices of dimension $m \times m$, $m \times n$ and $n \times n$ respectively, and O denotes a null matrix of dimension $n \times m$.

Multiplying two such matrices together we see that

$$D(g)D(g') \quad (= D(gg')) = \begin{pmatrix} A(g) & C(g) \\ O & B(g) \end{pmatrix}\begin{pmatrix} A(g') & C(g') \\ O & B(g') \end{pmatrix}$$

$$= \begin{pmatrix} A(g)A(g') & A(g)C(g') + C(g)B(g') \\ O & B(g)B(g') \end{pmatrix}$$

Thus

$$\begin{aligned} A(gg') &= A(g)A(g') \\ B(gg') &= B(g)B(g') \end{aligned} \tag{3.24}$$

(and also

$$C(gg') = A(g)C(g') + C(g)B(g'))$$

so that $\{A(g)\}$ and $\{B(g)\}$ constitute $[m]$ and $[n]$ representations of G respectively.

For finite groups it can be shown (see Maschke's theorem, below) that under equivalence C can be taken to be a null matrix, as it was in the case of C_3. The representation is then said to be *completely reducible* or *decomposable*, and we write

$$D(g) = A(g) \oplus B(g)$$

It may be that the representations A and B are themselves decomposable, in which case it is natural to continue the process, which will terminate when we reach the level of *irreducible representations*, namely representations which cannot be further reduced. While there is no limit to the number and dimensions of reducible representations, it turns out that the irreducible representations can be classified (by characters) and enumerated. A large part of the subsequent analysis will therefore be focused on these fundamental building blocks.

3.4 Groups Acting on Vector Spaces

In the above we have talked about representations as groups of matrices. For some purposes it is useful to become a little more abstract and talk about groups of linear transformations of vector spaces. These are realized as matrices only when a particular choice of basis is made, but this choice is far from unique. As we have already mentioned, the different matrix representations so obtained will be equivalent in the sense of equation (3.20). Thus by working at the 'intrinsic' or coordinate-free level we automatically include complete equivalence classes of matrix representations. The concept of reducibility also arises in a very natural way within this framework.

Axioms of a Vector Space

A *vector space* V (over the field of complex numbers) is a set of elements $\{\boldsymbol{v}\}$ with two operations defined.

The first is addition, '+', satisfying:

A0 (closure)

$$\boldsymbol{u} + \boldsymbol{v} \in V \qquad \forall \boldsymbol{u}, \boldsymbol{v} \in V \tag{3.25}$$

A1 (associativity)

$$\boldsymbol{u} + (\boldsymbol{v} + \boldsymbol{w}) = (\boldsymbol{u} + \boldsymbol{v}) + \boldsymbol{w} \tag{3.26}$$

A2 (existence of identity (null vector))

V contains the null element, denoted by $\mathbf{0}$, such that

$$\boldsymbol{u} + \mathbf{0} = \boldsymbol{u} \tag{3.27}$$

A3 (existence of inverse)

$\forall \boldsymbol{u} \in V, \exists$ an element, denoted by $-\boldsymbol{u}$, in V, such that

$$\boldsymbol{u} + (-\boldsymbol{u}) = \boldsymbol{0} \tag{3.28}$$

A4 (commutativity)

$$\boldsymbol{u} + \boldsymbol{v} = \boldsymbol{v} + \boldsymbol{u} \tag{3.29}$$

The above axioms can be summarized succinctly in the statement that V forms an Abelian group under the law of composition specified by '+'.

The other operation is that of multiplication by scalars (complex numbers), satisfying:

B0

$$a\boldsymbol{u} \in V \qquad \forall a \in \mathbb{C}, \boldsymbol{u} \in V \tag{3.30}$$

B1

$$a(\boldsymbol{u} + \boldsymbol{v}) = a\boldsymbol{u} + a\boldsymbol{v} \tag{3.31}$$

B2

$$(a + b)\boldsymbol{u} = a\boldsymbol{u} + b\boldsymbol{u} \tag{3.32}$$

B3

$$a(b\boldsymbol{u}) = (ab)\boldsymbol{u} \tag{3.33}$$

B4

$$1 \cdot \boldsymbol{u} = \boldsymbol{u} \tag{3.34}$$

These are all apparently innocuous, and even 'obvious', requirements, but they need to be checked in each case against the definition given of multiplication by scalars. Some further 'obvious' results follow as a direct consequence of B1–B4. For example,

$$(1 + 0)\boldsymbol{u} = \boldsymbol{u} \Rightarrow 0 \cdot \boldsymbol{u} = \boldsymbol{u} + (-\boldsymbol{u}) = \boldsymbol{0}$$

and then

$$(1 + (-1))\boldsymbol{u} = 0 \cdot \boldsymbol{u} = \boldsymbol{0} \Rightarrow (-1)\boldsymbol{u} = -\boldsymbol{u}$$

In the above, with a view to quantum mechanics, we have taken the scalars to be complex numbers, but a vector space can be defined for scalars belonging to any 'field', a mathematical construct of which both the real numbers and the complex numbers are examples.

Examples

1. Ordinary [3] vectors

Here the elements $\boldsymbol{u}$ are, say, position vectors, directed lengths in ordinary 3-dimensional space. Without reference to any axes we can define addition by the parallelogram law (figure 3.4). The null vector is a vector of zero length, and the inverse of a vector is a vector of the same length but in the opposite direction. Associativity and commutativity are indeed a property of vector addition. Multiplication by real scalars just means multiplying the length by that factor, and satisfies B1–B4, as you should check.

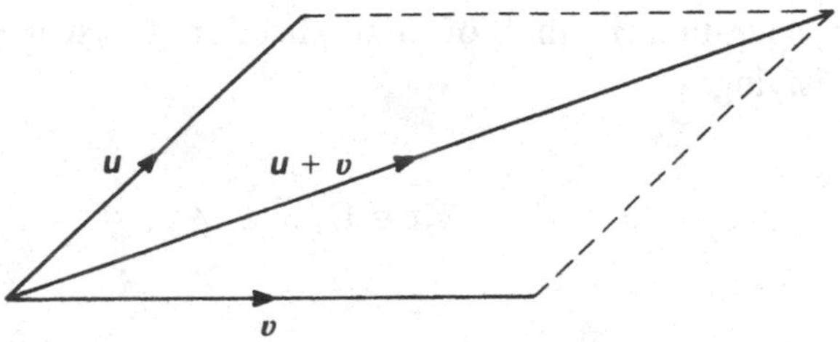

Figure 3.4 Addition of two vectors $\boldsymbol{u}$ and $\boldsymbol{v}$ by the parallelogram law.

An alternative way of describing 3-vectors is by their components, namely their projections on the standard vectors $\boldsymbol{i}, \boldsymbol{j}, \boldsymbol{k}$. That is, any vector $\boldsymbol{u}$ can be expressed uniquely as

$$\boldsymbol{u} = u_1\boldsymbol{i} + u_2\boldsymbol{j} + u_3\boldsymbol{k}$$

so that there is a 1:1 correspondence between the vector $\boldsymbol{u}$ and the triple of coordinates (u_1, u_2, u_3). Addition of two such triples is defined as

$$(u_1, u_2, u_3) + (v_1, v_2, v_3) = (u_1 + v_1, u_2 + v_2, u_3 + v_3) \quad (3.35)$$

i.e. addition of the individual components, which follows as a consequence of the previous definitions. Again, multiplication by scalars amounts to multiplication of the individual components:

$$a(u_1, u_2, u_3) = (au_1, au_2, au_3) \quad (3.36)$$

In this formulation there is nothing special about the three dimensions of space, and one can define a vector space of n-tuples in a precisely similar way.

2. Function space

Consider, for example, the space of complex functions f of the real variable x on the interval $0 \leqslant x \leqslant 1$, with $f(0) = f(1) = 0$. Addition of two functions is simply defined as ordinary addition of their values:

$$(f + g)(x) = f(x) + g(x) \tag{3.37}$$

and similarly for multiplication by (complex) scalars:

$$(af)(x) = af(x) \tag{3.38}$$

Such functions arise in the context of quantum mechanics as the wavefunctions appropriate to the infinite potential well $V(x) = 0$ in [0, 1], infinite outside this range. The energy eigenfunctions are of the form $u_n(x) = \sin(n\pi x)$, and under certain conditions any function f of the abovc type can be written as a superposition of these:

$$f(x) = \sum_{n=1}^{\infty} f_n u_n(x) \tag{3.39}$$

i.e. as a Fourier series. There is a 1:1 correspondence between the function f and its Fourier components $\{f_n\}$. So addition of functions and multiplication by scalars can be defined in terms of these components, just as we did for 3-vectors, with, however, the major difference that the number of components is in general infinite.

Returning now to the general case, let us give the formal definition of the dimension of a vector space. This involves the prior concepts of 'linear independence' and a 'basis'.

Definition
A set of vectors $\{\boldsymbol{e}_i\}$, $i = 1, \ldots, m$, is *linearly independent* if there is no non-trivial combination which yields the null vector:

$$\sum_{i=1}^{m} \lambda_i \boldsymbol{e}_i = \boldsymbol{0} \Rightarrow \lambda_i = 0 \qquad \forall i \tag{3.40}$$

For example, the unit vectors $\boldsymbol{i}$, $\boldsymbol{j}$ and $\boldsymbol{k}$ in ordinary [3] space, or equivalently the triples (1, 0, 0), (0, 1, 0) and (0, 0, 1), are linearly independent, whereas (−2, 1, 1), (1, −2, 1) and (1, 1, −2) are not.

Definition
A linearly independent set of vectors $\{\boldsymbol{e}_i\}$, $i = 1, \ldots, m$, forms a *basis* of a vector space V if they span the space, i.e. if any vector $\boldsymbol{u} \in V$ can be expressed as a linear superposition of the $\boldsymbol{e}_i$:

$$\boldsymbol{u} = \sum_{i=1}^{m} u_i \boldsymbol{e}_i \tag{3.41}$$

(From now on we will adopt the *summation convention* that repeated indices, such as the i above, are implicitly taken to be summed. In this convention the above equation would simply be written as $u = u_i e_i$.)

We have already seen two examples: the unit vectors $\boldsymbol{i}, \boldsymbol{j}, \boldsymbol{k}$ as a basis for ordinary [3] vectors, and the functions $u_n(x)$ as a basis for the functions f.

Definition
A vector space V is *n-dimensional* if it has a basis of n vectors.

Although it seems intuitively obvious, it is not completely trivial to prove that any two bases must have the same number of elements (see problem 3.5).

This definition conforms with our informal idea of dimensionality for finite-dimensional spaces: an *infinite-dimensional* vector space, such as a function space, is one in which there is no limit to the number of linearly independent vectors which can be found.

Linear Transformations

Definition
A map T: $V \to V$ is *linear* if

$$T(\alpha \boldsymbol{u} + \beta \boldsymbol{v}) = \alpha(T\boldsymbol{u}) + \beta(T\boldsymbol{v}) \tag{3.42}$$

for all $\boldsymbol{u}, \boldsymbol{v} \in V$; $\alpha, \beta \in \mathbb{C}$. That is, it is linear if the transform of any linear combination of vectors is equal to the same combination of the transforms.

A rather simple example is multiplication by a fixed scalar k, which satisfies (3.42) by virtue of B1 and B3. Less trivially, differentiation is a linear transformation in the vector space of differentiable functions.

Given a basis $\{\boldsymbol{e}_i\}$ of V, the map T has a concrete realization as a matrix D_{ij}. This is defined by giving the transforms of each of the basis vectors $\boldsymbol{e}_j$. Because the $\{\boldsymbol{e}_i\}$ form a basis, each such transform is a linear combination of the basis vectors, which we write in the form

$$T\boldsymbol{e}_j = D_{ij}\boldsymbol{e}_i \tag{3.43}$$

As already stated, we are now using the summation convention whereby a summation over the repeated index i is understood. The apparently 'wrong' order of the indices i, j is chosen so as to correspond to matrix multiplication of the coordinates of a general vector (equation (3.44) below).

Indeed, knowing the transforms of all the basis vectors we are in a position to write down the transform of a general vector $\boldsymbol{u} = u_j \boldsymbol{e}_j$, namely

$$\begin{aligned} T\boldsymbol{u} = T(u_j \boldsymbol{e}_j) &= u_j(T\boldsymbol{e}_j) \qquad \text{by linearity} \\ &= u_j(D_{ij}\boldsymbol{e}_i) \qquad \text{from (3.43)} \end{aligned}$$

So, writing $T\boldsymbol{u} = \boldsymbol{v} = v_i \boldsymbol{e}_i$ we can identify

$$v_i = D_{ij} u_j \tag{3.44}$$

Thus, thinking of the coordinates as a column vector, the transformation induced by T amounts to matrix multiplication by D: $v = Du$.

Similarity

The matrix D was derived from the linear transformation T by reference to a particular basis $\{\boldsymbol{e}_i\}$, which is far from unique. Referred to a different basis $\{\boldsymbol{f}_i\}$, say, we will obtain a different matrix D'. In analogy with equation (3.43) this will be defined by

$$T\boldsymbol{f}_j = D'_{ij}\boldsymbol{f}_i$$

and the new coordinates of $\boldsymbol{u}$ defined by writing $\boldsymbol{u} = u'_j \boldsymbol{f}_j$, will be transformed to

$$v'_i = D'_{ij} u'_j \tag{3.45}$$

or, in matrix form, $v' = D'u'$.

On the other hand we can relate u' to u and v' to v by expressing each old basis vector $\boldsymbol{e}_i$ as some linear superposition of the new basis vectors $\{\boldsymbol{f}_j\}$:

$$\boldsymbol{e}_i = S_{ji}\boldsymbol{f}_j \tag{3.46}$$

Thus

$$\boldsymbol{u} = u_i \boldsymbol{e}_i = u_i S_{ji} \boldsymbol{f}_j$$

so that

$$u'_j = u_i S_{ji} \tag{3.47}$$

or, in matrix form, $u' = Su$. Similarly $v' = Sv$. Altogether, then,

$$v' = Sv = S(Du) = SD(S^{-1}u')$$

Comparing with the matrix form of (3.45) we see that

$$D' = SDS^{-1} \tag{3.48}$$

substantiating the previously unsupported assertion that matrices representing the same linear transformation but referred to different bases are related by a similarity transformation. Moreover, we now see that the conjugating matrix S is the matrix which relates the two bases to each other according to equation (3.46). The existence of S^{-1} is assured by the invertibility of this relation. That is, we must be able to express the f_i in terms of the $\{e_j\}$.

G-Module

As stated at the beginning of this section, it is extremely useful to think of a representation of a group G as a homomorphism mapping G into the group of non-singular linear transformations of a vector space V. That is, every element g of G is mapped into a linear transformation $T(g)$ with the all-important property that

$$T(g'g) = T(g')T(g) \tag{3.49}$$

A vector space in which a group G acts in this way is called a *G-module*. As we have just seen, the linear operators can be realized as matrices in different ways depending on the basis chosen. In fact one set of linear operators $\{T(g)\}$ corresponds to an entire equivalence class of matrix representations, since, according to (3.48), $D'(g) = SD(g)S^{-1}$. Note that S, being the matrix relating the two different bases, is indeed independent of g, as required in the definition of equivalent representations.

Reducibility

In this language reducibility corresponds to the existence of a *submodule*, i.e. a subspace U of the vector space V which itself is closed under the action of the group:

$$\boldsymbol{u} \in U \subset V \Rightarrow T(g)\boldsymbol{u} \in U \tag{3.50}$$

In such a situation let $\{\boldsymbol{e}_i\}$, $i = 1, \ldots, m$, be a basis of U. We can extend this by vectors $\{\boldsymbol{e}_i\}$, $i = m + 1, \ldots, m + n$ (spanning a subspace W, say), to form a basis of V, which we take to have dimension $m + n$.

Relative to this basis the matrix $D(g)$ corresponding to the linear transformation $T(g)$ is given by

$$T(g)\boldsymbol{e}_j = D_{ij}(g)\boldsymbol{e}_i$$

The closure of U, namely the fact that the transform of a basis vector of

U has no component in W, is reflected in the statement that $D_{ij}(g) = 0$ if $j = 1, \ldots, m$ and at the same time $i = m + 1, \ldots, m + n$. That is, $D(g)$ takes the form

$$D(g) = \begin{array}{c} \\ m \\ n \end{array} \begin{array}{c} \begin{array}{cc} m & n \end{array} \\ \begin{pmatrix} A(g) & C(g) \\ 0 & B(g) \end{pmatrix} \end{array}$$

precisely as in (3.23).

For a finite group (also for a 'compact' group—see later) it turns out that $C(g)$ can be set to zero. The easiest way of showing this is to prove that all representations of a finite group are equivalent to *unitary* representations. To that end, and to define what is meant by 'unitary', we now need to add further structure by introducing the idea of a scalar (inner) product on a vector space: this is of course part and parcel of the framework of quantum mechanics.

3.5 Scalar Product; Unitary Representations; Maschke's Theorem

Scalar Product

The formal definition of a scalar product, abstracted from the familiar dot product of 3-vectors, is that it is a map $V \times V \to \mathbb{C}$ which assigns to each ordered pair $\boldsymbol{u}, \boldsymbol{v} \in V$ a *complex number* $(\boldsymbol{u}, \boldsymbol{v})$ with the following properties:

S1 (hermiticity)

$$(\boldsymbol{u}, \boldsymbol{v}) = (\boldsymbol{v}, \boldsymbol{u})^* \tag{3.51}$$

S2 (linearity)

$$(\boldsymbol{w}, \alpha\boldsymbol{u} + \beta\boldsymbol{v}) = \alpha(\boldsymbol{w}, \boldsymbol{u}) + \beta(\boldsymbol{w}, \boldsymbol{v}) \tag{3.52}$$

S3 (positivity)

$$(\boldsymbol{u}, \boldsymbol{u}) \geqslant 0 \tag{3.53}$$

with equality iff $\boldsymbol{u} = \boldsymbol{0}$.

(Note that (3.51) ensures that $(\boldsymbol{u}, \boldsymbol{u})$ is real, so that (3.53) makes sense.)

Examples are the usual scalar product $\boldsymbol{u} \cdot \boldsymbol{v}$ in [3] space, which in fact is real and therefore symmetric, and the overlap integral of wave mechanics: $(\psi, \varphi) := \int \psi^*(x)\varphi(x)\,d^3x$.

Given such a scalar product, we can define the *norm* of a vector $\boldsymbol{u}$ as $\|\boldsymbol{u}\| := (\boldsymbol{u}, \boldsymbol{u})^{1/2}$. A vector $\boldsymbol{u}$ is said to be *normalized* if it has unit norm. Borrowing from the language of ordinary [3] space, two vectors $\boldsymbol{u}$, $\boldsymbol{v}$ are said to be *orthogonal* if $(\boldsymbol{u}, \boldsymbol{v}) = 0$.

Orthonormal Basis

Given any basis $\{f_i\}$ of V it is possible to construct an *orthonormal* basis $\{\boldsymbol{e}_i\}$ satisfying

$$(\boldsymbol{e}_i, \boldsymbol{e}_j) = \delta_{ij} \tag{3.54}$$

by the 'Gram–Schmidt' procedure.

This goes as follows. We first construct $\boldsymbol{e}_1$ simply by normalizing $\boldsymbol{f}_1$:

$$\boldsymbol{e}_1 = \boldsymbol{f}_1/\|\boldsymbol{f}_1\|$$

Then we construct $\boldsymbol{e}_2$ from $\boldsymbol{f}_2$ by subtracting its 'component' along $\boldsymbol{e}_1$ to make it orthogonal to $\boldsymbol{e}_1$, and then normalizing:

$$\boldsymbol{e}_2 = (\boldsymbol{f}_2 - (\boldsymbol{e}_1, \boldsymbol{f}_2)\boldsymbol{e}_1)/\|\ldots\|$$

(You can check algebraically that this indeed satisfies $(\boldsymbol{e}_2, \boldsymbol{e}_1) = 0$, but it is also instructive to see how it works geometrically by drawing the vector diagram for the case when $\boldsymbol{e}_1$ and $\boldsymbol{f}_2$ are 3-vectors.) The next vector $\boldsymbol{e}_3$ is constructed from $\boldsymbol{f}_3$ by subtracting off its components along both $\boldsymbol{e}_1$ and $\boldsymbol{e}_2$, and then normalizing:

$$\boldsymbol{e}_3 = (\boldsymbol{f}_3 - (\boldsymbol{e}_1, \boldsymbol{f}_3)\boldsymbol{e}_1 - (\boldsymbol{e}_2, \boldsymbol{f}_3)\boldsymbol{e}_2)/\|\ldots\| \qquad \text{etc.}$$

If we have constructed such an orthonormal basis, the 'coordinates' u_i of a general vector $\boldsymbol{u}$, namely the coefficients which appear in the expansion $\boldsymbol{u} = u_i\boldsymbol{e}_i$, can be identified with the scalar products $(\boldsymbol{e}_i, \boldsymbol{u})$. Thus, taking the scalar product $(\boldsymbol{e}_j, \quad)$ with both sides of the equation $\boldsymbol{u} = (\Sigma_i)u_i\boldsymbol{e}_i$, the RHS is $(\Sigma_i)u_i\delta_{ij} = u_j$. By virtue of linearity and hermiticity, the scalar product of two arbitrary vectors $\boldsymbol{u}$, $\boldsymbol{v}$ can be expressed in terms of components as $(\boldsymbol{u}, \boldsymbol{v}) = (u_i\boldsymbol{e}_i, v_j\boldsymbol{e}_j) = u_i^*v_j(\boldsymbol{e}_i, \boldsymbol{e}_j) = u_i^*v_j\delta_{ij} = u_i^*v_i$, as in the familiar case of 3-vectors, where $\boldsymbol{a}\cdot\boldsymbol{b} = a_1b_1 + a_2b_2 + a_3b_3$.

Likewise, the matrix elements D_{ij} of a linear mapping T, defined as in equation (3.43) by $T\boldsymbol{e}_j = D_{ij}\boldsymbol{e}_i$, are given by $D_{ij} = (\boldsymbol{e}_i, T\boldsymbol{e}_j)$, while $(\boldsymbol{u}, T\boldsymbol{v})$ can be expressed as $(\boldsymbol{u}, T\boldsymbol{v}) = u_i^*D_{ij}v_j$, or $u^\dagger Dv$ in matrix notation.

Unitary Transformations

A linear transformation T acting in V is said to be *unitary* if

$$(T\boldsymbol{u}, T\boldsymbol{v}) = (\boldsymbol{u}, \boldsymbol{v}) \quad \forall \boldsymbol{u}, \boldsymbol{v} \in V \tag{3.55}$$

In the context of representations, the linear transformations $T(g)$ are all invertible. When T has an inverse, equation (3.55) is equivalent to

$$(T\boldsymbol{u}, \boldsymbol{v}) = (\boldsymbol{u}, T^{-1}\boldsymbol{v}) \tag{3.55$'$}$$

The corresponding relation for the matrix D_{ij} relative to an orthonormal basis $\{\boldsymbol{e}_i\}$ is obtained by taking $\boldsymbol{u}$ to be $\boldsymbol{e}_j$ and $\boldsymbol{v}$ to be $\boldsymbol{e}_i$, giving

$$D^*_{ij} = (D^{-1})_{ji} \tag{3.56}$$

or, in other words, $D^\dagger := (D^*)^{\mathrm{T}} = D^{-1}$, which defines a unitary matrix.

For completeness we note that a *Hermitian* linear transformation is one which satisfies instead

$$(T\boldsymbol{u}, \boldsymbol{v}) = (\boldsymbol{u}, T\boldsymbol{v})$$

Correspondingly its matrix referred to an orthonormal basis has the property that $D^\dagger = D$, the definition of a Hermitian matrix.

Complete Reducibility

We saw above that in coordinate-free language reducibility corresponded to the existence of a submodule U, closed under the action of the group: $\boldsymbol{u} \in u \Rightarrow T(g)\boldsymbol{u} \in U$. The subspace U had a basis $\{\boldsymbol{e}_i\}$, $i = 1, \ldots, m$, which we extended by vectors $\{\boldsymbol{e}_i\}$, $i = m + 1, \ldots, m + n$, to form a basis of the whole space V. Now that we have a scalar product, we can certainly choose this basis to be orthonormal, which means in particular that any vector $\boldsymbol{w}$ in the space W spanned by the additional vectors $\{\boldsymbol{e}_i\}$, $i = m + 1, \ldots, m + n$, is orthogonal to any $\boldsymbol{u} \in U$. That is, W is the *orthogonal complement* of U:

$$W = \{\boldsymbol{w} \in V | (\boldsymbol{w}, \boldsymbol{u}) = 0 \quad \forall \boldsymbol{u} \in U\} \tag{3.57}$$

In coordinate-free language, complete reducibility corresponds to the closure of W as well as U under the action of the group. This will always be the case if the $T(g)$ are unitary with respect to the scalar product we are using. For then, from (3.55′),

$$(T(g)\boldsymbol{w}, \boldsymbol{u}) = (\boldsymbol{w}, T^{-1}(g)\boldsymbol{u})$$

But the closure of U ensures that $T^{-1}(g)\boldsymbol{u} = T(g^{-1})\boldsymbol{u} = \boldsymbol{u}' \in U$. Hence

$$(T(g)\boldsymbol{w}, \boldsymbol{u}) = (\boldsymbol{w}, \boldsymbol{u}') = 0$$

showing that $T(g)w = w' \in W$, as required.

The closure of W is reflected at the matrix level in the vanishing of those matrix elements $D_{ij}(g)$ for which $j = m + 1, \ldots, m + n$ and at the same time $i = 1, \ldots, m$. That is, the submatrix $C(g)$ of equation (3.23) is the null matrix.

Maschke's Theorem

Maschke's theorem states that all reducible representations of a finite group are completely reducible (decomposable).

We have just shown that reducible *unitary* representations are decomposable, so all that remains to be proved is that for a finite (or indeed a compact) group it is always possible to choose a scalar product which makes the representation unitary.

This 'group-invariant' scalar product, which we denote by $\{\ ,\ \}$, is constructed from the original scalar product $(\ ,\ )$ in the following way:

$$\{\boldsymbol{v}, \boldsymbol{v}'\} := \frac{1}{[g]} \sum_{g \in G} (T(g)\boldsymbol{v}, T(g)\boldsymbol{v}') \tag{3.58}$$

where we are now using $\boldsymbol{v}$, $\boldsymbol{v}'$ to denote generic members of V. Now consider $\{T(h)\boldsymbol{v}, T(h)\boldsymbol{v}'\}$, where h is a group element $h \in G$. This is

$$\begin{aligned}\{T(h)\boldsymbol{v}, T(h)\boldsymbol{v}'\} &= \frac{1}{[g]} \sum_{g} (T(g)T(h)\boldsymbol{v}, T(g)T(h)\boldsymbol{v}') \\ &= \frac{1}{[g]} \sum_{g} (T(gh)\boldsymbol{v}, T(gh)\boldsymbol{v}')\end{aligned}$$

Writing $gh = g'$ and noting that $\Sigma_g = \Sigma_{g'}$ we have

$$\begin{aligned}\{T(h)\boldsymbol{v}, T(h)\boldsymbol{v}'\} &= \frac{1}{[g]} \sum_{g'} (T(g')\boldsymbol{v}, T(g')\boldsymbol{v}') \\ &= \{\boldsymbol{v}, \boldsymbol{v}'\}\end{aligned}$$

Thus by comparison with equation (3.55) we see that $T(h)$ is indeed unitary with respect to the scalar product $\{\ ,\ \}$.

(Note that the existence of the invariant sum over the group Σ_g was crucial to this construction. A compact group is a continuous group with an infinite number of elements for which nonetheless Σ_g can be replaced by a convergent group-invariant integration $\int d\mu(g)$.)

As far as matrix representations are concerned, $T(h)$ will be realized as a reducible unitary matrix $D'(h)$, say, if we choose a basis $\{f_i\}$, orthonormal with respect to the group-invariant scalar product, whose first m members span U. $D'(h)$ will be related to the original matrix

$D(h)$ by the similarity transformation $D'(h) = SD(h)S^{-1}$, where S is the matrix effecting the change of basis from the $\{f_i\}$ to the $\{e_i\}$. In other words, the representation D is equivalent to the reducible unitary representation D', which in fact is completely reducible.

Problems for Chapter 3

3.1 Show that the [3] representation of the group of rotations about the z axis is completely reducible when referred to $-(x + iy)/\sqrt{2}$, z, $(x - iy)/\sqrt{2}$ as basis functions. Calculate the character of the representation, showing that it is the same whether referred to the above spherical basis or to the Cartesian basis x, y, z.

3.2 Construct the [3] representation of D_3 in the basis x, y, z. Calculate the characters of all the elements of the group. They should be the same for elements in the same conjugacy class. Why?

3.3 Verify that

$$D(c) = \begin{pmatrix} 0 & 1 \\ -1 & -1 \end{pmatrix}$$

generates a [2] representation of C_3. Show that thc representation is irreducible over the field of real numbers.

3.4 Given a normal subgroup N of a group G, a representation $D^{G/N}$ of the quotient group G/N can be *lifted* to give a representation D^G of the full group G by the following definition:

$$D^G(g) := D^{G/N}(gN)$$

That is, each element of the group is assigned the matrix $D^{G/N}$ of the coset to which it belongs, as illustrated below.

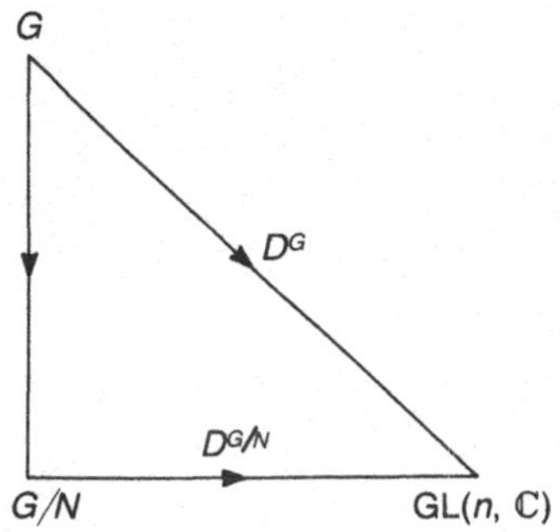

Verify that $D^G(g)$ indeed provides a representation of G, i.e. that $D^G(g_1)D^G(g_2) = D^G(g_1g_2)$.

3.5* Show that two bases of a finite-dimensional vector space must have the same number of elements.

3.6 The 'affine group' in $[n]$ is the group of rotations (by a matrix A) and translations (by a vector c), so that

$$x' = Ax + c$$

Find the composition law for two successive affine transformations and hence show that there is a 1:1 correspondence between this group and the group of $[n + 1]$ matrices of the form

$$\begin{pmatrix} A & c \\ 0 & 1 \end{pmatrix}.$$

(Note that these matrices, which are reducible, cannot be fully reduced unless $c = 0$, in which case we have a representation of the compact rotational subgroup rather than the full group, which is non-compact.)

4

Properties of Irreducible Representations

We have just seen that for finite or compact groups a representation is either irreducible or completely reducible. The matrices of a reducible representation can, by a similarity transformation, be cast in block diagonal form, in which the matrices along the diagonal are those of irreducible representations ('irreps', for short). In the present chapter we focus our attention on these fundamental building blocks, which turn out to have many remarkable properties. These all follow as consequences of the two lemmas due to Schur, which we will prove using the coordinate-free language of vector spaces developed in the last chapter. Once one has become accustomed to this language the proof is much simpler, and gives more insight, than the old method using matrices.

4.1 Schur's Lemmas

(i) The First Lemma

In matrix form this states that any matrix which commutes with all the matrices of an irreducible representation must be a multiple of the unit matrix, i.e.

$$BD(g) = D(g)B \quad \forall g \in G \Rightarrow B = \lambda \mathbb{1} \tag{4.1}$$

As a statement about linear operators it says that

$$\hat{B}T(g) = T(g)\hat{B} \quad \forall g \in G \Rightarrow \hat{B} = \lambda E \tag{4.1'}$$

where $\hat{B} : U \rightarrow U$ is the linear operator corresponding to B, and E is the identity map $\boldsymbol{u} \mapsto \boldsymbol{u}$.

To prove the identity in its second form, let $\boldsymbol{b}$ be an eigenvector of $\hat{B}$ with eigenvalue λ:

$$\hat{B}\boldsymbol{b} = \lambda \boldsymbol{b}$$

Then

$$\hat{B}(T(g)\boldsymbol{b}) = T(g)\hat{B}\boldsymbol{b} = T(g)\lambda\boldsymbol{b} = \lambda(T(g)\boldsymbol{b})$$

This means that $T(g)\boldsymbol{b}$ is also an eigenvector of $\hat{B}$, with the same eigenvalue λ. Now the space of such eigenvectors is, as you should check, a vector space, and forms a subspace of U. But according to the above result it is *closed* under the action of the group via the linear operators $\{T(g)\}$. In other words it is a *G-module*.

But now we invoke irreducibility, which means that the space U has no proper submodule. The only submodules allowed are (a) the whole space U or (b) the null vector $\mathbf{0}$.

The second case, (b), is in fact impossible, since every linear operator $\hat{B}$ has at least one proper eigenvector. In matrix form one has to solve the equation $Bb = \lambda b$, leading to the characteristic equation $\det(B - \lambda \mathbb{1}) = 0$. If we are working within the framework of complex numbers, this polynomial equation is guaranteed to have at least one root λ, with a corresponding eigenvector $\boldsymbol{b}$.

We are therefore left with case (a), which says that the space of eigenvectors of $\hat{B}$ with eigenvalue λ is in fact the whole space. So $\hat{B}\boldsymbol{u} = \lambda\boldsymbol{u}$ for *any* $\boldsymbol{u} \in U$, and $\hat{B} = \lambda E$.

(ii) The Second Lemma

This concerns two inequivalent irreducible representations D, D', and in matrix form states that

$$BD(g) = D'(g)B \quad \forall g \in G \Rightarrow B = \hat{O} \tag{4.2}$$

The corresponding linear operator $\hat{B}$ will in this case be a map $\hat{B} : U \to U'$ between the two vector spaces U and U' supporting the two representations, and the coordinate-free version of (4.2) is

$$\hat{B}T(g) = T'(g)\hat{B} \quad \forall g \in G \Rightarrow \hat{B} = \hat{O} \tag{4.2'}$$

where $\hat{O}$ is the linear operator which maps every vector in U onto the null vector in U': $\hat{O}\boldsymbol{u} = \mathbf{0}' \ \forall \boldsymbol{u} \in U$.

There are three separate cases to consider depending on the ordering of the dimensionalities n, n' of U, U' respectively.

(1) Suppose first that $n < n'$. Consider the action of $T'(g)\hat{B}$ on an arbitrary vector $\boldsymbol{u}$ of U. From (4.2'),

$$T'(g)\hat{B}\boldsymbol{u} = \hat{B}T(g)\boldsymbol{u}$$

Now since U is a G-module, $T(g)\boldsymbol{u}$ is a member of U. Thus

$$T'(g)(\hat{B}\boldsymbol{u}) \in \hat{B}U$$

which means that $\hat{B}U = \text{Im}\,\hat{B}$ is a *submodule* of U', as illustrated in figure 4.1.

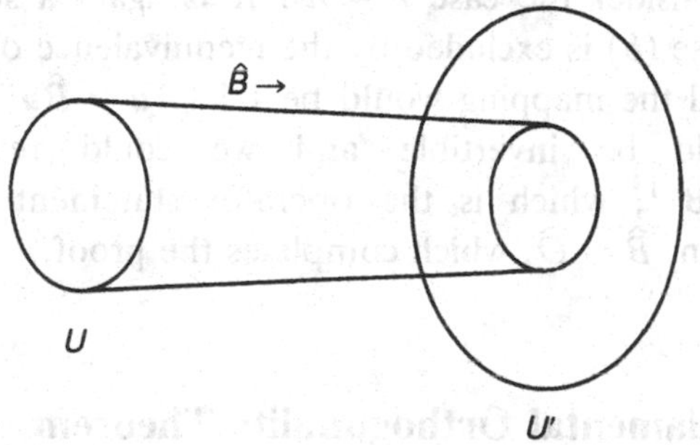

Figure 4.1 The image of the mapping $\hat{B}$: $U \to U'$ is a submodule of U'.

Again, however, U' is supposed to be irreducible, so $\hat{B}U$ must be either (a) the whole space U' or (b) the null vector $\mathbf{0}'$. Case (a) can be excluded because of our assumed inequality $n < n'$. As the image of U, the dimension m of $\hat{B}U$ cannot exceed that of U. So $m \leqslant n < n'$. This just leaves case (b), which means that $\hat{B} = \hat{O}$, as claimed.

(2) Now take $n > n'$. In this case we focus our attention on the *kernel* of the mapping, illustrated in figure 4.2, namely

$$K := \{\boldsymbol{k} \in U | \hat{B}\boldsymbol{k} = \mathbf{0}'\}$$

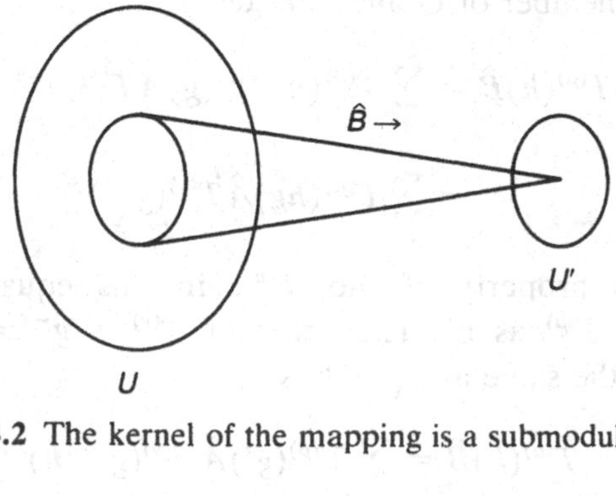

Figure 4.2 The kernel of the mapping is a submodule of U.

This space is a submodule of U, since

$$\hat{B}(T(g)\boldsymbol{k}) = T'(g)\hat{B}\boldsymbol{k} = \boldsymbol{0}'$$

Now we invoke the irreducibility of U: K must either be (a) the whole space U or (b) the null vector $\boldsymbol{0}$. In fact K must be non-trivial, since the dimensionality is reduced under $\hat{B}$. (The images $\hat{B}\boldsymbol{e}_i$ of the n vectors of a basis of U cannot all be linearly independent.) Thus we are left with case (a), which means that *every* vector $\boldsymbol{u} \in U$ satisfies $\hat{B}\boldsymbol{u} = \boldsymbol{0}'$, so that $\hat{B} = \hat{O}$ again.

(3) Finally consider the case $n = n'$. K is again a submodule of U, but this time case (b) is excluded by the inequivalence of D and D'. For if K were trivial the mapping would be 1:1 ($\hat{B}\boldsymbol{u} = \hat{B}\boldsymbol{u}' \Rightarrow \boldsymbol{u} - \boldsymbol{u}' \in K$). Then $\hat{B}$ would be invertible and we could rewrite (4.2′) as $T'(g) = \hat{B}T(g)\hat{B}^{-1}$, which is the operator statement of equivalence. Thus, once again, $\hat{B} = \hat{O}$, which completes the proof.

4.2 The Fundamental Orthogonality Theorem

Let U_ν and U_μ be two G-modules carrying inequivalent irreducible representations. For convenience we can take the labels μ, ν to be positive integers. Take an arbitrary linear mapping $\hat{A}$ from U_ν to U_μ and construct the operator

$$\hat{B} := \sum_g T^{(\mu)}(g)\hat{A}\,T^{(\nu)}(g^{-1}) \tag{4.3}$$

This involves the same sum over group elements that we used in §3.5. As it stands it applies to finite groups, but it can be extended to compact groups, when it is replaced by $\int \mathrm{d}\mu(g)$.

Let h be any member of G and consider

$$\begin{aligned} T^{(\mu)}(h)\hat{B} &= \sum_g T^{(\mu)}(h)T^{(\mu)}(g)\hat{A}\,T^{(\nu)}(g^{-1}) \\ &= \sum_g T^{(\mu)}(hg)\hat{A}\,T^{(\nu)}(g^{-1}) \end{aligned}$$

from the group property of the $T^{(\mu)}$. In this equation relabel the argument hg of $T^{(\mu)}$ as g'. Then that of $T^{(\nu)}$ is $g^{-1} = g'^{-1}h$, and we know that Σ_g is the same as $\Sigma_{g'}$. Thus

$$T^{(\mu)}(h)\hat{B} = \sum_{g'} T^{(\mu)}(g')\hat{A}\,T^{(\nu)}(g'^{-1}h)$$

$$= \sum_{g'} T^{(\mu)}(g')\hat{A}T^{(\nu)}(g'^{-1})T^{(\nu)}(h)$$

from the group property of the $T^{(\nu)}$. But the first three factors, summed over g', just give $\hat{B}$ again. So

$$T^{(\mu)}(h)\hat{B} = \hat{B}T^{(\nu)}(h) \tag{4.4}$$

Thus $\hat{B}$ satisfies the conditions of Schur's second lemma, and therefore $\hat{B} = \hat{O}$ unless $\mu = \nu$. In this latter case the two representations are one and the same. Then Schur's first lemma applies, giving $\hat{B} = \lambda E$.

We can express this succinctly in matrix form as

$$B := \sum_g D^{(\mu)}(g)AD^{(\nu)}(g^{-1}) = \lambda_A^{(\mu)}\delta^{\mu\nu}\mathbb{1} \tag{4.5}$$

The constant λ depends, as we have indicated, both on the label of the irreducible representation and on the choice of the matrix A, which so far is completely unspecified. Let us now take A to be a matrix whose entries are all zero except for $A_{rs} = 1$, which can be summarized as $A_{lm} = \delta_{lr}\delta_{ms}$. Accordingly we will rewrite $\lambda_A^{(\mu)}$ as $\lambda_{rs}^{(\mu)}$.

Then the ij matrix element of (4.5) reads

$$\sum_g D_{ir}^{(\mu)}(g)D_{sj}^{(\nu)}(g^{-1}) = \lambda_{rs}^{(\mu)}\delta^{\mu\nu}\delta_{ij} \tag{4.6}$$

$\lambda_{rs}^{(\mu)}$ can be found by setting $\nu = \mu$ and taking the trace, i.e. contracting with δ_{ij} (summation understood). This gives

$$\sum_g (D^{(\mu)}(g^{-1})D^{(\mu)}(g))_{sr} = n_\mu \lambda_{rs}^{(\mu)}$$

where n_μ is the dimensionality of $D^{(\mu)}$. The matrix $D^{(\mu)}(g^{-1})D^{(\mu)}(g)$, for each g, is just the unit matrix, with matrix elements δ_{rs}. Thus

$$[g]\delta_{rs} = n_\mu \lambda_{rs}^{(\mu)}$$

Substituting for $\lambda_{rs}^{(\mu)}$ in (4.5) we finally obtain the fundamental orthogonality relation for the matrices of irreducible representations:

$$\boxed{\sum_g D_{ir}^{(\mu)}(g)D_{sj}^{(\nu)}(g^{-1}) = \frac{[g]}{n_\mu}\delta^{\mu\nu}\delta_{ij}\delta_{rs}} \tag{4.7}$$

Restriction on the Number of Irreps

In view of the result proved at the end of the previous chapter, there is no loss of generality in taking the representations $D^{(\mu)}$ and $D^{(\nu)}$ to be unitary. In that case equation (4.7) can be rewritten as

$$\sum_g D_{ir}^{(\mu)}(g) D_{js}^{(\nu)*}(g) = \frac{[g]}{n_\mu} \delta^{\mu\nu}\delta_{ij}\delta_{rs} \tag{4.7'}$$

Let us first concentrate on a given irreducible representation $D^{(\mu)}$, and set $\nu = \mu$. For fixed i, r we can think of the set of objects $\{D_{ir}^{(\mu)}(g_1), D_{ir}^{(\mu)}(g_2), \ldots, D_{ir}^{(\mu)}(g_{[g]})\}$ as a $[g]$-dimensional column vector. Then the left-hand side of equation (4.7′) represents the complex scalar product of two vectors in this space, labelled by the pairs of indices (ir) and (js) respectively. Each of these indices can take any value from 1 to n_μ, the dimensionality of $D^{(\mu)}$. There are therefore n_μ^2 such vectors, and by (4.7′) they are all orthogonal to each other.

The same will be true for any other value of μ, μ' say, and moreover the vectors formed by the $D^{(\mu')}$ will be orthogonal to those formed by the $D^{(\mu)}$. Taking account of all possible values of μ we can form a total of $\Sigma_\mu n_\mu^2$ mutually orthogonal vectors. Now the number of such vectors cannot exceed the dimensionality of the space, namely $[g]$. We have therefore proved the important inequality

$$\sum_\mu n_\mu^2 \leqslant [g] \tag{4.8}$$

Since each n_μ must be at least 1, this means that the number of irreducible representations of a finite group is strictly limited. We shall see later on (§4.3) that the inequality is in fact an equality: $\Sigma_\mu n_\mu^2 = [g]$!

4.3 Orthogonality of Characters

Recall (§3.3) that the character of a representation D is the set $\{\chi(g)\}$, where $\chi(g)$ is the trace of the matrix $D(g)$. These traces have the following properties:

(i) χ is the same for equivalent representations, for which $D'(g) = SD(g)S^{-1}$.

(ii) χ is the same for conjugate elements, since $D(hgh^{-1}) = D(h)D(g)(D(h))^{-1}$.

(iii) If D is unitary, i.e. $D^{-1} = D^\dagger$, then

$$\chi(g^{-1}) = \mathrm{Tr}((D(g))^{-1}) = \mathrm{Tr}(D(g)^\dagger) = \chi^*(g) \tag{4.9}$$

In fact this is always true for finite or compact groups, since, as we have shown, any representation is equivalent to a unitary representation.

The orthogonality relation for characters is obtained by taking suitable traces of the fundamental orthogonality theorem of equation (4.7). Tracing over the indices i, r and s, j (i.e. multiplying by $\delta_{ir}\delta_{sj}$) gives

$$\sum_g \chi^{(\mu)}(g)\chi^{(\nu)}(g^{-1}) = \frac{[g]}{n_\mu}\,\delta^{\mu\nu}\delta_{ij}\delta_{ij}$$

Here $\delta_{ij}\delta_{ij} = \delta_{ii} = n_\mu$. The summation convention is always operative here, and the indices range from 1 to n_μ, the dimension of the irrep. Thus

$$\boxed{\frac{1}{[g]}\sum_g \chi^{(\mu)}(g)\chi^{(\nu)}(g^{-1}) = \delta^{\mu\nu}} \tag{4.10}$$

By virtue of (4.9) this can be cast in the alternative form

$$\frac{1}{[g]}\sum_g \chi^{(\mu)}(g)\chi^{(\nu)*}(g) = \delta^{\mu\nu} \tag{4.10$'$}$$

Up to the factor $[g]$, the left-hand side of this equation is just the usual complex scalar product of the two $[g]$-dimensional column vectors $(\chi^{(\mu)}(g_1), \chi^{(\mu)}(g_2), \ldots, \chi^{(\mu)}(g_{[g]}))$ and $(\chi^{(\nu)}(g_1), \chi^{(\nu)}(g_2), \ldots, \chi^{(\nu)}(g_{[g]}))$. So it is both convenient and illuminating to define the scalar product of two characters φ, χ by

$$\langle \varphi, \chi \rangle := \frac{1}{[g]}\sum_g \varphi(g)\chi(g^{-1}) = \langle \chi, \varphi \rangle \tag{4.11}$$

In this language (4.10) says that the characters of inequivalent irreducible representations are *orthonormal*: $\langle \chi^{(\mu)}, \chi^{(\nu)} \rangle = \delta^{\mu\nu}$.

We have noted in (ii) above that the characters of conjugate elements are equal. That means that all the elements of a conjugacy class will have the same character, so that the distinct characters may be labelled as χ_i, $i = 1, \ldots, k$, corresponding to the k conjugacy classes K_i.

Let k_i denote the number of elements in the conjugacy class K_i. Then the sum over g in (4.10′) can be rewritten as a sum over i:

$$\frac{1}{[g]}\sum_i k_i \chi_i^{(\mu)}\chi_i^{(\nu)*} = \delta^{\mu\nu} \tag{4.10$''$}$$

But this can now be interpreted as orthogonality of the vectors $k_i^{1/2}\chi_i^{(\mu)}$ (no summation) in k-dimensional space! And since there can be no more than k such vectors we have another inequality on the number r of different irreps. Namely, *the number of inequivalent irreducible representations is less than or equal to the number of conjugacy classes*:

$$r \leqslant k \tag{4.12}$$

Again, this is an inequality which turns out to be an equality. For remarkably the characters can be shown (e.g. Hamermesh, §3-7) to be orthogonal with respect to the index i as well in the sense that

$$\frac{1}{[g]}\sum_{\mu} k_i \chi_i^{(\mu)} \chi_j^{(\mu)*} = \delta_{ij} \tag{4.13}$$

This gives the inequality in the opposite direction, hence $r = k$.

Decomposition of Reducible Representation

For a finite or compact group any reducible representation is completely reducible into a direct sum of irreps, which means that the representation matrices can be cast in block diagonal form, the non-zero diagonal blocks being the matrices of irreducible representations. A given irrep may appear more than once in this decomposition: for example, a 5-dimensional representation could decompose into the trivial representation and two copies of the same 2-dimensional irrep. In the general case the decomposition is written

$$D = \sum_{\oplus} a_\nu D^{(\nu)} \tag{4.14}$$

where the non-negative integer a_ν denotes the number of times the irrep $D^{(\nu)}$ appears in the decomposition. Such decompositions feature prominently in physical applications.

How do we find the coefficients a_ν? Well, that is easily done using the orthogonality properties of the characters $\chi^{(\nu)}$.

Taking the trace of both sides of (4.14) for a general group element g we see that the character $\chi(g)$ of D decomposes as an ordinary arithmetic sum of the characters $\chi^{(\nu)}(g)$:

$$\chi(g) = \sum_{\nu} a_\nu \chi^{(\nu)}(g) \tag{4.15}$$

(As a matter of notation, χ is termed a compound character, the $\chi^{(\nu)}$ simple characters.) Multiply this equation by $\chi^{(\mu)}(g^{-1})$ and sum over g to obtain

$$\sum_{g} \chi^{(\mu)}(g^{-1})\chi(g) = \sum_{\nu} a_\nu \sum_{g} \chi^{(\mu)}(g^{-1})\chi^{(\nu)}(g)$$

But, using equation (4.10), the RHS is just $\Sigma_\nu a_\nu [g]\delta^{\mu\nu} = [g]a_\mu$. Thus

$$\boxed{a_\mu = \frac{1}{[g]}\sum_{g} \chi(g)\chi^{(\mu)}(g^{-1})} \tag{4.16}$$

In fact we did not need to go through this in detail: we could have used the abbreviated notation $\langle\ ,\ \rangle$ we introduced for the scalar product of two characters. Equation (4.15) gives a meaning to integer linear combinations of simple characters $\Sigma\, a_\nu \chi^{(\nu)}$. The object defined in (4.11) has the linearity property (3.52) required of a scalar product, and the $\chi^{(\nu)}$ are orthonormal with respect to this scalar product. Equation (4.16) is the equivalent of identifying the coefficients u_i of a vector $\boldsymbol{u}$ with the scalar products $(\boldsymbol{e}_i, \boldsymbol{u})$.

Thus the above derivation simply amounts to writing

$$\chi = \sum_\nu a_\nu \chi^{(\nu)}$$

and taking the scalar product $\langle \chi^{(\mu)}, \chi \rangle$ to obtain

$$a_\mu = \langle \chi^{(\mu)}, \chi \rangle \tag{4.16$'$}$$

which is the shorthand version of equation (4.16).

The Regular Representation

Recall (Cayley's theorem) that there is an isomorphism between the group G and a subgroup of the symmetric group $S_{[g]}$ provided by left multiplication. That is

$$gg_i = \sum_j D_{ji}(g) g_j \tag{4.17}$$

The $[g] \times [g]$ permutation matrices $D_{ji}(g)$ form a $[g]$-dimensional representation of the group, called the regular representation. Now gg_i is some single element g_l, say, of G, with $g_l \neq g_i$ except when $g = e$. Thus the matrix $D_{ji}(g)$ has only one non-zero element in each row and column. For $g \neq e$ all these elements are off-diagonal ($l \neq i$), while for $g = e$ they are all on the diagonal, and in fact $D_{ji}(e)$ is just the unit matrix, with elements δ_{ji}. For example, in C_3:

$$D(c) = \begin{pmatrix} 0 & 0 & 1 \\ 1 & 0 & 0 \\ 0 & 1 & 0 \end{pmatrix} \qquad D(e) = \begin{pmatrix} 1 & 0 & 0 \\ 0 & 1 & 0 \\ 0 & 0 & 1 \end{pmatrix}$$

Let us decompose the regular representation into its irreducible components:

$$D = \sum_\oplus a_\nu D^{(\nu)} \tag{4.18}$$

The coefficients a_μ are given by (4.16):

$$a_\mu = \frac{1}{[g]} \sum_g \chi(g)\chi^{(\mu)}(g^{-1})$$

But from the special form of the D_μ we can deduce that

$$\chi(g) = \begin{cases} 0 & g \neq e \\ [g] & g = e \end{cases} \tag{4.19}$$

Therefore

$$a_\mu = \chi^{(\mu)}(e) = n_\mu \tag{4.20}$$

the dimensionality of $D^{(\mu)}$.

Then, putting $g = e$ in

$$\chi(g) = \sum_\nu a_\nu \chi^{(\nu)}(g)$$

we obtain

$$[g] = \sum_\nu n_\nu \cdot n_\nu = \sum_\nu n_\nu^2 \tag{4.8$'$}$$

thus establishing that equation (4.8) is in fact an equality.

For $g \neq e$ we obtain instead

$$0 = \sum_\nu n_\nu \chi^{(\nu)}(g) \tag{4.21}$$

and the last two equations can be combined as

$$\frac{1}{[g]} \sum_\nu \chi^{(\nu)}(e)\chi^{(\nu)}(g) = \begin{cases} 0 & g \neq e \\ 1 & g = e \end{cases}$$

This is a special case of (4.13), since the class of the identity, K_1 say, contains the single element e and has $k_1 = 1$.

4.4 Construction of the Character Table

We have now developed the theory sufficiently far to enable us to determine the characters of the irreducible representations of the finite groups most commonly encountered in physical applications. The characters are most conveniently presented in the form of a table, the rows of which correspond to the different irreps, and the columns to the conjugacy classes of the group.

In the construction of this table the principal tools we use are:

(1) number of irreps = number of conjugacy classes: $r = k$;
(2) $\Sigma_\mu n_\mu^2 = [g]$;
(3) orthogonality: $\Sigma k_i \chi_i^{(\mu)} \chi_i^{(\nu)*} = [g]$;
(4) any other information we can come by (!).

In the last category is the fact that for 1-dimensional representations the characters are the same thing as the matrices, and hence the characters themselves must mimic the group multiplication properties.

In the case of an Abelian group all the irreps are in fact 1-dimensional. This follows from Schur's first lemma. For the matrix $D^{(\sigma)}(g)$ commutes with all the matrices of the representation $D^{(\sigma)}$ and is therefore a multiple of the unit matrix: $D^{(\sigma)} = \lambda_g^{(\sigma)} \mathbb{1}$. That is, all the matrices $D^{(\sigma)}(g)$ are diagonal. Thus if the dimension is greater than 1, the representation is reducible ⊗.

The above argument does not depend on the group being finite, and so applies to compact groups as well. For finite groups an alternative derivation uses (1), (2) and the fact that the conjugacy classes consist of single elements. Thus the number of classes, k, is equal to the order of the group, $[g]$, and (2) reads

$$\sum_{\mu=1}^{[g]} n_\mu^2 = [g]$$

whose only solution is $n_\mu = 1$ for all μ.

As a simple example of an Abelian group let us consider C_3.

Character Table of C_3

The table is a 3 × 3 array labelled by the three [1] irreps $D^{(1)} \ldots D^{(3)}$ and the three conjugacy classes consisting of the single elements e, c, c^2:

C_3	e	c	c^2
$D^{(1)}$			
$D^{(2)}$			
$D^{(3)}$			

As noted above, the characters of a 1-dimensional representation must mimic the group multiplication. In particular, $\chi(c^2) = (\chi(c))^2$ and $(\chi(c))^3 = \chi(c^3) = \chi(e) = 1$. $\chi(c)$ must therefore be one of the cube roots of unity, namely 1, $\omega := e^{2\pi i/3}$ or $\omega^2 = e^{4\pi i/3}$, and we can write the character table in the form

C_3	e	c	c^2
$D^{(1)}$	1	1	1
$D^{(2)}$	1	ω	ω^2
$D^{(3)}$	1	ω^2	ω

$D^{(1)}$ is the trivial representation, whereby each element is mapped onto unity. In crystallographic notation it is denoted by A, the letter used for 1-dimensional representations. Although $D^{(2)}$ and $D^{(3)}$ are also 1-dimensional, they are complex conjugates of each other. As we shall see later, the energy levels corresponding to these representations are the same in many physical situations (when there is no way of distinguishing between +i and −i). For this reason they are often thought of as forming a single 2-dimensional representation, denoted by E in crystallographic notation.

Thinking of the representations as mappings from C_3 into the group $GL(1, \mathbb{C})$, namely the non-zero complex numbers, we can identify the different kernels. These have to be normal subgroups of C_3. Since C_3 has no proper subgroups at all, the only possibilities are the whole group or the identity alone. For $D^{(1)}$ it is the first of these possibilities which is realized, while for $D^{(2)}$ and $D^{(3)}$ the kernel is just the unit element, which means that they are *faithful* representations, isomorphic to C_3 itself.

Let us immediately check orthonormality of the rows, equation (4.10′):

$$\langle \chi^{(1)}, \chi^{(2)} \rangle = \tfrac{1}{3}(1 + \omega^2 + \omega) = 0$$

by virtue of the factorization of $z^3 - 1$ into $(z - 1)(z^2 + z + 1)$. Similarly for $\langle \chi^{(1)}, \chi^{(3)} \rangle$ and $\langle \chi^{(2)}, \chi^{(3)} \rangle$. The normalization is assured by the fact that the characters all have modulus 1 (i.e. are unitary numbers, as befits a unitary representation!).

Finally let us use the character table to find out how the vector representation of equation (3.6), which acts on the components x, y and z, decomposes into irreps. The character is

$$\chi^V = (\chi^V(e), \chi^V(c), \chi^V(c^2)) = (3, 0, 0)$$

and we have to write this as

$$\chi^V = a_1\chi^{(1)} + a_2\chi^{(2)} + a_3\chi^{(3)}$$

By inspection it is easy to see that

$$\chi^V = \chi^{(1)} + \chi^{(2)} + \chi^{(3)}$$

Alternatively the coefficients are given by (4.16′) as

$$a_\nu = \langle \chi^V, \chi^{(\nu)} \rangle = \tfrac{1}{3}(3\chi^{(\nu)}(e)) = 1$$

Thus the vector representation D^V decomposes into the direct sum

$$D^V = D^{(1)} \oplus D^{(2)} \oplus D^{(3)} \tag{4.22}$$

It is not difficult to identify the coordinate combinations on which D^V acts irreducibly. We have already noted (§3.3) that the coordinate z is unchanged by any of the rotations of C_3, i.e. z forms the basis of the trivial representation $D^{(1)}$. As for $D^{(2)}$ and $D^{(3)}$, consider the transformation of the combinations $x \pm iy$. From equations (3.2) and (3.3) we have

$$\begin{aligned} x' \pm iy' &= x(\cos\theta \pm i\sin\theta) \pm iy(\cos\theta \pm i\sin\theta) \\ &= (x \pm iy)e^{\pm i\theta} \end{aligned}$$

Consider first the combination $x + iy$. The element c corresponds to $\theta = 2\pi/3$ and therefore gives $e^{i\theta} = \omega$. Similarly c^2 gives the multiplicative factor ω^2. Thus $x + iy$ forms the basis of the irrep $D^{(2)}$. Similarly $x - iy$ transforms irreducibly according to $D^{(3)}$.

All of this information, together with the crystallographic notation, is included in the final version of the character table, table 4.1.

As a second, less trivial, example let us consider D_3.

Table 4.1

	C_3	e	c	c^2	
A	$D^{(1)}$	1	1	1	z
E	$D^{(2)}$	1	ω	ω^2	$x + iy$
	$D^{(3)}$	1	ω^2	ω	$x - iy$

Character Table of D_3

Recall (§2.1) that the conjugacy classes are (e), (c, c^2) and (b, bc, bc^2), which we will denote by K_1, K_2 and K_3 respectively. There are three of them, and hence three irreducible representations. (2) therefore takes the form $n_1^2 + n_2^2 + n_3^2 = 6$. But there always exists the trivial representation $D^{(1)}(g) = 1$, with $n_1 = 1$, which leaves $n_2^2 + n_3^2 = 5$.

The only integer solution to this equation is $n_2 = 1$, $n_3 = 2$ (or vice versa, which is merely a matter of notation). This enables us to fill in the first column of the character table, since $\chi^{(\mu)}(e) = n_\mu$.

For the 1-dimensional representations the χ's must mimic the group structure. Thus $\chi(bc) = \chi(b)\chi(c)$. But $\chi(b) = \chi(bc) = \chi_3$, hence $\chi(c) = \chi_2 = 1$. Further $\chi(b)^2 = \chi(b^2) = \chi(e) = 1$, giving $\chi_3 = \pm 1$. The upper sign gives the trivial representation, leaving $\chi_3 = -1$ for $D^{(2)}$.

Using (1), (2) and (4) we have thus determined the character table to the extent shown below. The only unknown entries are $\chi_2^{(3)} = \alpha$ and $\chi_3^{(3)} = \beta$, say.

D_3	K_1	K_2	K_3
$D^{(1)}$	1	1	1
$D^{(2)}$	1	1	−1
$D^{(3)}$	2	α	β

We finally use orthogonality (3). Orthogonality of $\chi^{(3)}$ with $\chi^{(1)}$ and $\chi^{(2)}$ gives

$$2 + 2\alpha + 3\beta = 0$$

$$2 + 2\alpha - 3\beta = 0$$

which together determine $\beta = 0$, $\alpha = -1$. Here the numerical factors 2 and 3 in front of α and β are the multiplicities k_1 and k_2 of the respective conjugacy classes $K^{(1)}$ and $K^{(2)}$. For consistency you should also check the orthogonality of $\chi^{(1)}$ and $\chi^{(2)}$, and for good measure the normalization!

In the case of groups of rotations another notation is often used for the conjugacy classes which reflects both the number and nature of their members. As already remarked, conjugate rotations are necessarily rotations through the same angle, though the converse is not true. K_2 contains two members, both 3-fold rotations, which by themselves generate a C_3 subgroup. For this reason $K^{(2)}$ is written as $2C_3$. Similarly $K^{(3)}$ contains three 2-fold rotations, each of which generates a C_2 subgroup. Accordingly $K^{(3)}$ is written as $3C_2$. Finally $K^{(1)}$, the class of the identity, is often written as E.

We have already introduced the crystallographic notation for irreps. $D^{(1)}$ and $D^{(2)}$, being 1-dimensional, are labelled as A_1 and A_2 respectively, while the 2-dimensional representation $D^{(3)}$ is labelled as E.

As in the case of C_3, we can use the character table to determine how the vector representation acting on x, y, z breaks up into the different irreps. To this end we need the matrix of at least one of the b's. Without loss of generality we can take b to be a rotation about the x axis. The corresponding matrix will then be $D^V(b) = \mathrm{diag}(1, -1, -1)$, effecting the transformations $x' = x$, $y' = -y$, $z' = -z$. The trace of $D^V(b)$ is $\chi^V(b) = -1$. Thus

$$\chi^V = (3, 0, -1)$$

which decomposes as

$$\chi^V = \chi^{(2)} + \chi^{(3)}$$

Hence

$$D^V = D^{(2)} \oplus D^{(3)} = \mathrm{A}_2 \oplus \mathrm{E} \tag{4.23}$$

The basis functions of A_2 and E are not hard to find. As we have noted, z changes sign under b, but it is left invariant by c. Hence z forms the basis of A_2. The 2-dimensional representation E gives the transformations of x and y, but, unlike the case of C_3, and precisely because of the additional rotations b_i, cannot be further reduced.

All of this information is encapsulated in the final form of the character table, table 4.2.

Table 4.2

D_3	E	$2\mathrm{C}_3$	$3\mathrm{C}_2$	
A_1	1	1	1	
A_2	1	1	-1	z
E	2	-1	0	x, y

4.5 Direct Products of Representations and Their Decomposition

Suppose we have a system of two particles, for example atomic electrons, described by spatial wavefunctions $\psi_a(x)$, $\varphi_b(x)$ which transform according to the irreducible representations $D^{(\mu)}$, $D^{(\nu)}$ respectively of some symmetry group G. For atoms in a solid this could be one of the crystal point groups, while for free atoms it could be the full rotation group SO(3). The changes in the two wavefunctions induced, as in §3.2, by a transformation $g \in G$ are given by the respective matrices

$$\begin{aligned} \psi'_a(x) &= D^{(\mu)}_{ba}(g)\ \psi_b(x) \\ \varphi'_c(x) &= D^{(\nu)}_{dc}(g)\ \varphi_d(x) \end{aligned} \tag{4.24}$$

Then the product wavefunction $\psi_{ac}(x) := \psi_a(x)\varphi_c(x)$ transforms according to

$$\psi'_{ac}(x) = D^{(\mu)}_{ba}(g)D^{(\nu)}_{dc}(g)\psi_{bd}(x) \tag{4.25}$$

We can regard the product of matrices in (4.25) as a single matrix whose first index is the ordered pair $(bd) := B$ and whose second is the pair $(ac) := A$, thus defining the matrix of the direct product representation as

$$D^{(\mu\times\nu)}_{bd;ac}(g) := D^{(\mu)}_{ba}(g)D^{(\nu)}_{dc}(g) \tag{4.26}$$

so that (4.25) reads

$$\psi'_A(x) = D^{(\mu\times\nu)}_{BA}(g)\psi_B(x) \tag{4.27}$$

The representation of which the $D^{(\mu\times\nu)}_{bd;ac}$ are the matrices is itself written as $D^{(\mu)} \otimes D^{(\nu)}$. It is of course necessary to check that it is indeed a representation satisfying the fundamental requirement that

$$D^{(\mu\times\nu)}(g_1g_2) = D^{(\mu\times\nu)}(g_1)D^{(\mu\times\nu)}(g_2)$$

but this follows immediately from the corresponding property for $D^{(\mu)}$ and $D^{(\nu)}$.

If we are in a situation where as a first approximation we may neglect the interaction between the two electrons, then we really have a symmetry $G \times G$. That is, the system is invariant under *independent* rotations g_1, g_2 of the two electrons, and the appropriate wavefunctions are indeed just the products considered above, suitably symmetrized or antisymmetrized in order to conform with Fermi statistics. The matrix which implements such a transformation is

$$D^{(\mu\times\nu)}_{bd;ac}((g_1, g_2)) := D^{(\mu)}_{ba}(g_1)D^{(\nu)}_{dc}(g_2) \tag{4.28}$$

and it is easy to see that if $D^{(\mu)}$ and $D^{(\nu)}$ are irreducible representations of G, then $D^{(\mu)} \otimes D^{(\nu)}$, as so defined, is an irreducible representation of the *product group* $G \times G$. For suppose it were reducible. This would mean that there existed a proper invariant subspace $\forall g_1, g_2 \in G$. But in particular we could fix g_2 to be e, which would mean that there was a submodule of the vector space carrying the representation $D^{(\mu)}$, contrary to hypothesis.

However, when we take account of the interaction between the two electrons, the residual symmetry is reduced to the original group G, with the *same* rotation being applied to both electrons: $g_1 = g_2 = g$. In this situation the product representation will in general be reducible. The decomposition of $D^{(\mu)} \otimes D^{(\nu)}$ into its irreducible components

$$D^{(\mu)} \otimes D^{(\nu)} = \sum_{\oplus} a_\sigma D^{(\sigma)} \tag{4.29}$$

is called the *Clebsch–Gordan series* and is very important in physical applications of representation theory.

How in fact do we determine the coefficients a_σ? Well, by the machinery of characters that we have so painstakingly developed! The character of $D^{(\mu\times\nu)}(g)$ is, by definition,

$$\begin{aligned}\chi^{(\mu\times\nu)}(g) = D^{(\mu\times\nu)}_{AA}(g) &= D^{(\mu)}_{aa}(g)\, D^{(\nu)}_{cc}(g) \\ &= \chi^{(\mu)}(g)\chi^{(\nu)}(g).\end{aligned} \tag{4.30}$$

That is, the character of the product representation is just the product of the characters. Hence

$$a_\sigma = \langle \chi^{(\sigma)}, \chi^{(\mu)}\chi^{(\nu)} \rangle \tag{4.31}$$

Example: E ⊗ E in D_3

From table 4.2 the character of the irrep $D^{(3)}$ = E is (2, −1, 0). That of E ⊗ E is therefore (4, 1, 0). Then

$$a_1 = \tfrac{1}{6}(1\cdot 1\cdot 4 + 2\cdot 1\cdot 1 + 3\cdot 1\cdot 0) = 1$$

$$a_2 = \tfrac{1}{6}(1\cdot 1\cdot 4 + 2\cdot 1\cdot 1 + 3\cdot(-1)\cdot 0) = 1$$

$$a_3 = \tfrac{1}{6}(1\cdot 2\cdot 4 + 2\cdot(-1)\cdot 1 + 3\cdot 0\cdot 0) = 1$$

i.e.

$$\chi^{\mathrm{E}\otimes\mathrm{E}} = \chi_1 + \chi_2 + \chi_3$$

which could possibly have been seen by inspection. So the product representation E ⊗ E decomposes as

$$\mathrm{E} \otimes \mathrm{E} = \mathrm{A}_1 \oplus \mathrm{A}_2 \oplus \mathrm{E} \tag{4.32}$$

Problems for Chapter 4

4.1 Construct the matrices of the [2] representation of D_3 which acts on x, y and verify the fundamental orthogonality theorem for matrices.

4.2 If $D(g)$ is a representation of a group G, show that the matrices $D^*(g)$ also form a representation. This may or may not be equivalent to D. If the two representations are equivalent, so that $D^*(g) = C^{-1}D(g)C$, show that if D is irreducible then $CC^* = \lambda\mathbb{1}$. If further D is unitary show that $CC^\dagger = \mu\mathbb{1}$. Show that C may be redefined so that $\mu = 1$ and that C is then either symmetric or antisymmetric.

4.3 Show that the matrix $B_i^\nu := \Sigma_{g\in K_i} D^{(\nu)}(g)$, consisting of the sum of the matrices of an irreducible $[n_\nu]$ representation which correspond to the elements of the conjugacy class K_i, is a multiple of the identity: $B_i^\nu = \lambda_i^\nu\mathbb{1}$. The coefficient λ_i^ν is called the 'Dirac character' of the class K_i. By taking traces relate the Dirac character to the ordinary character χ_i^ν and hence obtain the dimensionality n_ν in terms of $[g]$, g_k and the λ_k^ν.

4.4 Let D be a general representation of a finite group G with characters χ_i. By considering its decomposition into irreducible representations show that

$$\sum_i g_i|\chi_i|^2 = [g]\sum_\mu a_\mu^2$$

and hence that the necessary and sufficient condition for irreducibility is

$$\sum_i g_i|\chi_i|^2 = [g]$$

4.5 Show that the character table for D_4 is

D_4	E	C_4^2	$2C_4$	$2C_2$	$2C_2'$
A_1	1	1	1	1	1
A_2	1	1	1	−1	−1
B_1	1	1	−1	1	−1
B_2	1	1	−1	−1	1
E	2	−2	0	0	0

4.6 Find the character table for the quaternion group Q, defined as gp$\{b, c\}$ with $c^4 = e$, $b^2 = c^2$, $bc = c^3b$. If you have done this correctly, it should turn out to be the same as that for D_4. Thus the characters cannot in general be relied on to distinguish between non-isomorphic groups of the same order.

4.7 Referring to problem 3.4, we can obtain characters of the group G by 'lifting' from those of the quotient group G/N:

$$\chi^G(g) = \chi^{G/N}(gN)$$

Consider the case of D_6. It has a normal subgroup C_2 (the centre) generated by c^3, and the structure of the factor group D_6/C_2 is D_3. Show that the characters of the irreps A_1, A_2, E_2 given in the table below can be lifted from those of the irreps of D_3 (table 4.2):

D_6	E	$2C_6$	$2C_6^2$	C_6^3	$3C_2$	$3C_2'$
A_1	1	1	1	1	1	1
A_2	1	1	1	1	−1	−1
B_1	1	−1	1	−1	1	−1
B_2	1	−1	1	−1	−1	1
E_1	2	1	−1	−2	0	0
E_2	2	−1	−1	2	0	0

4.8 Show that the character table for the tetrahedral group T is as follows (refer to problem 2.3 for the conjugacy classes):

T	E	$3C_2$	$4C_3$	$4C_3^2$
A	1	1	1	1
E_1	1	1	ω	ω^2
E_2	1	1	ω^2	ω
T	3	−1	0	0

Interpret the table in terms of mappings χ: $G \to \mathbb{C}$, establishing for each 1-dimensional representation the kernel K of the mapping and the way the cosets of K are mapped into $\mathbb{C}$.

4.9 Use the character table given in problem 4.5 to derive the Clebsch–Gordan series for the product representation $E \otimes E$ of D_4.

5

Physical Applications

In this chapter we apply the machinery so far developed to some problems in solid state and molecular physics. The first problem involves only classical mechanics, determining the restrictions placed on the macroscopic properties of a material by the symmetry group of its crystal structure. The second uses group theory to identify the normal modes of oscillation of a relatively simple molecule. This again is a classical calculation, but the discrete energy levels of the system follow immediately by quantizing the uncoupled oscillators which constitute the normal modes. The third example is purely quantum mechanical and is concerned with the determination of the splitting of the energy levels of atomic electrons by the reduced symmetry of a crystal environment.

5.1 Macroscopic Properties of Crystals

(i) Ferromagnetism and Ferroelectricity

Let us first consider the simple question of whether a crystal with a certain symmetry structure can possess a permanent magnetic dipole moment $\boldsymbol{M}$ or electric dipole moment $\boldsymbol{P}$.

The fundamental restriction on such quantities is that they should be invariant under the symmetry transformations of the *point group* P of the crystal, consisting of rotations and reflections. (These transformations, so called because they keep one point of the crystal fixed, involve large-scale displacements of other points of the crystal. Microscopic lattice translations, which together with P make up the full *space group* G, are not a relevant symmetry for such macroscopic quantities.)

A magnetic moment $\boldsymbol{M}$ is a vector quantity whose Cartesian components M_1, M_2, M_3 transform under rotations according to the [3] vector representation D^V:

$$M'_i = D^V_{ij}(g)M_j \tag{5.1}$$

Consider first the case when P does not include reflections but is a subgroup of the full rotation group. For a permanent dipole moment to exist which is invariant under the action of P we must be able to choose an $\boldsymbol{M}$ such that $\boldsymbol{M}' = \boldsymbol{M} \quad \forall g \in P$. In other words, *the representation D^V, when restricted to P, must contain the identity representation.*

It is obvious physically that for this to be the case the crystal must have a single invariant axis. For example, in a crystal with (trigonal) C_3 symmetry there is a well-defined direction in which the magnetic moment can point. But let us check this using characters.

Well, in the case of C_3 we have already calculated the decomposition of D^V, namely

$$D^V = D^{(1)} \oplus D^{(2)} \oplus D^{(3)} \tag{4.22}$$

In particular, the identity representation $D^{(1)}$ occurs just once, so there is one possible direction along which $\boldsymbol{M}$ can point.

If the symmetry were larger, for example D_3, this would no longer be possible, since there would be no invariant direction. This again is clear physically. C_3 is the symmetry of a triangle with *directed* sides, so there is a distinction between the $+z$ and $-z$ directions. However, this distinction is lost with D_3 because of the rotations b_ι which flip one into the other. As we have seen, the existence of these additional rotations is directly responsible for the different decomposition of D^V, which for D_3 is

$$D^V = A_2 \oplus E \tag{4.23}$$

and does not contain the identity representation A_1.

Magnetic and electric dipole moments have the same rotational properties but differ in their response to spatial reflections and inversions. Thus $\boldsymbol{P}$, like the electric field $\boldsymbol{E}$, is a true (polar) vector, with the property that $\boldsymbol{P} \to -\boldsymbol{P}$ when $\boldsymbol{x} \to -\boldsymbol{x}$. On the other hand, $\boldsymbol{M}$, like the magnetic induction $\boldsymbol{B}$, is an axial vector, remaining unchanged under spatial inversion. Thus crystals whose point groups include reflections may be able to support one or other of $\boldsymbol{P}$ and $\boldsymbol{M}$, but not both (see problem 5.1).

(ii) Conductivity Tensor σ_{ij}

In a non-isotropic medium such as a crystalline solid, the electric field $\boldsymbol{E}$ and the electric current density $\boldsymbol{j}$ may not be simply proportional but may be in different directions. In that case the most general linear relation between the two vectors is given by a conductivity *tensor* (which can also be thought of as a matrix) with two vector indices:

$$j_i = \sigma_{ij} E_j \tag{5.2}$$

The tensor σ_{ij} transforms like the product of the components of two vectors. For defining σ'_{ij} through

$$j'_i = \sigma'_{ij} E'_j \tag{5.3}$$

where $j'_i = D^V_{ik} j_k$ and $E'_j = D^V_{jl} E_l$, it is easy to show, using the orthogonality of the matrices D^V that

$$\sigma'_{ij} = D^V_{ik} D^V_{jl} \sigma_{kl} \tag{5.4}$$

Thus, in matrix form, $j' = D^V j$ and $E' = D^V E$, which can be inverted to give $j = (D^V)^{-1} j'$ and $E = (D^V)^{-1} E'$. Substituting in the matrix form of equation (5.2), $j = \sigma E$, and multiplying on the left by D^V gives $j' = D^V \sigma (D^V)^{-1} E'$. Comparing this with $j' = \sigma' E'$, we see that $\sigma' = D^V \sigma (D^V)^{-1}$. By virtue of the orthogonality of D^V, this is $\sigma' = D^V \sigma (D^V)^{\mathrm{T}}$, which is the matrix form of (5.4).

In technical terms this equation means that σ transforms according to the Kronecker product representation $V \otimes V$ (using the shorthand notation V for D^V). In fact we can say more than this. Time-reversal invariance (see e.g. Lax, §10.5) implies that σ_{ij} is *symmetric*.

In general we can decompose any 2-index tensor T_{ij} as $T_{ij} = T_{\{ij\}} + T_{[ij]}$, where $T_{\{ij\}} = \frac{1}{2}(T_{ij} + T_{ji})$ and $T_{[ij]} = \frac{1}{2}(T_{ij} - T_{ji})$. Its symmetric and antisymmetric parts $T_{\{ij\}}$ and $T_{[ij]}$ lie in invariant subspaces, transforming independently under rotations. The relevant transformation matrices, found by inserting the above decomposition into $T'_{ij} = D^V_{ik} D^V_{jl} T_{kl}$, are

$$D^{(V\otimes V)_\pm}_{ij,kl} = \tfrac{1}{2}[D^V_{ik} D^V_{jl} \pm D^V_{il} D^V_{jk}]$$

Correspondingly, the characters, obtained by contraction with $\delta_{ik}\delta_{jl}$, are

$$\begin{aligned} \chi^{(V\otimes V)_\pm}(g) &= \tfrac{1}{2}[D^V_{ii}(g) D^V_{jj}(g) \pm D^V_{ij}(g) D^V_{ji}(g)] \\ &= \tfrac{1}{2}[(\chi^V(g))^2 \pm \chi^V(g^2)] \end{aligned} \tag{5.5}$$

For the case of D_3 the relevant characters are displayed in table 5.1. The entry $\chi^{(V\otimes V)_+}(b)$, for example, is obtained as $\frac{1}{2}[(\chi^V(b))^2 + \chi^V(e)] = \frac{1}{2}[1 + 3]$.

The restriction on possible forms of the conductivity tensor is that they must be invariant under the transformations of $(V \otimes V)_+$ for g restricted to be a member of the point group P, in this case D_3. There will always be one such possible form, namely δ_{ij}, which is invariant

Table 5.1

D_3	(e)	(c, c^2)	(b, bc, bc^2)
V	3	0	−1
$(V \otimes V)_+$	6	0	2
$(V \otimes V)_-$	3	0	−1

under *all* rotations. In fact the invariance of δ_{ij} is just another way of expressing the fact that the matrices D^V are orthogonal. So the real question is whether there are any other forms which are invariant under the subgroup of rotations which form D_3.

This question can again be answered by use of the character table. In group theoretic terms it amounts to asking how many times the representation $(V \otimes V)_+$ contains the identity representation A_1. This is given by

$$\begin{aligned} a_1 &= \langle \chi^{(1)}, \chi^{(V\otimes V)_+} \rangle \\ &= \tfrac{1}{6}(1 \cdot 1 \cdot 6 + 2 \cdot 1 \cdot 0 + 3 \cdot 1 \cdot 2) = 2 \end{aligned}$$

Thus, apart from the unit matrix, there is one other independent symmetric matrix which is invariant under the rotations of D_3. It is easy to see that such a matrix is the diagonal matrix diag(0, 0, 1), or in tensor notation $\delta_{i3}\delta_{j3}$. It reflects the fact that the quadratic form z^2 is invariant, as well as $r^2 = x^2 + y^2 + z^2$. Thus the most general form of the matrix σ is a linear combination of the two:

$$\sigma = \begin{pmatrix} a & & \\ & a & \\ & & b \end{pmatrix}$$

On the other hand, for a crystal with cubic symmetry T the relevant characters are given in table 5.2. Now

$$a_1 = \tfrac{1}{12}(1 \cdot 6 + 3 \cdot 2 + 4 \cdot 0 + 4 \cdot 0) = 1$$

Thus such a crystal is macroscopically isotropic as far as electrical conductivity is concerned: $\sigma_{ij} = \sigma\delta_{ij}$.

Table 5.2

T	E	$3C_2$	$4C_3$	$4C_3^2$
V	3	−1	0	0
$(V \otimes V)_+$	6	2	0	0

In table 5.1 we included for completeness the character of the antisymmetric representation $(V \otimes V)_-$. Although not relevant for the conductivity tensor, it is interesting in its own right. As the observant reader may have noticed, the character of this representation is identical to that of V. This is no accident, but reflects the fact that the antisymmetric part of a 2-index tensor transforms under rotations like a vector. In the case when the tensor is actually formed from the product of the components of two vectors, $T_{ij} = a_i b_j$, this is just the familiar construction of the cross-product $\boldsymbol{a} \times \boldsymbol{b}$, with components $(a_2b_3 - a_3b_2, a_3b_1 - a_1b_3, a_1b_2 - a_2b_1)$.

In the case of conductivity σ_{ij}, a second-rank tensor, could be thought of as a matrix. However, tensors are more general than this: there exist tensors of higher rank which cannot be so regarded.

For example, in the phenomenon of piezoelectricity an electric polarization density P_i is induced by a mechanical stress, described by a second-rank tensor s_{ij}. The most general linear relation between them is then of the form

$$P_i = d_{ijk} s_{jk}$$

where d_{ijk} is a third-rank tensor, which transforms according to $V \otimes V \otimes V$.

5.2* Molecular Vibrations (H_2O)

(i) Small Oscillations and Normal Modes

Let us consider the small oscillations about its equilibrium position of a molecule consisting of N atoms. In the first instance we will treat the problem classically: the transition to quantum mechanics will be easily made at the end.

Denote by x_i, $i = 1, \ldots, 3N$, the displacements of the atoms from equilibrium, the first three entries giving the x, y and z displacements of the first atom, the next three giving those of the second atom and so on. The kinetic energy T of the system is of the form $\Sigma \frac{1}{2} m v^2$, i.e.

$$T = \tfrac{1}{2} \dot{x}_i M_{ij} \dot{x}_j$$

or in matrix notation

$$T = \tfrac{1}{2} \dot{x}^{\mathrm{T}} M \dot{x} \tag{5.6}$$

where M is the $3N \times 3N$ diagonal matrix $M = \mathrm{diag}(m_1, m_1, m_1, \ldots,$

m_N) in which $m_1, m_2, \ldots, m_N$ are the masses of the individual atoms.

Similarly the potential energy V is expressible in the form

$$V = \tfrac{1}{2}x^{\mathrm{T}}Kx \tag{5.7}$$

Without loss of generality K can be taken to be symmetric, since any antisymmetric part would not contribute to (5.7). The potential energy is quadratic in the displacements precisely because they are measured from the equilibrium position, which is a minimum of V.

Because M is diagonal we can simplify things by rescaling the coordinates, writing $y_1 = x_1\sqrt{m_1}, \ldots, y_{3N} = z_N\sqrt{m_N}$, i.e. $y = M^{1/2}x$. Equation (5.6) then reads

$$T = \tfrac{1}{2}\dot{y}^{\mathrm{T}}\dot{y} \tag{5.6$'$}$$

and (5.7) becomes

$$V = \tfrac{1}{2}y^{\mathrm{T}}K'y \tag{5.7$'$}$$

where $K' = M^{-1/2}KM^{-1/2}$.

The problem of small oscillations is solved if one can determine the *normal modes* of the system, which are patterns of displacements in which the x_i all vary coherently, with a common frequency. The most general solution of the equations of motion can be expressed as a linear superposition of these normal mode displacements.

In equations (5.6′) and (5.7′) let us look for a solution $y(t) = y(0)\cos\omega t$. Then $T = \tfrac{1}{2}\omega^2 y^{\mathrm{T}}(0)y(0)\sin^2\omega t$ and the total energy is

$$E = T + V = \tfrac{1}{2}y^{\mathrm{T}}(0)(K'\cos^2\omega t + \omega^2\mathbb{1}\sin^2\omega t)y(0) \tag{5.8}$$

In order for this to be constant we must have

$$K'y(0) = \omega^2 y(0) \tag{5.9}$$

which gives $E = \tfrac{1}{2}\omega^2 y^{\mathrm{T}}(0)y(0)$.

Equation (5.9) states that $y(0)$ is an eigenvector of the matrix K', with eigenvalue ω^2. The possible values of ω^2 are the roots of the characteristic equation $\det(K' - \omega^2\mathbb{1}) = 0$, or equivalently $\det(K - M\omega^2) = 0$. Let us label the distinct roots as ω_α^2, and the corresponding eigenvectors as $v^{(\alpha)}$, the normal mode displacements. It may well be that some eigenvalues are degenerate, i.e. that they are multiple roots of the characteristic equation. To cover that eventuality we will write the eigenvectors as $v_r^{(\alpha)}$, with an extra degeneracy label r.

It is a property of real symmetric matrices that eigenvectors corresponding to distinct eigenvalues are orthogonal in the sense that $v^{(\alpha)\mathrm{T}}v^{(\beta)} = 0$ for $\alpha \neq \beta$, and we might as well take them to be normalized. In the degenerate case the corresponding eigenvectors can be *chosen* to be orthonormal, i.e. $v_r^{(\alpha)\mathrm{T}} v_s^{(\alpha)} = \delta_{rs}$, by the real version of the Gram–Schmidt orthonormalization procedure discussed in §3.5.

Let us now write the general vector of scaled coordinates as a linear superposition of the form

$$y = \sum_{\alpha,r} q_{\alpha,r} v_r^{(\alpha)} \tag{5.10}$$

By virtue of the orthonormality of the $v^{(\alpha)}$ this substitution gives

$$T = \tfrac{1}{2} \sum_{\alpha,r} \dot{q}_{\alpha,r}^2 \tag{5.11}$$

Again

$$V = \tfrac{1}{2} \sum_{\alpha,r} \omega_\alpha^2 q_{\alpha,r}^2 \tag{5.12}$$

using orthonormality and the eigenvalue condition $K' v^{(\alpha)} = \omega_\alpha^2 v^{(\alpha)}$.

The dynamical system represented by equations (5.11) and (5.12) consists of a set of $3N$ independent harmonic oscillators. Classically the $q_{\alpha,r}$ have the time dependence $q_{\alpha,r}(t) = A_{\alpha,r} \cos(\omega_\alpha t + \varphi)$, and the energy of the configuration given by equation (5.10) is just

$$E = \tfrac{1}{2} \sum_{\alpha,r} \omega_\alpha^2 A_{\alpha,r}^2 \tag{5.13}$$

In quantum mechanics the energy levels of a simple harmonic oscillator with classical angular frequency ω are $E_n = (n + \tfrac{1}{2})\hbar\omega$. Thus the possible energy levels of our system are given by

$$E^{\mathrm{QM}} = \sum_{\alpha,r} (n_{\alpha,r} + \tfrac{1}{2})\hbar\omega_\alpha \tag{5.14}$$

In either case the solution of the problem involves the identification of the normal mode displacements $v_r^{(\alpha)}$ and finding the eigenfrequencies ω_α. This means diagonalizing the $3N \times 3N$ matrix K', a non-trivial task for even moderate values of N. However, when the dynamical system exhibits some kind of symmetry, group theory can be of great assistance in reducing the problem to manageable proportions.

Consider, then, the action of the symmetry group of the molecule on the displacements from equilibrium. Under a symmetry operation they will transform among themselves, being transferred from one atomic site to an equivalent one. The same is true for the scaled displacements y_i, since equivalent atoms have the same mass. The y_i thus form the basis for a (real) $3N$-dimensional representation D of the symmetry group G.

Moreover, both the kinetic energy and the potential energy are unchanged by a symmetry transformation $y \to y' = D(g)y$. The invariance of the kinetic energy is a consequence of the unitarity (orthogonality) of D, while that of the potential energy implies that K' is an invariant matrix, like σ in §5.1(ii), in the sense that

$$K' = D(g)^{\mathrm{T}} K' D(g) \tag{5.15}$$

or equivalently

$$[K', D(g)] = 0 \tag{5.15'}$$

As a consequence of (5.15′), the (scaled) normal mode displacements $v^{(\alpha)}$ form an invariant subspace under the action of the group. They are defined by the equation

$$K' v^{(\alpha)} = \omega_\alpha^2 v^{(\alpha)}$$

Multiplying on the left by $D(g)$ and using (5.15′), this becomes

$$K'(D(g)v^{(\alpha)}) = \omega_\alpha^2 (D(g)v^{(\alpha)})$$

showing that $D(g)v^{(\alpha)}$ is also a normal mode displacement corresponding to the same angular frequency ω_α.

Barring accidental degeneracy—for which there is almost always some deeper reason—this subspace will be irreducible. Thus each normal mode frequency ω_α will be in 1:1 correspondence with an irreducible representation $D^{(\alpha)}$, and the degeneracy of the level will be just the dimension n_α of this irrep. That is, for a given α the degeneracy label r introduced above will run from 1 to n_α.

So once again we need to find out how a reducible representation, in this case the $[3N]$ representation of the infinitesimal displacements, decomposes into irreducible components. This is most easily achieved by calculating the character χ of the representation and using the character table.

(ii) Example: The Water Molecule

(a) Decomposition of $D^{(3N)}$

The equilibrium configuration of the water molecule is illustrated in figure 5.1. It is left invariant by the following transformations, both of which exchange the positions of the two hydrogen atoms:

(1) a 2-fold rotation a about the y axis through O;
(2) a spatial reflection b in the y–z plane.

Together these generate the group $C_{2v} = \mathrm{gp}\{a, b\}$ with $a^2 = b^2 = (ab)^2$.

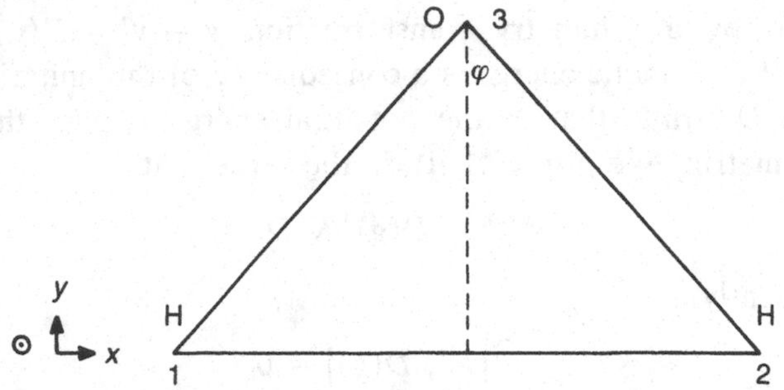

Figure 5.1 Equilibrium configuration of the water molecule H_2O.

As an abstract group it is isomorphic to D_2, but the physical meanings of the elements, and their effects on vectors, are different.

What is the effect of these transformations on the nine displacements $(x_1, y_1, \ldots, z_3)$? Consider first the rotation a. Acting on the components (x, y, z) of a general vector it reverses the signs of x and z. It also interchanges the two hydrogen atoms and therefore the suffixes 1 and 2. It thus induces the following changes:

$$\begin{aligned} a: \quad & (x_1, y_1, z_1) \to (-x_2, y_2, -z_2) \\ & (x_2, y_2, z_2) \to (-x_1, y_1, -z_1) \\ & (x_3, y_3, z_3) \to (-x_3, y_3, -z_3) \end{aligned} \tag{5.16}$$

From these transformations we can construct the 9×9 matrix $D^{(3N)}(a)$. In fact it consists mainly of zeros, with only the nine non-zero entries given by (5.16). Now we are trying to find the character $\chi^{(3N)}(a)$, which is the trace of this matrix, namely the sum of the diagonal elements. So for this purpose we can even ignore those elements which are off-diagonal. That means that we only need consider the *fixed points* of the transformation, for which the suffix remains unchanged. In the present case the fixed point is the position of the oxygen atom, with suffix 3.

Thus the trace of $D^{(3N)}(a)$ is the same as that of the submatrix which refers to the last three coordinates:

$$\chi^{(3N)}(a) = \mathrm{Tr}\begin{pmatrix} -1 & 0 & 0 \\ 0 & 1 & 0 \\ 0 & 0 & -1 \end{pmatrix} = -1 \tag{5.17}$$

Again, the reflection b interchanges the two hydrogen atoms, so for the purpose of calculating $\chi(b)$ we need only consider the fixed point O. Under b we have

$$b: \quad (x_3, y_3, z_3) \to (-x_3, y_3, z_3)$$

and hence

$$\chi^{(3N)}(b) = 1 \tag{5.18}$$

Finally ab is a reflection in the x–y plane, with $(x, y, z) \to (x, y, -z)$. All three atoms are fixed points under this transformation, so

$$\chi^{(3N)}(ab) = 3 \tag{5.19}$$

In order to determine the decomposition of $D^{(3N)}$ we append its character table to the character table of C_{2v}, table 5.3.

Table 5.3

C_{2v}	e	a	b	ab
A_1	1	1	1	1
A_2	1	1	−1	−1
B_1	1	−1	1	−1
B_2	1	−1	−1	1
$D^{(3N)}$	9	−1	1	3

Using equation (4.16) it is not difficult to establish that

$$D^{(3N)} = 3A_1 \oplus A_2 \oplus 2B_1 \oplus 3B_2 \tag{5.20}$$

(b) Elimination of translations and rotations

However, six of the normal modes are not vibrational modes at all. There are three which just correspond to uniform translations and another three to rotations of the system as a whole. Since no energy difference is produced by either type of transformation, they are in fact zero-frequency modes, with $\omega^2 = 0$. We would therefore like to subtract out these rather trivial motions so as to expose the true vibrational modes. In order to do so we need to determine which representations occurring in (5.20) correspond to uniform translations and rotations.

The normal mode pattern generated by a translation T_x in the x direction is $x_1 = x_2 = x_3 = q_{T_x}$, say. We have seen above that the x component of a vector changes sign under the action of both a and b. That is, it transforms according to the representation B_2. In a similar way it is easy to see that the patterns generated by translations T_y, T_z in the y and z directions transform according to the irreps A_1, B_1 respectively.

On the other hand the displacement $\delta \boldsymbol{r} = \delta\theta\, \boldsymbol{n} \times \boldsymbol{r}$ produced by an infinitesimal rotation is a pseudovector rather than a true vector, behaving oppositely with respect to spatial reflections. For example, the pattern generated by a rotation R_y about the y axis through O is $z_1 = -z_2 = q_{R_y}$, say, and all other displacements are zero. Referring to equation (5.16) we see that this pattern is left invariant under the action of a, which interchanges z_1 and $-z_2$. Under the action of b, $z_1 \rightarrow +z_2$, and so the pattern is reversed: $q_{R_y} \rightarrow -q_{R_y}$. Thus the normal mode corresponding to rotations about the y axis transforms according to the irrep A_2. Following a similar line of argument the reader should be able to show that q_{R_x} and q_{R_z} transform according to B_1 and B_2 respectively.

In table 5.4 we have repeated the character table of C_{2v}, incorporating the additional information we have just learned. The upshot is that the uniform translations and rotations account for the irreps $A_1 \oplus A_2 \oplus 2B_1 \oplus 2B_2$. Subtracting these from the RHS of equation (5.20) we arrive at the representation content of the true vibrational modes:

$$D^{(3N)}_{\text{vib}} = 2A_1 \oplus B_2 \tag{5.21}$$

Table 5.4

C_{2v}	e	a	b	ab		
A_1	1	1	1	1	T_y	
A_2	1	1	−1	−1		R_y
B_1	1	−1	1	−1	T_z	R_x
B_2	1	−1	−1	1	T_x	R_z

(c) Displacement patterns of vibrational modes

Our task is not yet completed, since we need to know the actual patterns of displacements which correspond to these remaining representations. The patterns transforming according to a given irrep $D^{(\sigma)}$ can be found by the application of the *projection operator*

$$P^{(\sigma)}_{ij} := \sum_{g} D^{(\sigma)*}_{ij}(g) T(g) \tag{5.22}$$

to the elements of a vector space V carrying a reducible representation which includes $D^{(\sigma)}$ in its decomposition. Here $T(g)$ is the linear operator acting in V induced by the group element g.

The claim is that for any $\boldsymbol{v} \in V$ and fixed j the vector $\boldsymbol{u}_i := P_{ij}^{(\sigma)}\boldsymbol{v}$ transforms according to the irrep $D^{(\sigma)}$, i.e. $T(g)\boldsymbol{u}_i = D_{ki}^{(\sigma)}(g)\boldsymbol{u}_i$. The proof involves manipulations very similar to those used in the derivation of the fundamental orthogonality theorem following equation (4.3). Thus

$$T(h)(P_{ij}^{(\sigma)})\boldsymbol{v} = \sum_g D_{ij}^{(\sigma)*}(g)T(h)T(g)\boldsymbol{v}$$

Writing $T(h)T(g) \equiv T(hg) = T(g')$, so that $g = h^{-1}g'$, changing the summation variable to g' and using the group property of the $D^{(\sigma)}$ gives

$$\begin{aligned} T(h)(P_{ij}^{(\sigma)})\boldsymbol{v} &= \sum_{g'} D_{ik}^{(\sigma)*}(h^{-1})D_{kj}^{(\sigma)*}(g')T(g')\boldsymbol{v} \\ &= D_{ki}^{(\sigma)}(h)\left(\sum_{g'} D_{kj}^{(\sigma)*}(g')T(g')\boldsymbol{v}\right) \end{aligned} \tag{5.23}$$

by virtue of the unitarity of $D^{(\sigma)}$. In other words

$$T(h)\boldsymbol{u}_i = D_{ki}^{(\sigma)}(h)\boldsymbol{u}_k \tag{5.24}$$

as required.

In the present case all the irreps are 1-dimensional. So the matrices $D^{(\sigma)}$ are identical to the characters, and the indices i, j are both equal to 1 and may as well be omitted. The expression for $P_{ij}^{(\sigma)}$ then simplifies to

$$P^{(\sigma)} = \sum_g \chi^{(\sigma)*}(g)T(g) \tag{5.22$'$}$$

According to equation (5.21) we need to find vectors which transform according to $D^{(1)} \equiv \mathrm{A}_1$ and $D^{(3)} \equiv \mathrm{B}_2$. Applying $P^{(1)}$ to $\boldsymbol{v}_{1x}$, a small displacement of atom 1 in the x direction, we obtain

$$\begin{aligned} P^{(1)}\boldsymbol{v}_{1x} &= (\chi^{(1)*}(e)T(e) + \chi^{(1)*}(a)T(a) + \chi^{(1)*}(b)T(b) \\ &\quad + \chi^{(1)*}(ab)T(ab))\boldsymbol{v}_{1x} \\ &= 1\cdot\boldsymbol{v}_{1x} + 1\cdot(-\boldsymbol{v}_{2x}) + 1\cdot(-\boldsymbol{v}_{2x}) + 1\cdot\boldsymbol{v}_{1x} \\ &= 2(\boldsymbol{v}_{1x} - \boldsymbol{v}_{2x}) \end{aligned} \tag{5.25}$$

This means a pattern of displacements in which $x_1 = -x_2 = q_1$, say, and all others are zero. Since $x_1 \rightarrow -x_2$ under the action of both a and b, we see that q_1 is indeed invariant under the action of the group. The motion of the atoms in this normal mode is illustrated in figure 5.2(a). It is clear that this is a pure vibrational mode since the position of the centre of mass is unchanged and the net angular momentum is zero.

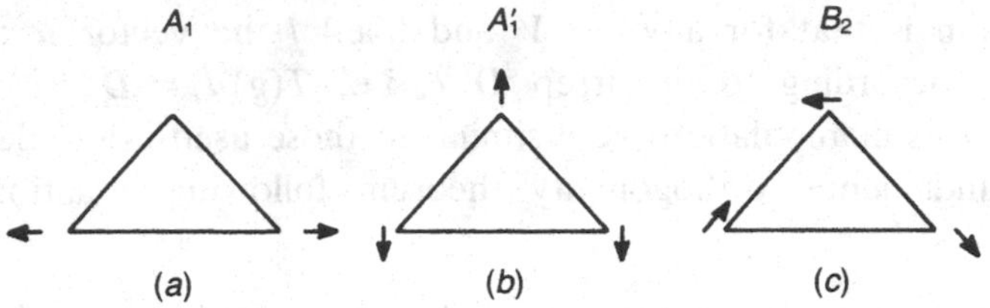

Figure 5.2 Displacement patterns in normal modes of vibration of the water molecule.

In a similar way one can show that

$$P^{(1)}\boldsymbol{v}_{1y} = 2(\boldsymbol{v}_{1y} + \boldsymbol{v}_{2y})$$
$$P^{(1)}\boldsymbol{v}_{3y} = 4\boldsymbol{v}_{3y} \tag{5.26}$$

Here there are two independent patterns: $y_1 = y_2$, and y_3. However, to form the true vibrational mode we must combine them in such a way that the centre of mass of the molecule is not displaced. The appropriate combination, $y_1 = y_2 = -(m_O/(2m_H))y_3 = q_2$, say, is illustrated in figure 5.2(b).

Turning now to the representation $D^{(3)} \equiv B_2$, the operator $P^{(3)}$ when applied to $\boldsymbol{v}_{1x}$, $\boldsymbol{v}_{1y}$ and $\boldsymbol{v}_{3x}$ projects out the following vectors:

$$P^{(3)}\boldsymbol{v}_{1x} = 2(\boldsymbol{v}_{1x} + \boldsymbol{v}_{2x})$$
$$P^{(3)}\boldsymbol{v}_{1y} = 2(\boldsymbol{v}_{1y} - \boldsymbol{v}_{2y}) \tag{5.27}$$
$$P^{(3)}\boldsymbol{v}_{3x} = 4\boldsymbol{v}_{3x}$$

In order to eliminate any net linear momentum, the two modes in the x direction must be combined as above to give $x_1 = x_2 = -(m_O/(2m_H))x_3$. However, this configuration carries net angular momentum, which must be cancelled by a judicious admixture of the other mode. Taking moments about O it is clear that that admixture must be just such as to make the displacements of the two hydrogen atoms directly along the sides of the equilibrium triangle, as shown in figure 5.2(c). Analytically that means that $y_1 = -y_2 = x_1 \cot\varphi = q_3$, say.

(d) Calculation of normal mode frequencies

This is the only part of the calculation which cannot be done by group theory alone: one needs to know the specific details of the potential V. What we have done, however, is to simplify the problem considerably. All that remains to be done to calculate the vibrational frequencies is to substitute into V the special patterns of displacements isolated in figure 5.2. In effect we have used representation theory to diagonalize a 9×9 matrix.

5.3 Raising of Degeneracy

(i) Degeneracy

Before considering the raising, or breaking, of degeneracy it is necessary to consider the concept of degeneracy itself and its relation to group theory. We are typically concerned with the eigenvalues and eigenvectors of a quantum Hamiltonian H_0 which is invariant under a group of symmetry transformations G.

In Dirac notation (Appendix A) the energy eigenvalue equation is

$$H_0|E^{(0)}\rangle = E^{(0)}|E^{(0)}\rangle \tag{5.28}$$

and the invariance of H_0 is expressed by

$$[H_0, U(g)] = 0 \tag{5.29}$$

where $U(g)$ is the unitary operator induced in the space of quantum mechanical states by the physical transformation g.

Because of the invariance of H_0 we have

$$\begin{aligned} H_0(U(g)|E^{(0)}\rangle) &= U(g)H_0|E^{(0)}\rangle \\ &= U(g)E^{(0)}|E^{(0)}\rangle \\ &= E^{(0)}(U(g)|E^{(0)}\rangle) \end{aligned} \tag{5.30}$$

That is, $U(g)|E^{(0)}\rangle$ is again an eigenstate of H_0 with the same eigenvalue $E^{(0)}$. Thus eigenstates $|E^{(0)}\rangle$ corresponding to a given eigenvalue $E^{(0)}$ transform among themselves under the action of the group. They thus form a submodule in the complete space of eigenvectors and provide the basis of a *representation* of G. It could be that there is only one eigenstate with the given eigenvalue, in which case we speak of a 'non-degenerate' level. The representation is then just the trivial representation. However, in many examples of physical interest there is more than one such eigenstate, in which case we speak of the level as being 'degenerate'.

In the latter case the action of the group on the space of degenerate states of the level induces an r-dimensional representation, where r is the number of degenerate eigenvectors. In general (i.e. barring 'accidental' degeneracy, which in fact can invariably be traced to some additional symmetry of the problem) there is no reason to expect smaller invariant subspaces, which means that the representation will be *irreducible*. Thus a given energy level $E^{(0)}$ will correspond to an irreducible representation $D^{(\mu)}$, say, of G, and the degeneracy r will be just the dimensionality n_μ of $D^{(\mu)}$. The level can be labelled as $E^{(0)}_{\alpha\mu}$, where α comprises other labels not connected with the group.

An example which immediately springs to mind is the energy spectrum of a particle moving in a central potential. The symmetry group is the 3-dimensional rotation group SO(3), whose irreducible representations, of dimension $2l + 1$, are labelled by the integer l associated with the angular part of the wave equation. The principal quantum number n, on the other hand, is associated with solutions of the radial equation. For a general potential $U(r)$ the levels $E^{(0)}_{nl}$ are distinct. However, in the most familiar problem of all, $U = -k/r$, there occurs the 'accidental' degeneracy $E^{(0)}_{nl} \rightarrow E^{(0)}_{n}$ with $l < n$, giving a degeneracy $r = n^2$. This additional degeneracy, which means that each level corresponds to a *reducible* representation of SO(3), arises from invariance of the $1/r$ potential under the larger group SO(4), as will be explained in §7.2.

(ii) Breaking of Degeneracy: $H_0 \rightarrow H_1$

Now suppose that the invariance group of the Hamiltonian is reduced by the introduction of an additional interaction which does not respect the full symmetry:

$$H_0 \rightarrow H_1 = H_0 + V \tag{5.31}$$

invariant under the subgroup $H \subset G$.

The new levels are now classified by irreps of H rather than G, and in order to find how a given level splits up we need to know how an irreducible representation of G decomposes into irreducible representations of H.

(It is important to realize that an irreducible representation of G does not in general remain irreducible when g is restricted to a subgroup. For irreducibility means that there are no invariant subspaces as g varies over all the members of G. But there could well be invariant subspaces under the weaker condition that g varies only over the subset H.)

If the relevant matrix elements of V are small compared with the separation of the original levels, we can identify where the new levels came from, as illustrated schematically in figure 5.3. Here μ, ν label irreps of G, while σ, τ denote irreps of H. We can easily find the pattern of splittings, though not the magnitude of the shifts, by our old friend the character table.

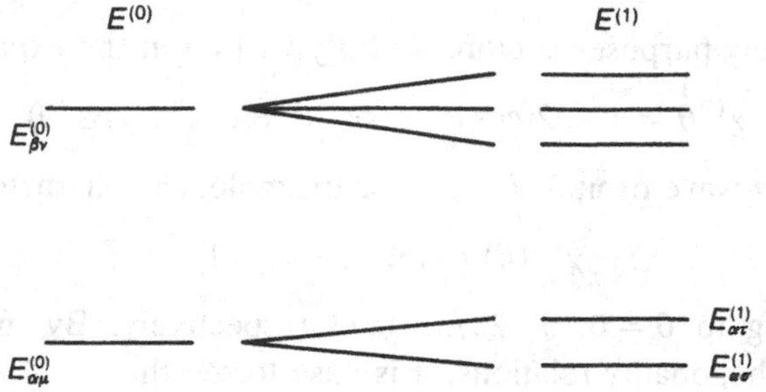

Figure 5.3 Splitting of degenerate energy levels by a perturbation V. Before the perturbation is added, the levels are classified by the irreps of G, the symmetry group of the original Hamiltonian H_0. After addition of V, the symmetry of $H_1 = H_0 + V$ is reduced to a subgroup H of G, and the levels are classified by the irreps of H.

Example: crystalline splitting of atomic energy levels

The symmetry of the atomic potential in, say, an alkali atom (or more generally a many-electron atom in the Hartree–Fock approximation; see §7.1) is reduced when the atom is situated in a crystal environment from the full rotational group SO(3) to the point group H of the crystal. Energy levels which were previously degenerate will accordingly split apart.

Consider, for example, a crystal whose symmetry group is the tetrahedral group T. Its character table (problem 4.8) is given in table 5.5.

Table 5.5

	T	E	$3C_2$	$4C_3$	$4C_3^2$
	A	1	1	1	1
E {	E_1	1	1	ω	ω^2
	E_2	1	1	ω^2	ω
	T	3	−1	0	0

We will have to anticipate a result from the next chapter, where it is shown that the character of SO(3) for orbital angular momentum l is

$$\chi^{(l)}(\theta) = \frac{\sin(l + \frac{1}{2})\theta}{\sin \frac{1}{2}\theta} \tag{5.32}$$

which for many purposes is more usefully written in the expanded form

$$\chi^{(l)}\theta = 1 + 2(\cos\theta + \cos 2\theta + \ldots + \cos l\theta) \tag{5.32'}$$

Taking a D-wave orbital ($l = 2$), for example, the character is

$$\chi^{(2)}(\theta) = (5, 1, -1, -1) \tag{5.33}$$

corresponding to $\theta = 0, \pi, 2\pi/3, 4\pi/3$ respectively. By inspection, or using the orthogonality relations, it is easy to see that

$$\chi^{(2)} = \chi^{(E_1)} + \chi^{(E_2)} + \chi^{(T)} \tag{5.34}$$

and hence

$$D^{(2)} = E_1 \oplus E_2 \oplus T \tag{5.35}$$

So the original 5-fold degeneracy is broken according to

$$5 \rightarrow 1 + 1 + 3$$

Actually there is an 'accidental' degeneracy in that the two levels E_1 and E_2 have the same energy. As usual this is not really accidental, but is a consequence of time-reversal invariance, or the reality of the perturbing potential V. Indeed, in many physical situations the two levels remain degenerate and for many purposes can be considered as a single level E. However, they will be split apart by a magnetic field.

Problems for Chapter 5

5.1 The point group C_{3v} is, among other things, the symmetry group of the ammonia molecule NH_3, which forms a right pyramid on an equilateral triangle base, as shown below:
It is generated by a 3-fold rotation c about the axis OD and a reflection σ_v in the plane OAD.

(i) Show that $C_{3v} \cong D_3$.
(ii) Construct the [3] representation provided by the transformation of a general vector $\boldsymbol{r}$.
(iii) Hence show that a crystal with this symmetry can possess a permanent electric dipole moment $\boldsymbol{P}$.

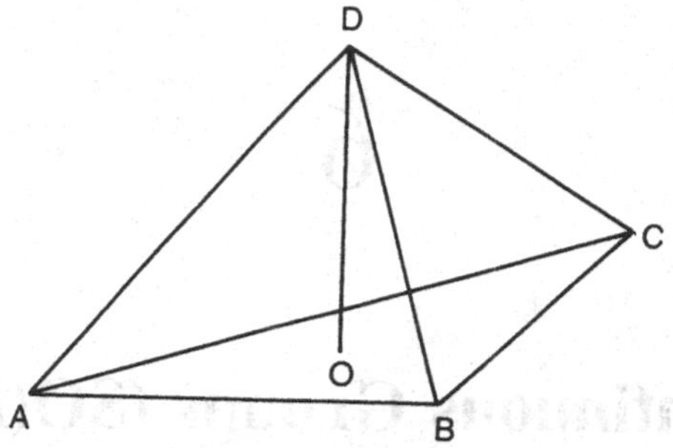

(iv) A magnetic dipole moment, like a magnetic field, is an *axial* vector, transforming under reflections with the opposite sign to a true (polar) vector. Show that such a crystal cannot possess a permanent magnetic moment.

5.2 Find the form of the conductivity tensor for a crystal with D_3 symmetry by explicitly finding the restrictions imposed on the elements σ_{ij} by the requirement that $\sigma = D(g)\sigma D(g)^{-1}$ (or $D(g)\sigma = \sigma D(g)$).

5.3 Find the normal mode frequencies and displacement patterns for small oscillations of the double pendulum consisting of masses m_1 and m_2 fixed on a weightless string at distances l and $2l$ respectively from the point of suspension.

5.4 Find how the 3-fold degeneracy of an $l = 1$ atomic orbital is broken when the SO(3) symmetry is reduced by a crystal environment to D_4. (The character table of D_4 is given in problem 4.5.)

5.5 An atom is located in a hexagonal crystal of point group D_6. Calculate how the $l = 1$ and $l = 2$ eigenstates of a central potential are split by the crystal potential, giving the new degeneracies. (Refer to problem 4.7 for the character table of D_6.)

The hexagonal symmetry is further broken, by a trigonal distortion, to the subgroup D_3. Does this produce any further splitting?

Finally, the introduction of a magnetic field along the original 6-fold axis reduces the symmetry to C_3. Without further calculation state how the previous degeneracy is affected.

6

Continuous Groups (SO(*N*))

So far we have restricted our attention to finite groups, consisting for the most part of rotations of finite order, i.e. through an angle $2\pi/n$. However, there are a large number of physical situations in which the relevant group is the full rotation group in 2 or 3 dimensions made up of rotations through an arbitrary angle.

These are *compact* groups, in which the angular parameters vary over finite ranges $[0, \pi)$ or $[0, 2\pi)$, and we can in fact adapt the previous machinery of characters, with the sum over group elements Σ_g being replaced by an appropriate integral. However, for continuous (Lie) groups it is also extremely useful to consider the *infinitesimal generators* of the group, which form a structure known as a *Lie algebra*. This is how the angular momentum operators J_i satisfying the commutation relations $[J_1, J_2] = i\hbar J_3$, etc. arise in the context of the [3] rotation group SO(3).

6.1 SO(2)

This is the group of proper rotations in 2 dimensions. These are all about the same axis, which we can take as the z axis, and are labelled by the angle φ, with $0 \leqslant \varphi \leqslant 2\pi$ (figure 6.1).

We have already worked out (see equation (3.4)) how the coordinates of a general point P transform under such a rotation. Restricting ourselves to the x–y plane, i.e. $z = 0$, the transformation is

$$\begin{pmatrix} x' \\ y' \end{pmatrix} = \begin{pmatrix} \cos\varphi & -\sin\varphi \\ \sin\varphi & \cos\varphi \end{pmatrix} \begin{pmatrix} x \\ y \end{pmatrix} \tag{6.1}$$

The 2×2 matrix

$$R(\varphi) = \begin{pmatrix} \cos\varphi & -\sin\varphi \\ \sin\varphi & \cos\varphi \end{pmatrix}$$

which implements the transformation is orthogonal, $R(\varphi)R^{\mathrm{T}}(\varphi) = \mathbb{1}$, and has unit determinant, $\det R(\varphi) = 1$. By virtue of these properties $R(\varphi)$ is a member of the group SO(2).

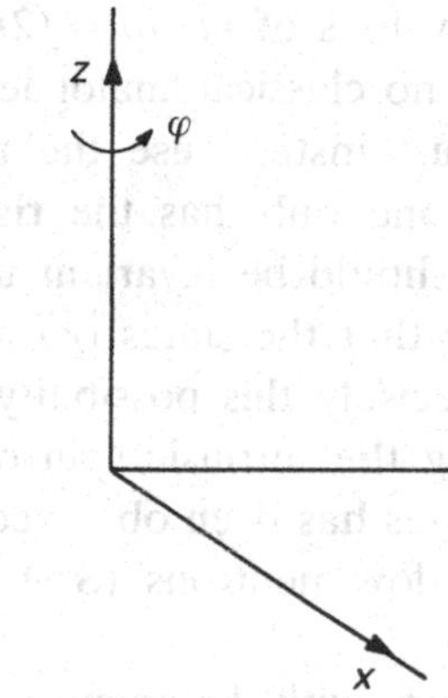

Figure 6.1 The SO(2) rotation $R(\varphi)$.

Strictly speaking the $R(\varphi)$ constitute a [2] *representation* of the abstract group, but since the kernel of the mapping is just the identity element ($\varphi = 0$) it is a faithful representation, isomorphic to the group itself.

The [2] representation is irreducible over the field of the real numbers, but reducible over the field $\mathbb{C}$ of complex numbers. We have already seen in fact (§4.4) that

$$\begin{pmatrix} x' + \mathrm{i}y' \\ x' - \mathrm{i}y' \end{pmatrix} = \begin{pmatrix} \mathrm{e}^{\mathrm{i}\varphi} & 0 \\ 0 & \mathrm{e}^{-\mathrm{i}\varphi} \end{pmatrix} \begin{pmatrix} x + \mathrm{i}y \\ x - \mathrm{i}y \end{pmatrix} \tag{6.2}$$

This makes perfect sense when translated back into polar coordinates (r, θ):

$$\begin{pmatrix} r'\mathrm{e}^{\mathrm{i}\theta'} \\ r'\mathrm{e}^{-\mathrm{i}\theta'} \end{pmatrix} = \begin{pmatrix} r\ \mathrm{e}^{\mathrm{i}(\theta+\varphi)} \\ r\ \mathrm{e}^{-\mathrm{i}(\theta+\varphi)} \end{pmatrix} \tag{6.2'}$$

i.e. $r' = r$, $\theta' = \theta + \varphi$. But again each of these [1] representations is faithful. They are 1×1 unitary matrices, and so the group can also be designated as U(1).

The group is clearly Abelian, with $R(\varphi)R(\varphi') = R(\varphi + \varphi')$, and hence its irreducible representations are all 1-dimensional. Its single-valued irreps are labelled by the integers:

$$D^{(m)}(\varphi) = \mathrm{e}^{-\mathrm{i}m\varphi} \tag{6.3}$$

with $m \in \mathrm{Z}$ in order that $D^{(m)}(\varphi + 2\pi) = D^{(m)}(\varphi)$.

In the context of quantum mechanics, however, it has to be recognized that the latter constraint is too restrictive. As we shall see in the case of SO(3) and the closely related group SU(2), representations do occur which have half-integral values of m: $m = (2n + 1)/2$. The point is that these states, which have no classical analogue, are not representable as wavefunctions: one must instead use the more general Dirac formalism. In this framework one only has the right to demand that bilinear objects such as $\langle \chi | \varphi \rangle$ should be invariant under a 2π rotation. This leaves open the possibility that the states $|\chi\rangle$ and $|\varphi\rangle$ change sign under such a rotation. It is precisely this possibility that is realized by the quantum states representing the intrinsic spin of fermions. The 4π periodicity of spin-$\frac{1}{2}$ fermion states has been observed directly by elegant interference experiments with slow neutrons (S A Werner *et al* 1975 *Physical Review Letters* **35** 1053).

Since SO(2) is a compact group, with the angular parameter φ varying over the finite range $0 \leqslant \varphi < 2\pi$, the group average $(1/[g])\Sigma_g$ which occurred for a finite group can be generalized, to $\int_0^{2\pi} \mathrm{d}\varphi/2\pi$. As is easily seen, this gives the correct orthonormality of the characters:

$$\langle \chi^{(m)}, \chi^{(m')} \rangle = \int_0^{2\pi} \frac{\mathrm{d}\varphi}{2\pi} \mathrm{e}^{\mathrm{i}m\varphi} \mathrm{e}^{-\mathrm{i}m'\varphi} = \delta_{mm'} \tag{6.4}$$

The Clebsch–Gordan series for the decomposition of the product of two irreps is very simple:

$$D^{(m)} \otimes D^{(m')} = D^{(m+m')} \tag{6.5}$$

as can be seen by inspection or use of (6.4).

We can use the characters to check the decomposition of the 2-dimensional defining representation:

$$\begin{aligned} a_m &= \frac{1}{2\pi} \int \mathrm{e}^{\mathrm{i}m\varphi}\, 2\cos\varphi\, \mathrm{d}\varphi \\ &= \frac{1}{2\pi} \int \mathrm{e}^{\mathrm{i}m\varphi} (\mathrm{e}^{\mathrm{i}\varphi} + \mathrm{e}^{-\mathrm{i}\varphi})\, \mathrm{d}\varphi \\ &= \delta_{m,-1} + \delta_{m1} \end{aligned}$$

Hence

$$R = D^{(1)} \oplus D^{(-1)}$$

as we found above.

The Infinitesimal Generator (J_z)

The generators of a continuous (Lie) group are introduced by considering elements infinitesimally near to the identity element. In the present case, for small φ we have an expansion of $R(\varphi)$ of the form

$$R(\varphi) = \mathbb{1} - \mathrm{i}\varphi X + \mathrm{O}(\varphi^2) \tag{6.6}$$

or equivalently

$$-\mathrm{i}X = \left.\frac{\mathrm{d}R(\varphi)}{\mathrm{d}\varphi}\right|_{\varphi=0} \tag{6.7}$$

X is called the *infinitesimal generator* of the SO(2) rotations, in the defining representation. In any other representation the corresponding generator is defined in a similar way.

In the defining representation, differentiation of the explicit expression for $R(\varphi)$ gives

$$-\mathrm{i}X = \begin{pmatrix} 0 & -1 \\ 1 & 0 \end{pmatrix} \tag{6.8}$$

Note that X is *Hermitian* (in fact $X = \sigma_2$, one of the Pauli matrices). This is a consequence of the unitarity (orthogonality) of R. Thus

$$\begin{aligned} \mathbb{1} = R^{\dagger}(\varphi)R(\varphi) &= (\mathbb{1} + \mathrm{i}\varphi X^{\dagger})(\mathbb{1} - \mathrm{i}\varphi X) \\ &= \mathbb{1} - \mathrm{i}\varphi(X - X^{\dagger}) + \mathrm{O}(\varphi^2) \end{aligned}$$

and hence $X = X^{\dagger}$. In a similar fashion the tracelessness of X is a consequence of the fact that R is a 'special' orthogonal matrix, with unit determinant.

The most important property of the infinitesimal generator is that the matrices of all proper rotations, i.e. those continuously connected to the identity (as opposed to the reflection $x \to x$, $y \to -y$), can be obtained by *exponentiation* of X, namely

$$R(\varphi) = \exp(-\mathrm{i}\varphi X) \tag{6.9}$$

The exponential of a matrix is given meaning as the exponential series, and in the present case this is easily evaluated, to verify (6.9). Thus

$$\exp(-\mathrm{i}\varphi X) = \sum_{0}^{\infty} (-\mathrm{i}\varphi)^n X^n$$

Since $X^2 = \mathbb{1}$, the series splits naturally into even and odd terms. The even terms are multiples of the unit matrix and form the series for $\cos\varphi$, while the odd terms all contain a single factor of $-\mathrm{i}X$ and sum to $-\mathrm{i}X\sin\varphi$. Thus

$$\exp(-\mathrm{i}\varphi X) = \mathbb{1}(\cos\varphi) - \mathrm{i}X(\sin\varphi)$$

which indeed is the matrix $R(\varphi)$.

The fundamental reason that R can be expressed in the form (6.9) is the Abelian composition law $R(\varphi)R(\varphi') = R(\varphi + \varphi')$. Differentiating this identity with respect to φ' and then setting $\varphi' = 0$ gives

$$\frac{\mathrm{d}R(\varphi)}{\mathrm{d}\varphi} = -\mathrm{i}XR(\varphi) \tag{6.10}$$

This is a generalization of the defining equation (6.7) to general φ. It is a matrix differential equation for $R(\varphi)$ whose formal solution is equation (6.9).

In a general representation D the corresponding generator is similarly defined as

$$-\mathrm{i}X = \left.\frac{\mathrm{d}D(\varphi)}{\mathrm{d}\varphi}\right|_{\varphi=0} \tag{6.11}$$

and again

$$D(\varphi) = \exp(-\mathrm{i}\varphi X)$$

For the irreducible representations, which are all 1-dimensional, the generators are rather trivial, namely $X^{(m)} = m$.

In wave mechanics the generator X is realized in a somewhat different way, as a differential operator $\hat{X}$ acting on wavefunctions $\psi(r, \theta)$. Recall that in general a physical rotation R induces a corresponding transformation in the space of wavefunctions (equation (3.10)). The transformation from ψ to ψ' defines a (unitary) operator $\hat{U}_R$ according to

$$\hat{U}_R\psi(\boldsymbol{x}) = \psi(R^{-1}\boldsymbol{x}) \tag{3.10'}$$

In the present case this reads

$$(1 - \mathrm{i}\varphi\hat{X})\psi(r, \theta) = \psi(r, \theta - \varphi) \tag{6.12}$$

for φ small. Hence we can identify $\mathrm{i}\hat{X} = \mathrm{d}/\mathrm{d}\theta$, i.e.

$$\hat{X} = -\mathrm{i}\frac{\mathrm{d}}{\mathrm{d}\theta} = \frac{\hat{J}_z}{\hbar} \tag{6.13}$$

where we recognize the usual operator for the z component of angular momentum. The wavefunctions of the irrep $D^{(m)}$ are $u_m(\theta) = e^{im\theta}$, satisfying $\hat{X}u_m = mu_m$.

6.2 SO(3) (SU(2))

This group is the group of all proper rotations in 3 dimensions. The defining representation consists of 3×3 matrices $R_{\boldsymbol{n}}(\varphi)$, say, which implement on the Cartesian coordinates x, y, z the effect of a rotation through φ about the axis given by the unit vector $\boldsymbol{n}$. As discussed at the end of §3.1 for the particular case $\boldsymbol{n} = (0, 0, 1)$, these matrices form a *faithful* representation of the abstract group, and so from the group theoretical point of view are synonymous with it. Again, as discussed there, the matrices R are orthogonal, in order to preserve scalar products, and have unit determinant, like the unit matrix, with which they are continuously connected. They thus form the special orthogonal group in 3 dimensions, SO(3). The relation of this group to SU(2), the group of special unitary matrices in 2 dimensions, will be elucidated in Chapter 8.

Infinitesimal Generators

Let us start again with rotations about the z axis, labelling the corresponding matrix as $R_3(\varphi)$. This is just equation (3.5) in another guise, or $R(\varphi)$ of SO(2) writ large, namely

$$R_3(\varphi) = \begin{pmatrix} \cos\varphi & -\sin\varphi & 0 \\ \sin\varphi & \cos\varphi & 0 \\ 0 & 0 & 1 \end{pmatrix} \tag{6.14}$$

The corresponding generator is given by

$$-\mathrm{i}X_3 = \left.\frac{\mathrm{d}R_3(\varphi)}{\mathrm{d}\varphi}\right|_{\varphi=0} = \begin{pmatrix} 0 & -1 & 0 \\ 1 & 0 & 0 \\ 0 & 0 & 0 \end{pmatrix} \tag{6.15}$$

which is just an enlarged version of equation (6.8). Again X_3 is Hermitian and traceless, reflecting respectively the orthogonality and unit determinant of R_3.

It is worth noting that the Cartesian matrix elements $(X_3)_{ij}$ of X_3 can be written in a compact form using the completely antisymmetric tensor ε_{ijk} and using the summation convention for repeated indices. The indices i, j, k each range from 1 to 3, and ε_{ijk} is fixed by the defining value $\varepsilon_{123} = +1$ and its property of antisymmetry under exchange of any two indices. Thus, for example, $\varepsilon_{213} = -1$ and $\varepsilon_{112} = 0$. In this notation we can write

$$(X_3)_{ij} = -\mathrm{i}\varepsilon_{ij3} \tag{6.15$'$}$$

Unlike the case of SO(2), that is not the end of the story. We need to know the matrices and generators about axes other than the z axis, and these will not commute with R_3, as any practitioner of the Rubik cube will know! The matrix R_1 is easily found to be

$$R_1(\varphi) = \begin{pmatrix} 1 & 0 & 0 \\ 0 & \cos\varphi & -\sin\varphi \\ 0 & \sin\varphi & \cos\varphi \end{pmatrix} \tag{6.16}$$

and correspondingly

$$-\mathrm{i}X_1 = \begin{pmatrix} 0 & 0 & 0 \\ 0 & 0 & -1 \\ 0 & 1 & 0 \end{pmatrix} \tag{6.17}$$

or in tensor form

$$(X_1)_{ij} = -\mathrm{i}\varepsilon_{ij1} \tag{6.17$'$}$$

Similarly

$$R_2(\varphi) = \begin{pmatrix} \cos\varphi & 0 & \sin\varphi \\ 0 & 1 & 0 \\ -\sin\varphi & 0 & \cos\varphi \end{pmatrix} \tag{6.18}$$

with

$$-\mathrm{i}X_2 = \begin{pmatrix} 0 & 0 & 1 \\ 0 & 0 & 0 \\ -1 & 0 & 0 \end{pmatrix} \tag{6.19}$$

or

$$(X_2)_{ij} = -\mathrm{i}\varepsilon_{ij2} \tag{6.19$'$}$$

Note that the three equations for the Cartesian matrix elements can be summarized in the single equation

$$(X_k)_{ij} = -\mathrm{i}\varepsilon_{ijk} \tag{6.20}$$

An alternative, more physical way of deriving the forms of the infinitesimal generators is to consider the actual displacement produced by a rotation through a small angle φ about a general axis $\boldsymbol{n}$. Referring to figure 6.2, the displacement is perpendicular to the plane defined by $\boldsymbol{r}$ and $\boldsymbol{n}$ and of magnitude $(r\sin\theta)\varphi$. This is summarized in the vector equation

$$\delta \boldsymbol{r} = \boldsymbol{n} \times \boldsymbol{r}\varphi$$

i.e.

$$\boldsymbol{r}' = \boldsymbol{r} + \varphi(\boldsymbol{n} \times \boldsymbol{r}) \tag{6.21}$$

or in tensor notation, where it is the ε symbol that gives the components of the cross-product,

$$\begin{aligned} r'_i &= r_i + \varphi\varepsilon_{ikj}n_k r_j \\ &= r_i - \varphi\varepsilon_{ijk}n_k r_j \end{aligned} \tag{6.22}$$

This can be written as

$$r'_i = (\delta_{ij} - \mathrm{i}\varphi\varepsilon_{ijk}n_k)r_j$$

showing that

$$\begin{aligned} (R_{\boldsymbol{n}}(\varphi))_{ij} &= \delta_{ij} - \mathrm{i}\varphi\varepsilon_{ijk}n_k \\ &= \delta_{ij} - \mathrm{i}\varphi(X_{\boldsymbol{n}})_{ij}, \text{ say} \end{aligned} \tag{6.23}$$

where

$$(X_{\boldsymbol{n}})_{ij} = -\mathrm{i}\varepsilon_{ijk}n_k = n_k(X_k)_{ij} \tag{6.24}$$

Thus

$$\boxed{X_{\boldsymbol{n}} = \boldsymbol{n}\cdot\boldsymbol{X}} \tag{6.25}$$

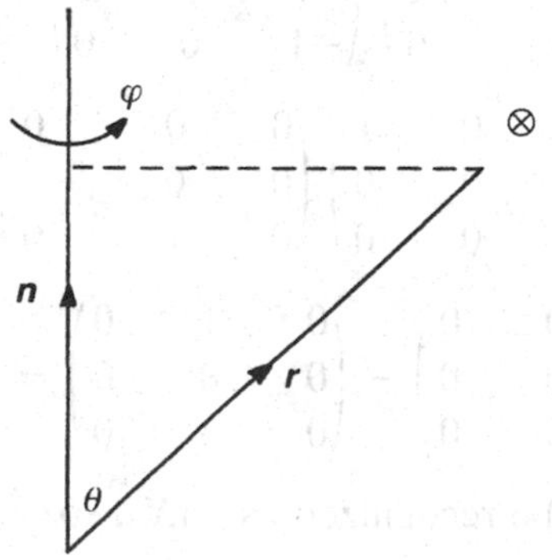

Figure 6.2 Displacement produced by a small rotation through φ about the axis $\boldsymbol{n}$.

This agrees with our previous derivation, and moreover shows that the matrix for a general rotation $R_{\boldsymbol{n}}(\varphi)$ is given by

$$\boxed{R_{\boldsymbol{n}}(\varphi) = \mathrm{e}^{-\mathrm{i}\boldsymbol{n}\cdot\boldsymbol{X}\varphi}} \tag{6.26}$$

Note that the general generator $X_{\boldsymbol{n}} = \boldsymbol{n}\cdot\boldsymbol{X}$ is also Hermitian (in fact pure imaginary and antisymmetric) and traceless, thus guaranteeing that $R_{\boldsymbol{n}}(\varphi)$ is orthogonal and has unit determinant.

Commutation Relations

The group is specified by its law of composition: how two rotations combine to make a third. The structure formed by the infinitesimal generators X_i is known as an algebra, which is in the first place a vector space, since (complex) linear combinations of the X_i are again generators. There is, however, an additional composition law given by commutation: for any two generators X, Y the commutator $[X, Y]$ is also a generator. (This composition law is non-associative, satisfying instead the Jacobi identity $[[X, Y], Z] + [[Y, Z], X] + [[Z, X], Y] = 0$.) In order to specify the algebra we therefore need to know how the generators commute with each other. We can find these commutation relations in (at least) two ways.

(*a*) Since the defining representation is a faithful representation of the group, we can simply calculate the commutation relations of the 3×3 matrices of equations (6.15), (6.17) and (6.19) which constitute the generators in that representation. Thus, for example,

$$-[X_1, X_2] = \begin{pmatrix} 0 & 0 & 0 \\ 0 & 0 & -1 \\ 0 & 1 & 0 \end{pmatrix} \begin{pmatrix} 0 & 0 & 1 \\ 0 & 0 & 0 \\ -1 & 0 & 0 \end{pmatrix}$$

$$- \begin{pmatrix} 0 & 0 & 1 \\ 0 & 0 & 0 \\ -1 & 0 & 0 \end{pmatrix} \begin{pmatrix} 0 & 0 & 0 \\ 0 & 0 & -1 \\ 0 & 1 & 0 \end{pmatrix}$$

$$= \begin{pmatrix} 0 & 0 & 0 \\ 1 & 0 & 0 \\ 0 & 0 & 0 \end{pmatrix} - \begin{pmatrix} 0 & 1 & 0 \\ 0 & 0 & 0 \\ 0 & 0 & 0 \end{pmatrix} = \begin{pmatrix} 0 & -1 & 0 \\ 1 & 0 & 0 \\ 0 & 0 & 0 \end{pmatrix}$$

The final matrix can be recognized as $-\mathrm{i}X_3$, so

$$[X_1, X_2] = \mathrm{i}X_3 \tag{6.27}$$

Similarly for the cyclic permutations of 1, 2, 3. In fact all the commutation relations can be nicely summarized, using the ε symbol, as

$$[X_i, X_j] = \mathrm{i}\varepsilon_{ijk}X_k \tag{6.28}$$

As an exercise in using the ε symbol the reader might like to rederive this relation using the tensor representation $(X_k)_{ij} = -\mathrm{i}\varepsilon_{ijk}$ rather than writing out 3×3 matrices explicitly. In order to do so you will need the extremely useful identity

$$\varepsilon_{ilk}\varepsilon_{jmk} = \delta_{ij}\delta_{lm} - \delta_{im}\delta_{lj} \tag{6.29}$$

which can be checked by taking particular values for the indices.

(*b*) The more physical way, which emphasizes the geometrical origin of the commutation relations, is to consider the composition of infinitesimal rotations.

More specifically we consider the conjugation $R_{n'}(\varphi) = S(\theta)R_n(\varphi)S^{-1}(\theta)$, where φ is infinitesimal. The structure of the group enters directly in the identification of the conjugate rotation as a rotation through the same angle φ about the rotated axis $\boldsymbol{n}' = S\boldsymbol{n}$.

In particular let us take $\boldsymbol{n}$ to be the x axis and S to be a rotation about the z axis, so that $S(\theta) = \exp(-\mathrm{i}X_3\theta)$. Referring to figure 6.3, the rotated axis $S\boldsymbol{n}$ has Cartesian components $(\cos\theta, \sin\theta, 0)$, so that $\boldsymbol{n}'\cdot\boldsymbol{X} = X_1\cos\theta + X_2\sin\theta$. Thus

$$\mathrm{e}^{-\mathrm{i}X_3\theta}\mathrm{e}^{-\mathrm{i}X_1\varphi}\mathrm{e}^{\mathrm{i}X_3\theta} = \mathrm{e}^{-\mathrm{i}(X_1\cos\theta + X_2\sin\theta)\varphi} \tag{6.30}$$

Expanding each exponential to first order in φ gives

$$\mathrm{e}^{-\mathrm{i}X_3\theta}(1 - \mathrm{i}X_1\varphi)\mathrm{e}^{\mathrm{i}X_3\theta} = 1 - \mathrm{i}(X_1\cos\theta + X_2\sin\theta)\varphi$$

Thus

$$\mathrm{e}^{-\mathrm{i}X_3\theta}X_1\mathrm{e}^{\mathrm{i}X_3\theta} = X_1\cos\theta + X_2\sin\theta \tag{6.31}$$

Differentiating with respect to θ and then setting $\theta = 0$ we obtain

$$-\mathrm{i}[X_3, X_1] = X_2$$

which is one of the commutation relations (6.28). The others can obviously be obtained in a similar manner.

Since the conjugacy classes of rotations are all rotations through the same angle about different axes, they can be labelled simply by that angle, say φ. Equally the characters are just a function of φ. In the defining 3-dimensional representation we have

$$\chi^{[3]}(\varphi) = 2\cos\varphi + 1 \tag{6.32}$$

as can be verified directly in the special cases R_1, R_2, R_3.

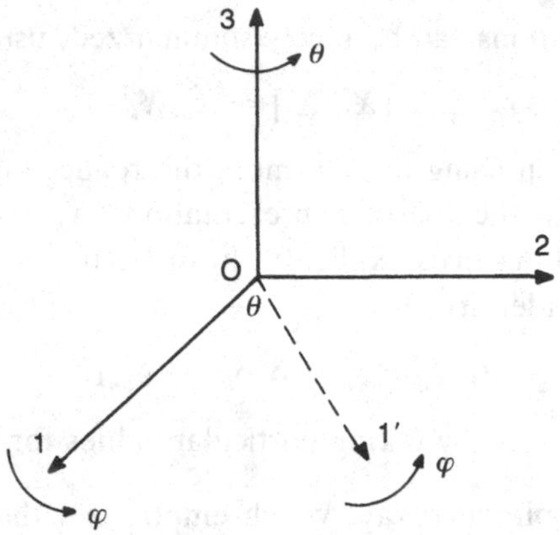

Figure 6.3 $R_1(\varphi)$ is a rotation through φ about the axis 01. The conjugate rotation $S_3(\theta)R_1(\varphi)(S_3(\theta))^{-1}$ is a rotation through the same angle φ about the axis 01′.

Irreducible Representations of SO(3)

Since we can obtain finite rotations by exponentiation, the problem of finding irreps of the *group* is (essentially) the same as that of finding irreps of the *algebra*, i.e. to find irreducible matrices which satisfy the commutation relations (6.28).

This is a standard problem in quantum mechanics, repeated for completeness in Appendix B. One is looking for the eigenvalues and eigenvectors of the angular momentum operators $\hat{J}_i$, which up to a factor $\hbar$ can be identified with the $\hat{X}_i$: $\hat{J} = \hbar\hat{X}$. The result is that the irreps are labelled by the eigenvalue $j(j+1)\hbar^2$ of the total angular momentum operation $\hat{J}^2$. Within the representation specified by j, the matrix elements of the generators X_i are given (in Dirac notation) by

$$\begin{aligned} \langle jm'|X_3|jm\rangle &= m\delta_{m'm} \\ \langle jm'|X_+|jm\rangle &= [(j-m)(j+m+1)]^{1/2}\,\delta_{m',m+1} \\ \langle jm'|X_-|jm\rangle &= [(j+m)(j-m+1)]^{1/2}\,\delta_{m',m-1} \end{aligned} \tag{6.33}$$

Here $X_\pm$ are the linear combinations $X_\pm = X_1 \pm \mathrm{i}X_2$. The eigenvalue m of X_3 ranges from $+j$ to $-j$ in integer steps, thus taking on $2j+1$ values in all.

For the case $j = 1$ these are just the matrices (6.15), (6.17) and (6.19) in disguise, i.e. referred to the spherical basis $-(1/\sqrt{2})(x+\mathrm{i}y)$, z, $(1/\sqrt{2})(x-\mathrm{i}y)$ rather than the Cartesian basis.

The general analysis of the irreps of the algebra defined in equation (6.28) requires $2j+1$, the multiplicity, or dimensionality of the representation, to be an integer. This, however, can be satisfied by j being half-integral as well as integral. In the context of orbital angular momentum, which has a classical analogue and periodicity 2π, we must take j ($=l$) integral. Thus strictly speaking it is only for integral values of j that the generators exponentiate to give a representation of SO(3), with $D(2\pi) = 1$. That is why we inserted the caveat 'essentially' in the introduction to this subsection. The half-integral representations cannot be realized as differential operators acting on wavefunctions, which would inevitably have a periodicity of 2π, but only as matrices. When exponentiated they give representations of the 'covering' group SU(2) rather than SO(3) itself, as discussed in Chapter 8.

Characters

In the $[2j + 1]$ representation labelled by j, m, the matrix for X_3 is (cf. (6.33))

$$X_3 = \mathrm{diag}(j, j-1, \ldots, -j+1, -j) \tag{6.34}$$

This diagonal matrix exponentiates trivially to give

$$R_3(\varphi) = \mathrm{e}^{-\mathrm{i}X_3\varphi} = \mathrm{diag}(\mathrm{e}^{-\mathrm{i}j\varphi}, \mathrm{e}^{-\mathrm{i}(j-1)\varphi}, \ldots, \mathrm{e}^{\mathrm{i}(j-1)\varphi}, \mathrm{e}^{\mathrm{i}j\varphi}) \tag{6.35}$$

with trace

$$\chi^{(j)}(\varphi) = \mathrm{e}^{-\mathrm{i}j\varphi} + \mathrm{e}^{-\mathrm{i}(j-1)\varphi} + \ldots + \mathrm{e}^{\mathrm{i}(j-1)\varphi} + \mathrm{e}^{\mathrm{i}j\varphi} \tag{6.36}$$

This is just a geometric series of $2j + 1$ terms with constant ratio $\mathrm{e}^{-\mathrm{i}\varphi}$, so that

$$\chi^{(j)}(\varphi) = \mathrm{e}^{-\mathrm{i}j\varphi}\,\frac{1 - \mathrm{e}^{\mathrm{i}(2j+1)\varphi}}{1 - \mathrm{e}^{\mathrm{i}\varphi}}$$

Writing the first factor as $\mathrm{e}^{-\mathrm{i}(j+1/2)\varphi}/\mathrm{e}^{-\mathrm{i}\varphi/2}$ this becomes

$$\chi^{(j)}(\varphi) = \frac{\mathrm{e}^{-\mathrm{i}(j+1/2)\varphi} - \mathrm{e}^{\mathrm{i}(j+1/2)\varphi}}{\mathrm{e}^{-\mathrm{i}\varphi/2} - \mathrm{e}^{\mathrm{i}\varphi/2}}$$

or finally

$$\boxed{\chi^{(j)}(\varphi) = \frac{\sin(j + \frac{1}{2})\varphi}{\sin \frac{1}{2}\varphi}} \tag{6.37}$$

In the first instance this is the character of a rotation through φ about the z axis, but as discussed above it is also the character of a rotation through φ about any other axis.

In fact, although equation (6.37) is the most compact form for the character, the original extended form (6.36) is often of more practical use. In the case of orbital angular momentum, when $j = l$ is integral, the complex exponentials pair up to give

$$\chi^{(l)}(\varphi) = 1 + 2\sum_{m=1}^{l} \cos(m\varphi) \tag{6.38}$$

while for half-integral spin, when $j = s = n + \frac{1}{2}$,

$$\chi^{(s)}(\varphi) = 2\sum_{m=0}^{n} \cos(m + \tfrac{1}{2})\varphi \tag{6.39}$$

Orthogonality

Since we now have a continuous group we need to know what the group-invariant integration is which replaces the group average $(1/[g])\Sigma_g \ldots$ of a finite group. More accurately we wish to replace $(1/[g])\Sigma_i k_i \ldots$, writing the sum as a sum over conjugacy classes with weight k_i, the number of elements in the conjugacy class K_i. In the continuous case the conjugacy classes are labelled by the angle of rotation φ, irrespective of the orientation of the axis of rotation. Thus we expect to make the replacement $(1/[g])\Sigma_g \ldots \to \int d\mu(\varphi) \ldots$, where $d\mu(\varphi)$ is a suitable integration measure.

The conditions this measure must satisfy are in the first place $\int d\mu(\varphi) = 1$, in order that it be a group *average*. More importantly it must have the property of group invariance in the form $\int d\mu(\varphi) \ldots = \int d\mu(\varphi') \ldots$, where in the second measure every rotation $R_n(\varphi)$ has been multiplied by a constant rotation S, say, to give $R_{n'}(\varphi')$. In the proofs of the theorems on characters we made frequent use of the equivalence of $\Sigma_g \ldots$ and $\Sigma_{g'} \ldots$, where $g' = hg$, and the measure for the continuous case must behave analogously.

The measure $d\mu(\varphi)$ which satisfies these conditions is shown in Appendix C to be $d\mu(\varphi) = (1 - \cos\varphi)\, d\varphi/2\pi$. So for SO(3) the replacement is

$$\frac{1}{[g]}\sum_g \to \int_0^{2\pi} \frac{d\varphi}{2\pi}(1 - \cos\varphi) \tag{6.40}$$

As a check let us calculate the 'scalar product' of two characters $\chi^{(j)}$ and $\chi^{(j')}$, where j and j' can each be either integral or half-integral:

$$\begin{aligned}\langle \chi^{(j)}, \chi^{(j')} \rangle &= \int \frac{d\varphi}{2\pi}\, 2\sin^2 \tfrac{1}{2}\varphi\, \frac{\sin(j + \frac{1}{2})\varphi}{\sin\frac{1}{2}\varphi}\, \frac{\sin(j' + \frac{1}{2})\varphi}{\sin\frac{1}{2}\varphi} \\ &= \int \frac{d\varphi}{2\pi}[\cos(j - j')\varphi - \cos(j + j' + 1)\varphi] \\ &= \delta_{jj'}\end{aligned} \tag{6.41}$$

as required.

The Clebsch–Gordan Series

We recall that the Clebsch–Gordan series gives the decomposition of the direct product of two irreps of the group into a sum of irreps with certain multiplicities. This is the same as the problem of 'addition of angular momentum' in quantum mechanics, where one is dealing with the algebra of the operators $\boldsymbol{J}$ rather than the group itself. The well-known result, reproduced in Appendix B, is that the addition of two angular momenta with eigenvalues j_1 and j_2 gives all j values spaced integrally between $j_1 + j_2$ and $|j_1 - j_2|$. In terms of the group this means that

$$D^{(j_1)} \otimes D^{(j_2)} = \sum_{|j_1-j_2|}^{j_1+j_2} \oplus D^{(j)} \tag{6.42}$$

Let us prove this directly using characters, taking $j_1 \geqslant j_2$ for definiteness. Then

$$\chi^{(j_1)}(\varphi)\,\chi^{(j_2)}(\varphi) = \frac{e^{i(j_1+1/2)\varphi} - e^{-i(j_1+1/2)\varphi}}{2i\sin\frac{1}{2}\varphi} \sum_{m=-j_2}^{j_2} e^{im\varphi}$$

using a judicious choice of expressions for $\chi^{(j_1)}$ and $\chi^{(j_2)}$. Taking the first numerator inside the summation we get

$$\chi^{(j_1)}(\varphi)\chi^{(j_2)}(\varphi) = \frac{1}{2i\sin\frac{1}{2}\varphi} \sum_{m=-j_2}^{j_2} (e^{i(j_1+m+1/2)\varphi} - e^{-i(j_1-m+1/2)\varphi})$$

Because the summation over m is over an even range we can change $-m$ to m in the second exponent. Having done this, m and j_1 appear in the single combination $j := j_1 + m$. The generic term of the series is just $2i\sin(j + \frac{1}{2})\varphi$, with j running from $j_1 - j_2$ to $j_1 + j_2$. Thus

$$\chi^{(j_1)}(\varphi)\chi^{(j_2)}(\varphi) = \sum_{j=j_1-j_2}^{j_1+j_2} \frac{\sin(j+\frac{1}{2})\varphi}{\sin\frac{1}{2}\varphi}$$

$$= \sum_{j_1-j_2}^{j_1+j_2} \chi^{(j)}(\varphi) \tag{6.43}$$

which confirms (6.42).

6.3 Clebsch–Gordan Coefficients

Corresponding to the block diagonal decomposition of the matrices of the direct product representation $D^{(j_1)} \otimes D^{(j_2)}$ there is a similar decomposition of the vectors, in this case the quantum mechanical state vectors $|j_1m_1\rangle|j_2m_2\rangle \equiv |j_1m_1;\ j_2m_2\rangle$, on which they act. That is, these vectors, which are eigenstates of the individual angular momentum operators $\hat{\boldsymbol{J}}_1^2$, $\hat{J}_{1z}$, $\hat{\boldsymbol{J}}_2^2$ and $\hat{J}_{2z}$, can be written as a linear superposition of eigenstates still of $\hat{\boldsymbol{J}}_1^2$ and $\hat{\boldsymbol{J}}_2^2$, but now of the total angular momentum operators $\hat{J}^2$ and $\hat{J}_z$ instead of $\hat{J}_{1z}$ and $\hat{J}_{2z}$:

$$|j_1m_1;\ j_2m_2\rangle = \sum_{j,m} C(j_1j_2j;\ m_1m_2m)|(j_1j_2)jm\rangle \tag{6.44}$$

Here the summation on j runs from $|j_1 - j_2|$ to $j_1 + j_2$, in accordance with (6.42), and $m \equiv m_1 + m_2$ runs, as always, from $-j$ to j. The labels j_1 and j_2 are inert in this decomposition.

The Clebsch–Gordan (CG) coefficients C can be found from the structure of the group, but they involve more detailed information than can be obtained from the characters alone. Unlike the characters, they are dependent on the precise choice of basis. In the case of SO(3) the CG coefficients can be taken as *real* by a suitable phase convention, but that does not exhaust the ambiguity. No physical result can depend on this choice of phase, but the reader should beware of mixing different conventions when consulting tables or books on the subject. In this book we will always calculate any required coefficients explicitly.

Equation (6.44) is a transformation between the two bases $\{|(j_1)m_1;\ (j_2)m_2\rangle\}$ and $\{|(j_1j_2)jm\rangle\}$. For fixed j_1, j_2 each basis contains $(2j_1 + 1)(2j_2 + 1)$ vectors. They are each orthogonal, since the labels are the distinct eigenvalues of Hermitian operators, and all conventions agree in making them orthonormal:

$$\langle(j_1)m_1';\ (j_2)m_2'|(j_1)m_1;\ (j_2)m_2\rangle = \delta_{m_1m_1'}\delta_{m_2m_2'} \tag{6.45}$$

$$\langle(j_1j_2)j'm'|(j_1j_2)jm\rangle = \delta_{jj'}\delta_{mm'} \tag{6.46}$$

Using (6.46) we can write $C(j_1j_2j;\ m_1m_2m)$ as the scalar product

$$C(j_1j_2j;\ m_1m_2m) = \langle(j_1)m_1;\ (j_2)m_2|(j_1j_2)jm\rangle \tag{6.47}$$

Since it implements a transformation between two orthonormal bases, C, considered as a matrix whose first index is the pair (m_1m_2) and second index the pair (jm), is unitary; in fact orthogonal since we have chosen it to be real. That is

$$\sum_{m_1,m_2} C(j_1j_2j';\ m_1m_2m')C(j_1j_2j;\ m_1m_2m) = \delta_{jj'}\delta_{mm'} \tag{6.48}$$

reflecting the orthogonality of the second basis and the completeness of the first. Equally

$$\sum_{j,m} C(j_1j_2j;\, m_1'm_2'm)C(j_1j_2j;\, m_1m_2m) = \delta_{m_1m_1'}\delta_{m_2m_2'} \tag{6.49}$$

reflecting the orthogonality of the first and the completeness of the second.

Using (6.48) we can invert the decomposition (6.44) to write

$$|(j_1j_2)jm\rangle = \sum_{m_1,m_2} C(j_1j_2j;\, m_1m_2m)|(j_1)m_1;\, (j_2)m_2\rangle \tag{6.50}$$

which is perhaps the more familiar expansion, whereby states of definite total angular momentum are constructed out of the eigenstates of the individual angular momentum operators. For fixed m the sum over m_1, m_2 is actually only a single sum because of the additivity property $m_1 + m_2 = m$, so that an alternative notation for the CG coefficient is just $C(j_1j_2j;\, m_1\; m - m_1)$, omitting the last entry entirely. For a given j the way the series is constructed is first to find that linear combination which corresponds to the state $|(j_1j_2)jj\rangle$ with the maximum possible m value. This state is characterized by the condition $\hat{J}_+|(j_1j_2)jj\rangle = 0$, where $\hat{J}_+$ is the total raising operator: $\hat{J}_+ = (\hat{J}_+)_1 + (\hat{J}_+)_2$. Having found this 'top' state, the combinations corresponding to the other m values, $m \geqslant -j$, can be generated by successive application of the total lowering operator $\hat{J}_-$.

Example: $D^{(1/2)} \otimes D^{(1/2)} = D^{(1)} \oplus D^{(0)}$

For brevity let us simply write the states as $|m_1;\, m_2\rangle$ and $|j,\, m\rangle$, suppressing the inactive labels $j_1 = \frac{1}{2}$ and $j_2 = \frac{1}{2}$. The possible values of j are just 1 and 0. It is trivial to make the identification

$$|1,\, 1\rangle = |\tfrac{1}{2};\, \tfrac{1}{2}\rangle \tag{6.51}$$

since this is the only state with $m = 1$. However, it is worth noting that we have already made a phase choice: the sign could have been minus!

In operating on equation (6.51) to generate the other states $|1\; m\rangle$ we adopt a sort of schizophrenic attitude. That is, on the left-hand side we forget how $\hat{J}_+$ is made up and just use equation (6.33) in the form

$$\hat{J}_-|j,\, m\rangle = \hbar[(j + m)(j - m + 1)]^{1/2}|j,\, m-1\rangle \tag{6.52}$$

(which again involves a choice of phase!). On the right-hand side, however, we remember the decomposition of $\hat{J}_-$, which strictly speaking should be written as $\hat{J}_- = (\hat{J}_-)_1 \otimes \mathbb{1}_2 + \mathbb{1}_1 \otimes (\hat{J}_-)_2$, and operate on the states $|j_1,\, m_1\rangle$, $|j_2,\, m_2\rangle$ individually, again using (6.52).

This procedure gives

$$\sqrt{2}\,|1,\, 0\rangle = |\tfrac{1}{2};\, -\tfrac{1}{2}\rangle + |-\tfrac{1}{2};\, \tfrac{1}{2}\rangle \tag{6.53}$$

which we note is correctly normalized, and again

$$|1, -1\rangle = |-\tfrac{1}{2}; -\tfrac{1}{2}\rangle \tag{6.54}$$

In order to have $m = 0$, the state $|0, 0\rangle$ must be some linear combination $\alpha|\tfrac{1}{2}, -\tfrac{1}{2}\rangle + \beta|-\tfrac{1}{2}, \tfrac{1}{2}\rangle$. Implementing the condition $\hat{J}_+|0, 0\rangle = 0$ by using the individual operators on the right-hand side fixes $\alpha = -\beta$. So normalizing, and again choosing a phase, we can take

$$\sqrt{2}|0, 0\rangle = |\tfrac{1}{2}; -\tfrac{1}{2}\rangle - |-\tfrac{1}{2}; \tfrac{1}{2}\rangle \tag{6.55}$$

Note that this is orthogonal to $|1, 0\rangle$, as it should be.

Equations (6.51) and (6.53)–(6.55) implicitly give the Clebsch–Gordan coefficients $C(\tfrac{1}{2}\tfrac{1}{2}j;\ m_1m_2m)$. By comparison with (6.50) we can identify

$$\begin{aligned} C(\tfrac{1}{2}\tfrac{1}{2}1;\ \tfrac{1}{2}\tfrac{1}{2}1) &= C(\tfrac{1}{2}\tfrac{1}{2}1;\ -\tfrac{1}{2}-\tfrac{1}{2}-1) &&= 1 \\ C(\tfrac{1}{2}\tfrac{1}{2}1;\ \tfrac{1}{2}-\tfrac{1}{2}0) &= C(\tfrac{1}{2}\tfrac{1}{2}1;\ -\tfrac{1}{2}\tfrac{1}{2}0) &&= 1/\sqrt{2} \\ C(\tfrac{1}{2}\tfrac{1}{2}0;\ \tfrac{1}{2}-\tfrac{1}{2}0) &= -\ C(\tfrac{1}{2}\tfrac{1}{2}0;\ -\tfrac{1}{2}\tfrac{1}{2}0) &&= 1/\sqrt{2} \end{aligned} \tag{6.56}$$

all others zero.

The decomposition (6.50) affords a means of constructing the rotation matrices† $D^{(j)}_{m'm}(R)$ for larger values of j by systematically working upwards from the simplest non-trivial representation $j = \tfrac{1}{2}$. That is, if the states $|jm\rangle$ are expressed as linear combinations of states $|j_1m_1\rangle$, $|j_2m_2\rangle$ of lower angular momentum, the rotation matrices $D^{(j)}_{m'm}(R)$, which are the matrix elements of the unitary operator $U(R)$ induced in the space of quantum states by the physical rotation R, can be correspondingly expressed as

$$\begin{aligned} D^{(j)}_{m'm}(R) &= \langle jm'|U(R)|jm\rangle \\ &= \sum_{m_1,m_1'}\sum_{m_2'm_2} C(j_1j_2j;\ m_1'm_2'm)C(j_1j_2j;\ m_1m_2m) \\ &\quad\times \langle j_1m_1';\ j_2m_2'|U(R)|j_1m_1;\ j_2m_2\rangle \\ &= \sum_{m_1,m_1'}\sum_{m_2'm_2} C(j_1j_2j;\ m_1'm_2'm)C(j_1j_2j;\ m_1m_2m) \\ &\quad\times D^{(j_1)}_{m_1'm_1}(R)D^{(j_2)}_{m_2'm_2}(R) \end{aligned} \tag{6.57}$$

A general rotation R can be parametrized by the three Euler angles φ, θ, ψ, which in the first place refer to rotations through φ, θ and then ψ about successively transformed axes. However, as the reader is invited to show in problem 6.2, it has an alternative expression in terms of

†More frequently written as $D^j_{m'm}(R)$.

rotations about the fixed coordinate axes. Namely, R can be written as the product

$$R(\varphi, \theta, \psi) = R_3(\varphi)R_2(\theta)R_3(\psi) \tag{6.58}$$

where $R_i(\alpha)$ is a rotation through α about the ith axis. Since X_3 is diagonal, the matrix $D(R_3(\alpha)) \equiv \exp(-\mathrm{i}X_3\alpha)$ can be obtained immediately, as in equation (6.35). Thus the only non-trivial factor in $D(R)$ is the matrix corresponding to $R_2(\theta)$, the so-called *reduced* matrix $d(\theta) := D(R_2(\theta)) = \exp(-\mathrm{i}X_2\theta)$.

Using the Clebsch–Gordan coefficients of equation (6.56) it is not difficult to derive the reduced rotation matrix for $j = 1$ from that for $j = \frac{1}{2}$ (see problems 6.3 and 6.4).

Tensor Operators and the Wigner–Eckart Theorem

In relating the QM formalism to physical measurements we invariably have to evaluate quantities of the nature $\langle\psi|O|\varphi\rangle$. Typically $|\varphi\rangle$, $|\psi\rangle$ are the eigenstates of an unperturbed Hamiltonian H_0, which is invariant under some symmetry. In the present context we are concerned with a rotationally symmetric Hamiltonian, invariant under the rotations of SO(3). A great deal of information about these matrix elements can then be extracted from the statement that the operator O transforms in a certain way under the rotation group.

As the simplest example consider a scalar operator S which is actually invariant with respect to SO(3) rotations. This means that

$$U(R)SU(R)^{-1} = S \tag{6.59}$$

or equivalently

$$[X, S] = 0 \tag{6.60}$$

From the latter equation we can immediately derive the *selection rules*, that $\langle j'm'|S|j\,m\rangle$ vanishes unless $j = j'$ and $m = m'$. Moreover, this matrix element is independent of m, so that

$$\langle j'm'|S|jm\rangle = N_j\delta_{jj'}\delta_{mm'} \qquad \text{(no summation)} \tag{6.61}$$

The condition $j = j'$ is a consequence of the vanishing of the commutator $[S, X^2]$. For then

$$\begin{aligned} j(j+1)\langle j'\,m'|S|j\,m\rangle &= \langle j'\,m'|SX^2|jm\rangle \\ &= \langle j'\,m'|X^2S|jm\rangle \\ &= j'(j'+1)\langle j'\,m'|S|jm\rangle \end{aligned}$$

where to obtain the last line we have operated to the left (see Appendix A) with the Hermitian operator X^2. Thus

$$(j - j')(j + j' + 1)\langle j'\, m'|S|j\, m\rangle = 0 \tag{6.62}$$

The second factor is always positive. Thus either $j = j'$ or

$$\langle j'\, m'|S|j\, m\rangle = 0.$$

Similarly the condition $m = m'$ follows from the fact that $[S, X_3] = 0$. Finally the fact that the matrix element depends on j only and not on m is a consequence of the vanishing of $[S, X_\pm]$. Thus

$$\begin{aligned}\sqrt{[(j - m + 1)(j + m)]}\langle jm|S|jm\rangle &= \langle jm|SX_+|j,m - 1\rangle \\ &= \langle jm|X_+S|j,m - 1\rangle \\ &= \sqrt{[(j + m)(j - m + 1)]} \\ &\quad \times \langle j,m - 1|S|j,m - 1\rangle\end{aligned}$$

where again we have operated X_+ to the left, remembering that $(X_+)^\dagger = X_-$.

A similar procedure can be adopted to find the selection rules for a vector operator V, or higher tensor operators. However, the procedure rapidly becomes cumbersome, and a more powerful method is required. This is provided by the Wigner–Eckart theorem, which introduces the idea of a tensor operator T^j_m and, using the machinery already developed, shows that the m dependence of its matrix elements is completely determined by an appropriate Clebsch–Gordan coefficient.

Definition

An (irreducible) *tensor operator* T^j_m is an operator with the property that it transforms under rotations according to

$$U(R)T^j_m U(R)^{-1} = D^j_{m'm}(R)T^j_{m'} \tag{6.63}$$

Here $U(R)$ is the unitary operator induced in the space of quantum states by the rotation R. Equation (6.63) should be compared to the transformation of a state vector $|jm\rangle$, for which

$$U(R)|jm\rangle = D^j_{m'm}(R)|jm'\rangle \tag{6.64}$$

In this notation the scalar operator S previously considered is a T^0_0, transforming according to the trivial representation $j = 0$. A vector operator V is an operator with *spherical* components T^1_m, $m = \pm 1, 0$. As a particular example the generators $\boldsymbol{X}$ themselves form a vector operator. This is effectively how we found their commutation relations in the previous section, equation (6.30) *et seq*.

To spell this out in a little more detail, let $R = R_3(\theta)$ and consider the transformation of $X_{+1} \equiv (-1/\sqrt{2})(X_1 + \mathrm{i}X_2)$. This is

$$e^{-iX_3\theta}X_{+1}\,e^{iX_3\theta} = (-1/\sqrt{2})[X_1\cos\theta + X_2\sin\theta + i(X_2\cos\theta - X_1\sin\theta)]$$
$$= X_{+1}(\cos\theta - i\sin\theta) = X_{+1}e^{-i\theta}$$
$$= D^1_{11}(R_3(\theta))X_{+1} \tag{6.65}$$

as required. The crucial elements in this derivation were the commutation relations $[X_3, X_1] = iX_2$ and $[X_3, X_2] = -iX_1$. The other commutator $[X_1, X_2]$ would enter if we were to examine the transformation of X_{+1} under a rotation $R_2(\theta)$ (try it!).

In this particular case we have established the equivalence of (6.63), which refers to the spherical components of $\boldsymbol{X}$, and the corresponding commutation relations in their Cartesian form $[X_i, X_j] = i\varepsilon_{ijk}X_k$. These commutation relations, then, are just the statement that $\boldsymbol{X}$ is a vector operator. For a general vector operator, indeed, the equivalent statement to (6.63) is

$$[X_i, V_j] = i\varepsilon_{ijk}V_k \tag{6.66}$$

Consider now the matrix element of a tensor operator T^J_M sandwiched between angular momentum eigenstates $\langle j'\ m'|T^J_M|j\,m\rangle$. First we note that the state $T^J_M|j\,m\rangle$ transforms according to the direct product representation $D^{(J)} \otimes D^{(j)}$. Thus

$$U(R)(T^J_M|j\,m\rangle) = U(R)T^J_M U(R)^{-1}U(R)|j\,m\rangle$$
$$= D^J_{M'M}(R)D^j_{m'm}(R)(T^J_{M'}|j\,m'\rangle) \tag{6.67}$$

Hence we can project out eigenstates $|j'\ m'\rangle$ by using the appropriate Clebsch–Gordan coefficients:

$$|j'\ m'\rangle \propto \sum_{m,M} C(Jjj';\ Mmm')T^J_M|j\,m\rangle \tag{6.68}$$

The constant of proportionality in this equation can be a function of the three j values, but cannot depend on the m's, since this would destroy the required transformation properties. Thus we can write

$$N_{Jjj'}|j'\ m'\rangle = \sum_{m,M} C(Jjj';\ Mmm')T^J_M|j\,m\rangle \tag{6.68'}$$

which, using the orthogonality of the CG coefficients, can be inverted to give

$$T^J_M|j\,m\rangle = N_{Jjj''}\sum_{j'',m''} C(Jjj'';\ Mmm'')|j''\,m''\rangle \tag{6.69}$$

Taking the scalar product with $\langle j'\ m'|$, which on the right-hand side picks out the term with $j'' = j'$, $m'' = m'$, we arrive at

$$\langle j'\ m'|T^J_M|j\,m\rangle = N_{Jjj'}C(Jjj';\ Mmm') \tag{6.70}$$

The constant $N_{Jjj'}$ is called the 'reduced' or 'double-bar' matrix element, usually denoted by $\langle j'\|T^J\|j\rangle$. In this notation (6.70) reads

$$\boxed{\langle j'\,m'|T^J_M|j\,m\rangle = C(Jjj';\,Mmm')\langle j'\|T^J\|j\rangle} \tag{6.70'}$$

which is the Wigner–Eckart theorem. The scalar case dealt with earlier (equation (6.61)) can now be seen as a trivial example of this theorem.

Examples

Dipole selection rules

Radiative electric dipole transitions in atoms are governed by the matrix elements of $(\boldsymbol{E}\cdot)\boldsymbol{x}$, viz.: $\langle n'l'm'|\boldsymbol{x}|n\,l\,m\rangle$, with the transition rate proportional to their squared moduli. Thus only those sets of quantum numbers for which the matrix element is non-zero will correspond to allowed transitions: otherwise the transitions are 'forbidden', but may in fact proceed through higher multipole operators.

From the Wigner–Eckart theorem we know that

$$\langle n'l'm'|x_M|n\,l\,m\rangle = C(1ll';\,Mmm')\langle n'l'\|\boldsymbol{x}\|n\,l\rangle \tag{6.71}$$

Hence we have the selection rules:

(i) $\Delta l \equiv l' - l = \pm 1, 0$ (this follows from the CG series alone);
(ii) $\Delta m \equiv m' - m = 0$ for $\boldsymbol{E}$ along the z axis (in which case $\boldsymbol{E}\cdot\boldsymbol{x} \propto x_0$), $\Delta m = \pm 1$ for $\boldsymbol{E}$ along the x or y axis (when $\boldsymbol{E}\cdot\boldsymbol{x}$ involves x_{+1} and x_{-1}).

In fact it turns out that $\Delta l = 0$ is not allowed. This is because $\boldsymbol{x}$ is not just any old vector, but is related to $\boldsymbol{L}$ by $\boldsymbol{L} = \boldsymbol{x} \times \boldsymbol{p}$, and thus satisfies $\boldsymbol{x}\cdot\boldsymbol{L} = 0$.

Consider, then, $0 = \langle l\,m|\boldsymbol{x}\cdot\boldsymbol{L}|l\,m\rangle$. Expressed in terms of spherical components, the scalar product of any two vectors $\boldsymbol{A}$ and $\boldsymbol{B}$ is $\boldsymbol{A}\cdot\boldsymbol{B} = \Sigma_M(-1)^M A_M B_{-M}$. Thus, inserting a complete set of states,

$$0 = \sum_M (-1)^M \sum_{l',m'} C(1ll';\,Mmm')C(1l'l;\,-Mm'm)\langle l\|\boldsymbol{x}\|l'\rangle\langle l'\|\boldsymbol{L}\|l\rangle$$

Now $\langle l'\|\boldsymbol{L}\|l\rangle$ is zero unless $l' = l$, and the m values are additive. So the above equation reduces to

$$\begin{aligned} 0 &= \langle l\,m|\boldsymbol{x}\cdot\boldsymbol{L}|l\,m\rangle \\ &= \left(\sum_M (-1)^M C(1ll;\,Mm)C(1ll;\,-M,\,m+M)\right)\langle l\|\boldsymbol{x}\|l\rangle\langle l\|\boldsymbol{L}\|l\rangle \end{aligned} \tag{6.72}$$

Now the reduced matrix element $\langle l\|\boldsymbol{L}\|l\rangle$ is certainly not zero, nor is the sum over CG coefficients, since they both appear as factors in the exactly analogous reduction of $\langle lm|\boldsymbol{L}^2|lm\rangle$. Hence the remaining factor $\langle l||\boldsymbol{x}||l\rangle$ must be zero.

An alternative and somewhat easier proof of this latter selection rule invokes the parity of $\boldsymbol{x}$, namely its behaviour under spatial inversions, which lie outside the proper rotation group SO(3). Since $\boldsymbol{x}$ is odd under parity, its matrix elements between states of the same parity must vanish.

Landé g-factor in l–s coupling

Consider the magnetic moment of an electron in a hydrogenic state with orbital angular momentum l and spin $\frac{1}{2}$. (More generally one could consider a multi-electron state with total orbital angular momentum L and total spin S, as in §7.1.) The states are classified by the eigenvalues j of the total angular momentum $\boldsymbol{J} = \boldsymbol{L} + \boldsymbol{S}$.

In calculating the magnetic moment $\boldsymbol{\mu}$ arising from such a configuration, the problem arises that the gyromagnetic ratios for orbital and spin angular momenta differ by a factor of 2:

$$\boldsymbol{\mu}_l = (e/2m_e)\boldsymbol{L} \tag{6.73a}$$

whereas

$$\boldsymbol{\mu}_s = (e/m_e)\boldsymbol{S} \tag{6.73b}$$

We therefore have to evaluate the expectation value

$$\begin{aligned} \boldsymbol{\mu} \equiv \langle \boldsymbol{\mu}_l + \boldsymbol{\mu}_s \rangle &= (e/2m_e)\langle \boldsymbol{L} + 2\boldsymbol{S} \rangle \\ &= (e/2m_e)\langle \boldsymbol{J} + \boldsymbol{S} \rangle \end{aligned} \tag{6.74}$$

The matrix element of $\boldsymbol{J}$ is easy to evaluate, since the states are eigenstates of $\boldsymbol{J}$; however, we also need those of $\boldsymbol{S}$. Like $\boldsymbol{x}$, $\boldsymbol{S}$ is a vector operator, so by writing down the analogues of (6.72) for $\langle jm|\boldsymbol{S}\cdot\boldsymbol{J}|jm\rangle$ and $\langle jm|\boldsymbol{J}^2|jm\rangle$ and taking the ratio we have

$$\frac{\langle jm|\boldsymbol{S}\cdot\boldsymbol{J}|jm\rangle}{\langle jm|\boldsymbol{J}^2|jm\rangle} = \frac{\langle j||\boldsymbol{S}\cdot\boldsymbol{J}||j\rangle}{\langle j||\boldsymbol{J}^2||j\rangle} = \frac{\langle j||\boldsymbol{S}||j\rangle}{\langle j||\boldsymbol{J}||j\rangle} \tag{6.75}$$

Note that this method only serves to evaluate the diagonal reduced matrix elements $\langle j||\boldsymbol{S}||j\rangle$. For $\langle j'||\boldsymbol{S}||j\rangle$ with $j' \neq j$, equation (6.75) would just give 0/0.

Again, using the Wigner–Eckart theorem for $\langle jm|S_M|jm\rangle$ and $\langle jm|J_M|jm\rangle$ and taking the ratio gives

$$\frac{\langle jm|\boldsymbol{S}_M|jm\rangle}{\langle jm|\boldsymbol{J}_M|jm\rangle} = \frac{\langle j||\boldsymbol{S}||j\rangle}{\langle j||\boldsymbol{J}||j\rangle} \tag{6.76}$$

Putting the two together we arrive at

$$\langle jm|\boldsymbol{S}|jm\rangle = \langle jm|\boldsymbol{J}|jm\rangle \frac{\langle j||\boldsymbol{S}\cdot\boldsymbol{J}||j\rangle}{\langle j||\boldsymbol{J}^2||j\rangle} \tag{6.77}$$

Equation (6.77) is the group theoretic justification of the vector model, according to which **S** and **L** precess about the conserved vector **J**, and all that matters in the expectation value is the component of **S** along **J** (figure 6.4).

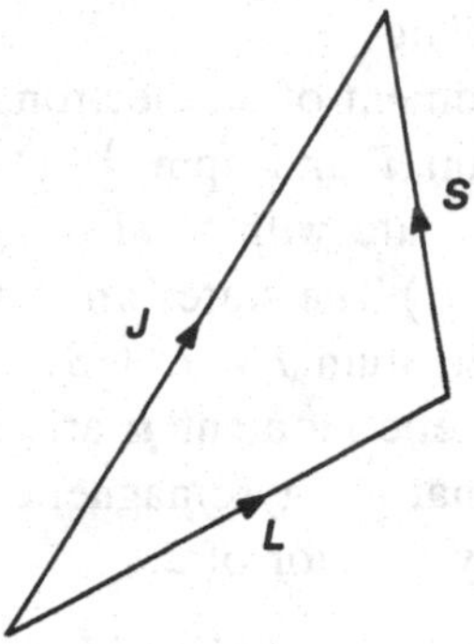

Figure 6.4 The vector model for the addition of spin **S** and orbital angular momentum **L**.

The calculation is completed by noting that

$$\boldsymbol{S}\cdot\boldsymbol{J} = \tfrac{1}{2}(\boldsymbol{S}^2 + \boldsymbol{J}^2 - \boldsymbol{L}^2) \tag{6.78}$$

so that

$$\langle j||\boldsymbol{S}\cdot\boldsymbol{J}||j\rangle = \tfrac{1}{2}\hbar^2[s(s+1) + j(j+1) - l(l+1)]$$

Hence

$$\boldsymbol{\mu} = (e/2m_e)g\langle \boldsymbol{J}\rangle$$

where

$$g = 1 + \frac{s(s+1) + j(j+1) - l(l+1)}{2j(j+1)} \tag{6.79}$$

For a single electron j can only be $l + \frac{1}{2}$ or $l - \frac{1}{2}$, and g reduces to

$$g = \begin{cases} (j + \frac{1}{2})/j & \text{for } j = l + \frac{1}{2} \\ (j + \frac{1}{2})/(j+1) & \text{for } j = l - \frac{1}{2} \end{cases} \tag{6.80}$$

in the two cases.

Problems for Chapter 6

6.1 The group O_2 ($\cong C_{\infty v}$) contains, in addition to rotations $R(\varphi)$ about the z axis, reflections σ_v in a plane containing the z axis.

If S is the reflection in the x–z plane show that $SR(\varphi)S^{-1} = R(-\varphi)$, so that the group is no longer Abelian. Show that for $m \neq 0$ it is necessary to combine the irreps $D^{(m)}$ and $D^{(-m)}$ of SO(2) in order to provide an irreducible representation of the larger group, and write down the characters in such a representation. What happens in the case $m = 0$?

6.2 A general rotation is often described by the Euler angles:

$$R(\varphi, \theta, \psi) = \mathrm{e}^{-\mathrm{i}X''_3\psi}\, \mathrm{e}^{-\mathrm{i}X'_2\theta}\, \mathrm{e}^{-\mathrm{i}X_3\varphi}$$

i.e. a rotation through φ about the z axis, followed by a rotation through θ about the new y axis, and finally a rotation through ψ about the new z axis (which is in the direction with polar angles (θ, φ)).

Show, using the meaning of conjugacy, that $R(\varphi, \theta, \psi)$ can be expressed in terms of rotations about the *fixed* axes as

$$R(\varphi, \theta, \psi) = \mathrm{e}^{-\mathrm{i}X_3\varphi}\, \mathrm{e}^{-\mathrm{i}X_2\theta}\, \mathrm{e}^{-\mathrm{i}X_3\psi}$$

6.3 Show that

$$d^{1/2}_{m'm}(\theta) = \begin{pmatrix} \cos\frac{1}{2}\theta & -\sin\frac{1}{2}\theta \\ \sin\frac{1}{2}\theta & \cos\frac{1}{2}\theta \end{pmatrix}$$

given that the generator X_2 of rotations about the y axis is represented by the Pauli matrix $\frac{1}{2}\sigma_2$ (equation (8.1)).

6.4 From the product representation $D^{(1/2)} \otimes D^{(1/2)}$ construct the matrix $d^1_{m'm}(\theta) := D^1_{m'm}(R_2(\theta))$ and check that this is indeed the appropriate matrix of the vector representation referred to $-(1/\sqrt{2})(x + \mathrm{i}y)$, z, $(1/\sqrt{2})(x - \mathrm{i}y)$ as basis.

6.5 Show that the reduced matrix element $\langle j\|\boldsymbol{J}\|j\rangle$ appearing in

$$\langle jm'|J_M|jm\rangle = C(1\,jj;\, Mmm')\langle j\|\boldsymbol{J}\|j\rangle$$

is given by

$$\langle j\|\boldsymbol{J}\|j\rangle = \hbar\sqrt{[j(j+1)]}$$

(Hint: take $m' = j$ and $M = 0$ or $M = 1$ in the above relation and find the relevant Clebsch–Gordan coefficient by explicitly constructing the combination $\alpha|jj\rangle|1, 0\rangle + \beta|j, j-1\rangle|1, 1\rangle$ which transforms as $|j\,j\rangle$.)

6.6 In a strong crystal field the electronic states of an atom are classified according to the point group P rather than SO(3). Hence the dipole selection rules analogous to $\Delta l = \pm 1, 0$ will arise from constructing the Clebsch–Gordan series for $D^{(\nu)} \times D^{[3]}$, where $D^{[3]}$ is the [3] representation afforded by x, y, z. Show that for the crystal symmetry C_{3v} (problem 5.1) $D^{[3]} = A_1 \oplus E$, generated by z and (x, y) respectively, and hence find the dipole selection rules for radiation polarized (i) in the z direction and (ii) in the x or y direction.

6.7 Find the selection rules for dipole radiation between levels classified by the irreducible representations of T (problem 4.8).

7

Further Applications

7.1 Energy Levels of Atoms in Hartree–Fock Scheme

The procedure in this method, which is the basis of the periodic table, is to take as a trial wavefunction for an n-electron atom a product of single-electron wavefunctions, which in fact has to be antisymmetrized because of Fermi statistics:

$$\psi(\boldsymbol{r}_1, s_1; \boldsymbol{r}_2, s_2; \ldots; \boldsymbol{r}_n, s_n) = A(\varphi_1(\boldsymbol{r}_1, s_1)\varphi_2(\boldsymbol{r}_2, s_2) \ldots \varphi_n(\boldsymbol{r}_n, s_n)) \tag{7.1}$$

where A denotes antisymmetrization with respect to the n pairs of indices $\rho_i := (\boldsymbol{r}_i, s_i)$. The product state of equation (7.1) is known as a *configuration*: apart from the antisymmetrization, it represents n independent electrons in the single-particle states, or *orbitals*, described by the wavefunctions φ_i.

Requiring the expectation value of the energy $\langle\psi|H|\psi\rangle$ to be a minimum with respect to variations of the single-electron orbitals then leads (e.g. Heine, §12) to the following set of coupled equations for the φ_i:

$$\left[-\hbar^2\nabla_i^2/2m - Ze_{\mathrm{M}}^2/r_i + \sum_{j\neq i}\left(\int \varphi_j^*(\rho_j)(e_{\mathrm{M}}^2/r_{ij})\varphi_j(\rho_j)\,\mathrm{d}^3\boldsymbol{r}_j\right)\right]\varphi_i(\rho_i)$$

$$-\left[\sum_{j\neq i}\left(\int \varphi_j^*(\rho_j)(e_{\mathrm{M}}^2/r_{ij})\varphi_i(\rho_j)\mathrm{d}^3\boldsymbol{r}_j\right)\varphi_j(\rho_i)\right] = E_i\varphi_i(\rho_i) \tag{7.2}$$

Here $e_{\mathrm{M}}^2 = e^2/4\pi\varepsilon_0$, $r_i = |\boldsymbol{r}_i|$ and $r_{ij} = |\boldsymbol{r}_i - \boldsymbol{r}_j|$. In the first line we have the usual Coulomb interaction with the nucleus, together with the expectation value of the electrostatic repulsion of the other electrons. The second line represents an exchange term. Since r_{ij} does not involve the spin, only those states with spin quantum number s_j identical to that of φ_i will contribute.

In order to make the problem tractable, the calculational procedure is to *average* the interaction terms over the direction of $\boldsymbol{r}_i$, thus making them spherically symmetric. The equations are then iterated until they converge on a self-consistent set of wavefunctions φ_i.

As far as group theory is concerned, this means that at this level of approximation the equations are invariant under *separate* rotations of the coordinates $\boldsymbol{r}_1, \boldsymbol{r}_2, \ldots, \boldsymbol{r}_n$, so that the symmetry group is $(\mathrm{SO}(3))^n$. In fact, since we have so far ignored spin–orbit coupling, each individual SO(3) factor here is actually a product $\mathrm{SO}(3)_l \times \mathrm{SO}(3)_s$ acting respectively on the orbital and spin degrees of freedom. The total degeneracy of the n-particle state described by orbitals of angular momentum l_i is therefore $\{\Pi_{i=1}^{n}[2(2l_i+1)]\}/n!$, except that in general some combinations will be excluded by the requirement of antisymmetry.

The large symmetry of the self-consistent field potential V_{scf} is an artefact of the simplifying approximations that have been made: the true interaction potential V_{actual} is only invariant when the coordinates $\boldsymbol{r}_i$ are subjected to the *same* rotation. At the next level of approximation, when the true interaction is taken into account by perturbation theory in the difference $V_{\mathrm{actual}} - V_{\mathrm{scf}}$, the symmetry group is therefore a single SO(3), or at the stage when we are still neglecting spin–orbit coupling, a product $\mathrm{SO}(3)_L \times \mathrm{SO}(3)_S$. The energy levels, called *terms*, are classified according to irreps of this residual group and labelled by the *total* angular momentum L and spin S (see figure 7.1). When spin–orbit coupling, proportional to $\boldsymbol{L}\cdot\boldsymbol{S}$, is finally taken into account, the invariance group is truly reduced to a single SO(3) and the terms split further into levels labelled by the total angular momentum J, which is now the only good quantum number.

Example: Carbon

Carbon has six electrons, which occur in the configuration $(1s)^2(2s)^2(2p)^2$. That is, the orbitals consist of two S-waves ($l = 0$) with different spatial wavefunctions and a P-wave ($l = 1$). Each of these spatial states can accommodate two electrons, provided that they are in an antisymmetric spin combination, i.e. with their spins opposed to give total spin zero. For most purposes we can neglect the S-state electrons, which form closed shells of lower energy, and concentrate on the valence electrons. Thus we have two indistinguishable electrons, to be assigned to $2 \times 3 = 6$ possible states. Because of Fermi statistics, two electrons cannot occupy the same state, so the number of different possibilities is just $\frac{1}{2}(6^2 - 6) = 15$.

As explained above, when we go from the self-consistent field potential to the non-averaged potential, the levels are split into terms labelled by the total angular momentum numbers L and S. In the present case L is the result of the addition of two individual orbital angular momenta with $l = 1$. From the Clebsch–Gordan series the possible values are therefore $L = 0$, 1 or 2, or in spectroscopic notation S, P or D. As far as the spin is concerned, we are combining two spin-$\frac{1}{2}$ systems, which can give $S = 0$ or 1, the singlet and triplet states respectively $(2S + 1 = 1, 3)$.

However, we still have to impose overall antisymmetry on the resulting states. It is clear from (6.56) that the triplet state is symmetric under interchange of the spin labels, while the singlet state is antisymmetric. In a similar fashion one finds that the $L = 0$ and $L = 2$ states are symmetric and $L = 1$ antisymmetric under interchange of the orbital angular momentum labels. Hence the allowed possibilities are $S = 0$ with $L = 0$, 2 and $S = 1$ with $L = 1$. In the spectroscopic notation ^{2S+1}L used in figure 7.1 these are denoted by ^{1}S, ^{1}D and ^{3}P respectively.

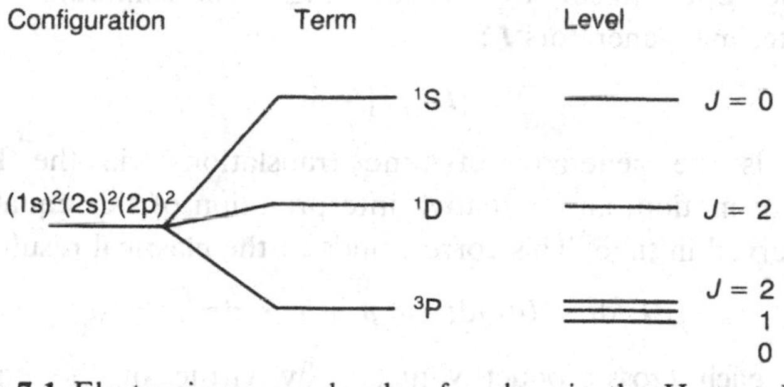

Figure 7.1 Electronic energy levels of carbon in the Hartree–Fock scheme.

Let us check that we have not lost any states in this rearrangement! The term ^{1}S has degeneracy 1 only. ^{1}D has degeneracy 1×5, while ^{3}P has $3 \times 3 = 9$. Altogether the total degeneracy is $1 + 5 + 9 = 15$, in agreement with the original count.

The final splitting due to the spin–orbit interaction breaks the degeneracy of the ^{3}P term. The possible values of the total angular momentum J arising from the combination of a spin $S = 1$ and an orbital angular momentum $L = 1$ are $J = 2$, 1 or 0. The other terms are not split, since with $S = 0$ the value of J was already fixed.

7.2 'Accidental' Degeneracy of the H Atom and SO(4)

As we have mentioned several times before, most 'accidental' degeneracies are the result of some deeper symmetry of the problem which is not apparent at first sight. The spectrum of the non-relativistic hydrogen atom is a case in point: the Hamiltonian for a general central potential has SO(3) symmetry because of its rotational invariance. However, for the particular case of a $1/r$ potential there is a further symmetry which enlarges the symmetry group to SO(4), which is locally isomorphic to SO(3) × SO(3). From this symmetry one can not only explain the additional degeneracy of the hydrogen atom, but even deduce the spectrum!

The rotational invariance of the Hamiltonian

$$H = \boldsymbol{p}^2/2m - e_{\mathrm{M}}^2/r \tag{7.3}$$

is expressed operationally by the vanishing of the commutator of H with the infinitesimal generators $\boldsymbol{L}$:

$$[\boldsymbol{L}, H] = 0 \tag{7.4}$$

Since H is the generator of time translations via the Heisenberg equation of motion, an alternative interpretation of this equation is that $\boldsymbol{L}$ is conserved in time. This corresponds to the classical result

$$\mathrm{d}\boldsymbol{L}/\mathrm{d}t = (\mathrm{d}\boldsymbol{r}/\mathrm{d}t) \times \boldsymbol{p} + \boldsymbol{r} \times \mathrm{d}\boldsymbol{p}/\mathrm{d}t = 0 \tag{7.5}$$

in which each cross-product vanishes by virtue of the equations of motion

$$\begin{aligned} \mathrm{d}\boldsymbol{p}/\mathrm{d}t &= -(e_{\mathrm{M}}^2/r^3)\boldsymbol{r} \\ \mathrm{d}\boldsymbol{r}/\mathrm{d}t &= \boldsymbol{p}/m \end{aligned} \tag{7.6}$$

Conservation of the angular momentum, a vector quantity, means that the classical motion lies in a plane, whose normal is defined by $\boldsymbol{L}$. Now a classical bound orbit, in general an ellipse, also has another fixed direction, namely that of the major axis. So one might suspect that there was another conserved vector which pointed in that direction. There is indeed such a vector, the *Runge–Lentz* vector

$$\boldsymbol{M} := \boldsymbol{v} \times \boldsymbol{L} - e_{\mathrm{M}}^2 \boldsymbol{r}/r \tag{7.7}$$

with

$$\begin{aligned} \mathrm{d}\boldsymbol{M}/\mathrm{d}t &= (\mathrm{d}\boldsymbol{v}/\mathrm{d}t) \times \boldsymbol{L} - e_{\mathrm{M}}^2 \mathrm{d}\hat{\boldsymbol{r}}/\mathrm{d}t \\ &= -(e_{\mathrm{M}}^2/mr^3)\boldsymbol{r} \times \boldsymbol{L} - (e_{\mathrm{M}}^2/mr)[\boldsymbol{p} - \boldsymbol{r}(\boldsymbol{r}\cdot\boldsymbol{p})/r^2] \\ &= 0 \end{aligned} \tag{7.8}$$

It is perpendicular to $\boldsymbol{L}$, $\boldsymbol{L}\cdot\boldsymbol{M} = 0$, and so is a vector in the plane of the orbit. To justify our assertion that it is in the direction of the semi-major axis, consider the scalar product with $\boldsymbol{r}$:

$$\boldsymbol{r}\cdot\boldsymbol{M} = (\boldsymbol{r} \times \boldsymbol{v})\cdot\boldsymbol{L} - e_{\mathrm{M}}^2 r = L^2/m - e_{\mathrm{M}}^2 r$$

Denoting by θ the angle between $\boldsymbol{r}$ and $\boldsymbol{M}$, this can be written as

$$r(e_{\mathrm{M}}^2 + M\cos\theta) = L^2/m \tag{7.9}$$

It is hoped that this will be recognized as the polar equation of an elliptic orbit. The eccentricity is found by computing $\boldsymbol{M}^2$ (which is zero for a circle).

Problem
Show that

$$M^2 = (2/m)L^2 E + e_{\mathrm{M}}^4 \tag{7.10}$$

The moral of this classical result is that one could solve the equations of motion by spotting the conserved vector $\boldsymbol{M}$, in addition to $\boldsymbol{L}$. An analogous vector can similarly be defined in the quantum problem, where all of the above properties are closely paralleled.

In seeking the quantum mechanical version of (7.7) one would certainly replace $\boldsymbol{v}$ by $\boldsymbol{p}/m$, since it is $\boldsymbol{p}$ rather than $\boldsymbol{v}$ which is the canonical dynamical variable. However, at the operator level we face an ordering ambiguity: $\boldsymbol{p}$ and $\boldsymbol{L}$ do not commute, so $\boldsymbol{p} \times \boldsymbol{L}$ is not equal to $-\boldsymbol{L} \times \boldsymbol{p}$. But for this very reason neither of these operators is separately Hermitian. In order to obtain a Hermitian operator we must adopt the symmetrized definition

$$\boldsymbol{M} = (\boldsymbol{p} \times \boldsymbol{L} - \boldsymbol{L} \times \boldsymbol{p})/2m - e_{\mathrm{M}}^2 \boldsymbol{r}/r \tag{7.11}$$

It is by no means a trivial matter to calculate $[H, \boldsymbol{M}]$. In principle all it involves is repeated use of the canonical commutation relation $[p_i, r_j] = \delta_{ij}$, but it takes more than a few lines. So we will leave it as an exercise for the dedicated reader to verify the result

$$[H, \boldsymbol{M}] = 0 \tag{7.12}$$

which is the quantal analogue of equation (7.8).

Since they comprise a vector, the components of $\boldsymbol{M}$ have the standard commutation relations with the angular momentum generators $\boldsymbol{L}$

$$[L_i, M_j] = i\hbar\varepsilon_{ijk}M_k \tag{7.13}$$

which closely follow the mutual commutation relations of the L_i

$$[L_i, L_j] = i\hbar\varepsilon_{ijk}L_k \tag{7.14}$$

To complete these commutation relations we must also calculate $[M_i, M_j]$. This involves even more algebra than the computation of $[H, \boldsymbol{M}]$, but eventually gives the surprisingly simple result

$$[M_i, M_j] = i\hbar\varepsilon_{ijk}(-2H/m)L_k \tag{7.15}$$

Thus the six operators $\boldsymbol{L}$ and $\boldsymbol{M}$ close under commutation, forming a larger algebra than SO(3). The structure of this algebra can be exposed by a number of redefinitions. First let us define

$$\boldsymbol{M}' := \boldsymbol{M}(-2H/m)^{-1/2} \tag{7.16}$$

Though it may look a little peculiar, the function $(-2H/m)^{-1/2}$ occurring on the right is in fact perfectly well-defined. In general a function $f(H)$ of a Hermitian operator H is defined by its action $f(H)|\lambda_i\rangle = f(\lambda_i)|\lambda_i\rangle$ on the complete set of eigenstates $\{|\lambda_i\rangle\}$. Since the spectrum of H is negative, $-2E_n/m$ is a positive quantity, and we will take the positive square root. This rescaling of $\boldsymbol{M}$ leaves equation (7.13) unaltered, but removes the factor of $-2H/m$ from equation (7.15).

A further redefinition,

$$\begin{aligned} \boldsymbol{J}^{(1)} &:= \tfrac{1}{2}(\boldsymbol{L} + \boldsymbol{M}') \\ \boldsymbol{J}^{(2)} &:= \tfrac{1}{2}(\boldsymbol{L} - \boldsymbol{M}') \end{aligned} \tag{7.17}$$

leads to the decoupled commutation relations

$$\begin{aligned} [J_i^{(1)}, J_j^{(1)}] &= i\hbar\varepsilon_{ijk}J_k^{(1)} \\ [J_i^{(2)}, J_j^{(2)}] &= i\hbar\varepsilon_{ijk}J_k^{(2)} \\ [J_i^{(1)}, J_j^{(2)}] &= 0 \end{aligned} \tag{7.18}$$

The operators $\boldsymbol{J}^{(1)}$ and $\boldsymbol{J}^{(2)}$ each separately satisfy an SO(3) algebra, and they commute with each other. The algebra of (7.18) can thus be characterized as SO(3) × SO(3) or SU(2) × SU(2). Alternatively it can be considered as the algebra of SO(4) generated by the operators

$$L_{\mu\nu} := x_\mu p_\nu - x_\nu p_\mu \tag{7.19}$$

where μ and ν run from 0 to 3 and x_μ, p_ν satisfy the canonical commutation relations $[x_\mu, p_\nu] = i\hbar\delta_{\mu\nu}$. Breaking these commutators up into $L_1 := L_{23}$, etc. and $K_1 = L_{01}$, etc. gives precisely the commutation relations of $\boldsymbol{L}$, $\boldsymbol{M}'$ above.

In addition to the algebra, however, we have further relations between $\boldsymbol{M}$, $\boldsymbol{L}$ and H. Just as in the classical case, $\boldsymbol{L}$ and $\boldsymbol{M}$ are orthogonal:

$$\boldsymbol{L}\cdot\boldsymbol{M} = \boldsymbol{M}\cdot\boldsymbol{L} = 0 \tag{7.20}$$

Moreover (more algebra!), one can show that

$$\boldsymbol{M}^2 = (2H/m)(\boldsymbol{L}^2 + \hbar^2) + e_{\mathrm{M}}^4 \tag{7.21}$$

the quantum analogue of (7.10), to which it reduces in the limit $\hbar \to 0$.

From the algebra (7.18) we know that the possible eigenvalues of $(\boldsymbol{J}^{(1)})^2$ and $(\boldsymbol{J}^{(2)})^2$ in units of $\hbar^2$ are $j_1(j_1+1)$ and $j_2(j_2+1)$ respectively, where j_1 and j_2 can be integral or half-integral. However, since $\boldsymbol{L} = \boldsymbol{J}^{(1)} + \boldsymbol{J}^{(2)}$ and $\boldsymbol{M}' = \boldsymbol{J}^{(1)} - \boldsymbol{J}^{(2)}$, equation (7.20) imposes the condition $j_1 = j_2 = j$, say.

The only equation we have not used so far is (7.21). Multiplying by $(-2H/m)^{-1}$ this becomes

$$-\tfrac{1}{2}me_{\mathrm{M}}^4 H^{-1} = (\boldsymbol{M}')^2 + \boldsymbol{L}^2 + \hbar^2 \tag{7.22}$$

But $(\boldsymbol{M}')^2 + \boldsymbol{L}^2 = 2(\boldsymbol{J}^{(1)2} + \boldsymbol{J}^{(2)2}) = 4j(j+1)\hbar^2$ on the above states. Thus for these states the energy eigenvalues E are determined as

$$\begin{aligned} -\tfrac{1}{2}me_{\mathrm{M}}^4 E^{-1} &= [4j(j+1) + 1]\hbar^2 \\ &= (2j+1)^2\hbar^2 \end{aligned} \tag{7.23}$$

In other words, we can write

$$E_n = \frac{-me_{\mathrm{M}}^4}{2\hbar^2 n^2} \tag{7.24}$$

which is precisely the spectrum of hydrogen, with the principal quantum number n now identified as the integer $2j+1$.

As for the degeneracy, E_n is independent of l, but for a given value of n the range of l is restricted. This is where the Clebsch–Gordan series comes in! Thus $\boldsymbol{L}$ is obtained as the addition of the two 'angular momentum' operators $\boldsymbol{J}^{(1)} + \boldsymbol{J}^{(2)}$, with $j_1 = j_2 = j$. The combined angular momentum quantum number therefore ranges in integer steps from $j_1 + j_2 = 2j$ down to $j_1 - j_2 = 0$. In terms of n, then, $l \leqslant n-1$, which is the standard result.

7.3* The Partial Wave Expansion; Unitarity

(i) Scattering Theory

In this subsection we give a lightning review of scattering theory, in the restricted context of the scattering of a non-relativistic spinless particle off a short-range central potential $V(r)$. In the absence of absorptive effects, V is real and consequently the probability current is conserved, giving rise to the optical theorem (equation (7.40)). In more formal terms this result follows from the unitarity of the S matrix, defined by equation (7.44). We defer discussion of the partial wave expansion to the following subsection, where the 'phase shifts' are introduced naturally as a consequence of unitarity and conservation of angular momentum.

There are two ways of looking at a scattering process. The first follows the time development of an individual wave packet as it passes through the scattering region. With a short-range potential (one that falls off faster than $1/r$, e.g. the Yukawa potential $e^{-\mu r}/r$) this region is effectively limited, and before and after the interaction the wave packet behaves like a free particle. The second way of thinking about the problem is to consider the steady state situation where a continuous stream of particles is scattered by the potential. It is like looking at the overall configuration of the spray from a hose (the Euler formulation of hydrodynamics) rather than concentrating on the motion of the individual molecules (the Lagrange formulation).

Adopting the second point of view, we are concerned with the characteristics of scattering solutions of the time-independent Schrödinger equation

$$[-(\hbar^2/2m)\nabla^2 + V(r)]\psi(\boldsymbol{r}) = E\psi(\boldsymbol{r}) \tag{7.25}$$

In particular we can look for solutions, labelled $\psi_{\boldsymbol{k}}^{\text{in}}(\boldsymbol{r})$, which at large distances have the form of a plane wave $e^{i\boldsymbol{k}\cdot\boldsymbol{r}}$ together with a spherical wave e^{ikr}/r, modulated by some angular function $f(\theta)$, where θ is the angle between $\boldsymbol{k}$ and $\boldsymbol{r}$:

$$\psi_{\boldsymbol{k}}^{\text{in}}(\boldsymbol{r}) \sim e^{i\boldsymbol{k}\cdot\boldsymbol{r}} + f(\theta)e^{ikr}/r \tag{7.26}$$

The reason for the nomenclature is that if a wave packet is formed of such solutions it will behave in the far past like the same wave packet formed of just the incoming plane wave part $e^{i\boldsymbol{k}\cdot\boldsymbol{r}}$, with the spherical waves interfering destructively. For large positive times, however, the spherical waves will interfere constructively, to give an outgoing scattered wave packet.

The interpretation of this solution is that the *scattering amplitude* $f(\theta)$ gives a measure of the amount of scattering in the direction θ relative to $\boldsymbol{k}$. A more precise statement can be made by considering the probability current

$$\boldsymbol{j} := -\frac{i\hbar}{2m}(\psi^*\nabla\psi - \psi\nabla\psi^*) \tag{7.27}$$

The contribution to $\boldsymbol{j}$ from the spherical part of $\psi_{\boldsymbol{k}}^{\text{in}}(\boldsymbol{r})$ is easily calculated as

$$\boldsymbol{j}^{\text{scatt}} = (\hbar k/m)|f(\theta)|^2\,\hat{\boldsymbol{r}}/r^2 \tag{7.28}$$

The flux scattered into the area $r^2\,\mathrm{d}\Omega$ given by the element of solid angle $\mathrm{d}\Omega$ is therefore

$$F^{\text{scatt}} = (\hbar k/m)|f(\theta)|^2\,\mathrm{d}\Omega \tag{7.29}$$

Dividing by the incoming flux per unit area, which is $\hbar k/m$, gives the *cross-section*

$$\mathrm{d}\sigma = |f(\theta)|^2\,\mathrm{d}\Omega \tag{7.30}$$

We have introduced the scattering amplitude $f(\theta)$ through the asymptotic form (7.26) of $\psi_{\boldsymbol{k}}^{\text{in}}(\boldsymbol{r})$, which is a solution of the Schrödinger equation (7.25) with $E = \hbar^2k^2/2m$. This differential equation can usefully be transformed into an integral equation of the form

$$\psi_{\boldsymbol{k}}^{\text{in}}(\boldsymbol{r}) = \mathrm{e}^{i\boldsymbol{k}\cdot\boldsymbol{r}} + \int \mathrm{d}^3\boldsymbol{r}'\,G(\boldsymbol{r}-\boldsymbol{r}')U(\boldsymbol{r}')\psi_{\boldsymbol{k}}^{\text{in}}(\boldsymbol{r}') \tag{7.31}$$

where $U := 2mV/\hbar^2$. Here $G(\boldsymbol{r}-\boldsymbol{r}')$ is a *Green function* of the operator $(\nabla^2 + k^2)$. That is, it satisfies the equation

$$(\nabla^2 + k^2)G(\boldsymbol{r}) = \delta^3(\boldsymbol{r}) \tag{7.32}$$

It is easily verified that

$$G(\boldsymbol{r}) = -\frac{1}{4\pi}\frac{\mathrm{e}^{ikr}}{r} \tag{7.33}$$

is a solution of (7.32), and moreover it incorporates the outgoing boundary conditions which characterize $\psi_{\boldsymbol{k}}^{\text{in}}(\boldsymbol{r})$.

Substitution of $G(\boldsymbol{r}-\boldsymbol{r}')$ into (7.31) gives

$$\psi_{\boldsymbol{k}}^{\text{in}}(\boldsymbol{r}) = \mathrm{e}^{i\boldsymbol{k}\cdot\boldsymbol{r}} - \frac{1}{4\pi}\int \mathrm{d}^3\boldsymbol{r}'\,\frac{\mathrm{e}^{ik|\boldsymbol{r}-\boldsymbol{r}'|}}{|\boldsymbol{r}-\boldsymbol{r}'|}\,U(\boldsymbol{r}')\psi_{\boldsymbol{k}}^{\text{in}}(\boldsymbol{r}') \tag{7.34}$$

To find the asymptotic form of $\psi_{\boldsymbol{k}}^{\text{in}}(\boldsymbol{r})$, i.e. $f(\theta)$, we expand the exponent in powers of r'/r, noting that r' is effectively limited because of the short-range nature of the potential. This gives

$$\psi_k^{\text{in}}(\boldsymbol{r}) = \mathrm{e}^{\mathrm{i}\boldsymbol{k}\cdot\boldsymbol{r}} - \frac{\mathrm{e}^{\mathrm{i}kr}}{4\pi r}\int \mathrm{d}^3\boldsymbol{r}'\mathrm{e}^{-\mathrm{i}\boldsymbol{k}'\cdot\boldsymbol{r}'}U(\boldsymbol{r}')\psi_k^{\text{in}}(\boldsymbol{r}') \tag{7.35}$$

where we have written $\boldsymbol{k}' = k\hat{\boldsymbol{r}}$, the scattered wavevector. Thus we can identify $f(\theta)$ as

$$f(\theta) = -\frac{1}{4\pi}\int \mathrm{d}^3\boldsymbol{r}'\,\mathrm{e}^{-\mathrm{i}\boldsymbol{k}'\cdot\boldsymbol{r}'}U(\boldsymbol{r}')\psi_k^{\text{in}}(\boldsymbol{r}') \tag{7.36}$$

which can be written in Dirac notation as

$$f(\theta) = -(2\pi)^2 m\hbar\langle\boldsymbol{p}'|V|\boldsymbol{p}\rangle^{\text{in}} \tag{7.37}$$

where $|\boldsymbol{p}'\rangle$ is a plane wave state with momentum $\boldsymbol{p}' = \hbar\boldsymbol{k}'$ normalized according to $\langle\boldsymbol{p}'|\boldsymbol{p}\rangle = \delta^3(\boldsymbol{p}' - \boldsymbol{p})$ and $|\boldsymbol{p}\rangle^{\text{in}}$ is the state with wavefunction $\psi_k^{\text{in}}(\boldsymbol{r})$ consisting of a plane wave with momentum $\boldsymbol{p} = \hbar\boldsymbol{k}$ and outgoing spherical waves.

Note that the scattering amplitude involves the solution $|\boldsymbol{p}\rangle^{\text{in}}$ of the full Schrödinger equation. If, however, the potential is in some sense weak, an expansion can be made in terms of V. In the lowest-order approximation, the *first Born approximation*, one simply replaces $|\boldsymbol{p}\rangle^{\text{in}}$ by the plane wave state $|\boldsymbol{p}\rangle$ to give

$$f(\theta) \simeq -(2\pi)^2 m\hbar\langle\boldsymbol{p}'|V|\boldsymbol{p}\rangle \tag{7.38}$$

a result also obtainable from time-dependent perturbation theory.

The hermiticity of the Hamiltonian H, i.e. the reality of V, has important implications for scattering. One immediate consequence is that the probability current $\boldsymbol{j}$ of equation (7.27) is conserved: $\boldsymbol{\nabla}\cdot\boldsymbol{j} = 0$. More generally, the *Wronskian* $\boldsymbol{W} := \psi^*\boldsymbol{\nabla}\varphi - \varphi\boldsymbol{\nabla}\psi^*$ of two solutions ψ and φ with the same energy E is divergenceless. Hence by Gauss's theorem the flux of $\boldsymbol{W}$ out of the surface of a large sphere is zero. Applying this result to two in-states with wavefunctions $\psi_k^{\text{in}}(\boldsymbol{r})$ and $\psi_{k'}^{\text{in}}(\boldsymbol{r})$ ($\boldsymbol{k}'^2 = \boldsymbol{k}^2$) and using their asymptotic forms as in (7.26), one arrives, after some algebra, at the *unitarity relation*

$$\mathrm{Im}\, f(\theta_{\text{sc}}) = \frac{k}{4\pi}\int \mathrm{d}\Omega\, f^*(\theta')f(\theta) \tag{7.39}$$

Here θ and θ' are the angles between $\hat{\boldsymbol{r}}$ and $\boldsymbol{k}$, $\boldsymbol{k}'$ respectively, while θ_{sc} is the scattering angle between $\boldsymbol{k}$ and $\boldsymbol{k}'$. An important special case of this relation, obtained by taking $\boldsymbol{k} = \boldsymbol{k}'$, is the *optical theorem*:

$$\mathrm{Im}\, f(0) = \frac{k}{4\pi}\int \mathrm{d}\Omega\, |f(\theta)|^2 \tag{7.40}$$

which relates the imaginary part of the forward scattering amplitude to the total elastic cross-section.

In order to appreciate the full significance of the unitarity relation, and indeed the reason for the nomenclature, one has to go over to Dirac notation and adopt a somewhat more formal approach to scattering theory. In this language the Green function of (7.33) can be written up to a factor as $(H_0 - E - \mathrm{i}\varepsilon)^{-1}$, where H_0 is the free Hamiltonian $\hat{\boldsymbol{p}}^2/2m$. That is, the Schrödinger equation

$$(H_0 - E)|\boldsymbol{p}\rangle^{\mathrm{in}} = -V|\boldsymbol{p}\rangle^{\mathrm{in}} \qquad (7.25')$$

has the formal solution

$$|\boldsymbol{p}\rangle^{\mathrm{in}} = |\boldsymbol{p}\rangle - (H_0 - E - \mathrm{i}\varepsilon)^{-1}|\boldsymbol{p}\rangle^{\mathrm{in}} \qquad (7.34')$$

where the infinitesimal imaginary part $-\mathrm{i}\varepsilon$, which tells one how to treat the singularity at $E' = E$, gives the correct outgoing boundary conditions. Indeed, changing the sign of this term gives the Green function appropriate for 'out' states, states consisting of a plane wave and *incoming* spherical waves, which satisfy the integral equation

$$|\boldsymbol{p}\rangle^{\mathrm{out}} = |\boldsymbol{p}\rangle - (H_0 - E + \mathrm{i}\varepsilon)^{-1}|\boldsymbol{p}\rangle^{\mathrm{out}} \qquad (7.41)$$

Equation (7.34′) is just a rewrite of the more explicit (7.34). However, in the formal approach we can write the alternative equation (the *Lippman–Schwinger* equation)

$$|\boldsymbol{p}\rangle^{\mathrm{in}} = |\boldsymbol{p}\rangle - (H - E - \mathrm{i}\varepsilon)^{-1}V|\boldsymbol{p}\rangle \qquad (7.42)$$

and a similar equation for $|\boldsymbol{p}\rangle^{\mathrm{out}}$. Here $(H - E - \mathrm{i}\varepsilon)^{-1}$ is the Green function of the full Hamiltonian $H = H_0 + V$. It is not clear what we have gained thereby, since we only know this Green function if we can solve the problem exactly. But by manipulating (7.42) and the corresponding equation for $|\boldsymbol{p}\rangle^{\mathrm{out}}$ it is not too difficult to relate the scattering amplitude $f(\theta)$ as given by (7.38) to the overlap of the states $|\boldsymbol{p}\rangle^{\mathrm{in}}$ and $|\boldsymbol{p}'\rangle^{\mathrm{out}}$, namely

$$^{\mathrm{out}}\langle\boldsymbol{p}'|\boldsymbol{p}\rangle^{\mathrm{in}} = \delta^3(\boldsymbol{p} - \boldsymbol{p}') - 2\pi\mathrm{i}\delta(E - E')\langle\boldsymbol{p}'|V|\boldsymbol{p}\rangle^{\mathrm{in}} \qquad (7.43)$$

We are now in a position to understand the significance of the unitarity relation. The two sets of states $\{|\boldsymbol{p}\rangle^{\mathrm{in}}\}$ and $\{|\boldsymbol{p}\rangle^{\mathrm{out}}\}$ are the eigenstates of a Hermitian Hamiltonian, but with different boundary conditions. They each form a complete set of scattering states and may be taken as orthonormal. So there must be a unitary transformation between the two orthonormal bases, which is called the scattering operator S:

$$|\boldsymbol{p}\rangle^{\text{in}} = S|\boldsymbol{p}\rangle^{\text{out}} \tag{7.44}$$

The left-hand side of equation (7.43) is then just the matrix element $S(\boldsymbol{p}', \boldsymbol{p}) := {}^{\text{out}}\langle \boldsymbol{p}'|S|\boldsymbol{p}\rangle^{\text{out}}$.

As we have just said, S must be a *unitary* operator, satisfying $S^{\dagger}S = \mathbb{1}$. This equation can be made explicit by sandwiching it between states ${}^{\text{out}}\langle \boldsymbol{p}'|$ and $|\boldsymbol{p}\rangle^{\text{out}}$ and inserting a complete set of states $\int d^3\boldsymbol{p}''{}^{\text{out}}|\boldsymbol{p}''\rangle\langle \boldsymbol{p}''|^{\text{out}}$. After some manipulation one arrives precisely at equation (7.39).

(ii) Partial Waves

So far we have made no use of the rotational invariance of the central potential $V(r)$. That invariance is best exploited by going over from the momentum states $|\boldsymbol{p}\rangle$ to states $|plm\rangle$. These have the same energy $E = p^2/2m$ but are eigenstates of the orbital angular momentum operators $\boldsymbol{L}^2$, L_z rather than momentum. They are conveniently normalized according to

$$\langle p'l'm'|plm\rangle = (1/p^2)\delta(p' - p)\delta_{l'l}\delta_{m'm} \tag{7.45}$$

in which $(1/p^2)\delta(p' - p)$ is the radial δ function appropriate to spherical polar coordinates, where $\int d^3\boldsymbol{p}$ becomes $\int p^2\, dp\, d\Omega$.

The expansion of a plane wave state $|\boldsymbol{p}\rangle$ in terms of the angular momentum states takes the form

$$|\boldsymbol{p}\rangle = \sum_{l,m} Y^*_{lm}(\hat{\boldsymbol{p}})|plm\rangle \tag{7.46}$$

In group theoretic terms the coefficients $Y^*_{lm}(\hat{\boldsymbol{p}})$ in this expansion are Clebsch–Gordan coefficients expressing the states of a reducible representation of the rotation group in terms of those of the irreducible representations labelled by l. The $Y_{lm}(\theta, \varphi)$ are the *spherical harmonics*, which in wave mechanics are the eigenfunctions of the differential operators $\boldsymbol{L}^2$ and L_z. As far as the present section is concerned, their most important properties are their orthonormality,

$$\int Y^*_{l'm'}(\theta, \varphi)Y_{lm}(\theta, \varphi)\, d\Omega = \delta_{l'l}\delta_{m'm} \tag{7.47}$$

and the group property,

$$\sum_m Y^*_{lm}(\theta_1, \varphi_1)Y_{lm}(\theta_2, \varphi_2) = [(2l + 1)/4\pi]P_l(\cos\theta_{12}) \tag{7.48}$$

where $P_l(\cos\theta)$ is the Legendre polynomial and θ_{12} is the angle between the two unit vectors with polar angles (θ_1, φ_1) and (θ_2, φ_2).

The proof of these relations would take us too far afield, and we refer to standard books on the rotation group (e.g. Brink and Satchler). We should, however, mention that Y_{lm} and P_l are both special cases of the rotation matrices, with

$$\begin{aligned} Y_{lm}(\theta, \varphi) &= N_l D^l_{m0}(\varphi, \theta, \psi) \\ P_l(\cos\theta) &= D^l_{00}(\varphi, \theta, \psi) \end{aligned} \tag{7.49}$$

where $N_l := [(2l+1)/4\pi]^{1/2}$. Equation (7.47) is then a special case of the orthogonality property of the rotation matrices, which is the continuum analogue of equation (4.7), while (7.48) is a consequence of the group property and unitarity of the D's. The inverse of equation (7.46), namely

$$|plm\rangle = \int d\Omega\, Y_{lm}(\hat{\boldsymbol{p}})|\boldsymbol{p}\rangle \tag{7.50}$$

is a special case of the construction of angular momentum helicity states given in Chapter 9.

A similar decomposition of the 'in' and 'out' scattering states can also be made, though not without a word of explanation. Although they are eigenstates of the complete Hamiltonian, they are labelled by their plane wave part, which is a free particle specification identifying the form of a wave packet at times well before or well after its interaction with the scattering centre. The angular momentum decomposition refers to this free particle label. Thus we can write

$$|\boldsymbol{p}\rangle^{\text{in}} = \sum_{l,m} Y^*_{lm}(\hat{\boldsymbol{p}})|plm\rangle^{\text{in}} \tag{7.51}$$

and similarly for the out-states.

We now come to the crucial point. The S operator is an operator determined by the rotationally invariant operators H_0 and V, and is therefore itself rotationally invariant. In group theoretic terms it is a scalar operator, and the Wigner–Eckart theorem implies that

$$^{\text{out}}\langle p'l'm'|S|plm\rangle^{\text{out}} = (1/p^2)\delta(p'-p)\delta_{l'l}\delta_{m'm}R_l(p) \tag{7.52}$$

say, with the reduced matrix element $R_l(p)$ being independent of m. In this angular momentum basis where the matrix is diagonal, the unitarity of the operator S, $S^\dagger S = \mathbb{1}$, reduces to the statement that $R_l(p)$ is a unitary number, i.e. $|R_l(p)| = 1$. It can therefore be parametrized by a real number, the *phase shift* $\delta_l(p)$, as

$$R_l(p) = \exp(2i\delta_l(p)) \tag{7.53}$$

The reason for the nomenclature will become apparent later.

In polar coordinates equation (7.43) becomes

$$^{\text{out}}\langle \boldsymbol{p}'|S|\boldsymbol{p}\rangle^{\text{out}} = (1/p^2)\delta(p'-p)[\delta(\Omega'-\Omega) + (\mathrm{i}k/2\pi)f(\theta)] \tag{7.54}$$

where we have used equation (7.37) and re-expressed $\delta(E'-E)$ as $mp\delta(p'-p)$. In the angular momentum basis this gives R_l as

$$R_l = 1 + 2\mathrm{i}kf_l \tag{7.55}$$

where f_l is the partial wave coefficient of $f(\theta)$, defined by the expansion

$$f(\theta) = \sum_l (2l+1)P_l(\cos\theta)f_l \tag{7.56}$$

But from (7.53) f_l can be expressed in terms of the phase shift δ_l as

$$f_l = \frac{1}{k}\,\mathrm{e}^{\mathrm{i}\delta_l}\sin\delta_l \tag{7.57}$$

and therefore satisfies

$$\mathrm{Im}\, f_l = k|f_l|^2 \tag{7.58}$$

The unitarity equation (equation (7.39)) and the optical theorem then follow rather simply using equation (7.56) and this expression of partial wave unitarity.

Apart from being a convenient way of implementing unitarity, the partial wave expansion (7.56) is also very useful in the context of low-energy scattering. There it is not too difficult to show that the phase shifts δ_l behave for small k as k^{2l+1}, or more accurately as $(ka)^{2l+1}$, where a is some measure of the range of the potential. (Recall that our whole discussion applies only to short-range potentials.) Thus for energies such that ka is small only the first few phase shifts are significant. This can be understood from a semiclassical argument illustrated in figure 7.2. A particle approaching the scattering centre O with momentum p and *impact parameter* b, the perpendicular distance from O to the asymptotic straight-line trajectory, has orbital angular momentum $L \simeq l\hbar = pb$. Thus $b \simeq l/k$. If this is very much greater than a, i.e. if $l \gg ka$, the particle does not come within the range of the potential and is not scattered. For low energies the whole scattering process can be described in terms of a few real parameters, the low-order phase shifts.

It remains to be explained why the angles δ_l introduced in equation (7.53) are known as phase shifts. The nomenclature arises from the asymptotic form of ψ^{in} when expressed in partial waves. To this end we need the partial wave expansion of $e^{i\boldsymbol{k}\cdot\boldsymbol{r}}$:

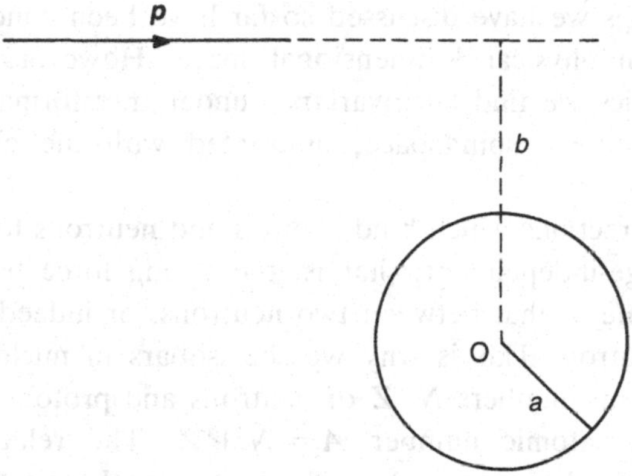

Figure 7.2 Impact parameter b compared with the range of the potential, a.

$$e^{i\boldsymbol{k}\cdot\boldsymbol{r}} = \sum_l (2l+1) i^l j_l(kr) P_l(\cos\theta) \tag{7.59}$$

(see e.g. Schiff, §19). Here $j_l(kr)$ is the spherical Bessel function, whose asymptotic behaviour for large r is

$$\begin{aligned} j_l(kr) &\sim (1/kr)\sin(kr - \tfrac{1}{2}l\pi) \\ &= (1/2ikr)((-i)^l e^{ikr} - i^l e^{-ikr}) \end{aligned} \tag{7.60}$$

In the expression for ψ^{in} this is added to

$$e^{ikr} f(\theta)/r = \sum_l (2l+1)(1/2ikr) e^{ikr}(e^{2i\delta_l} - 1) P_l(\cos\theta) \tag{7.61}$$

When the two are combined we find after some algebra that

$$\psi_{\boldsymbol{k}}^{\text{in}}(\boldsymbol{r}) \sim \sum_l (2l+1) i^l (1/kr) e^{i\delta_l} \sin(kr - \tfrac{1}{2}l\pi + \delta_l) P_l(\cos\theta) \tag{7.62}$$

Thus compared with the asymptotic form of $j_l(kr)$, the phase of the sine function has been shifted by an amount δ_l.

As a final remark, the exponential form of (7.62), namely

$$\psi_{\boldsymbol{k}}^{\text{in}}(\boldsymbol{r}) \sim \sum_l (2l+1)(1/2ikr)(e^{2i\delta_l} e^{ikr} - (-1)^l e^{-ikr}) P_l(\cos\theta) \tag{7.63}$$

clearly exhibits partial wave unitarity: for each value the incoming and outgoing fluxes, proportional to $|(-1)^l|^2$ and $|e^{2i\delta_l}|^2$ respectively, exactly balance.

7.4 Isotopic Spin; πN Scattering

Most of the groups we have discussed so far have been concerned with transformations in physical 3-dimensional space. However, in nuclear and particle physics we find an invariance under transformations in an *internal* space, isotopic spin space, associated with the charge of a particle.

The strong interactions which bind protons and neutrons together in a nucleus are charge-independent; that is, the strong force between two protons is the same as that between two neutrons, or indeed between a proton and a neutron. This is why we see isobars in nuclear physics; nuclei with different numbers N, Z of neutrons and protons respectively, but the same atomic number $A = N + Z$. The relatively small difference in masses between such nuclei is due to the electromagnetic interaction, which *does* distinguish between protons and neutrons.

This invariance under the interchange p $\leftrightarrow$ n, known as charge symmetry, does not exhaust the symmetry of the strong interactions. They are in fact invariant under the larger symmetry group of rotations in a space known as isotopic spin space†, as we shall now explain.

The proton and neutron are thought of as the upper and lower components of a 2-component object N = (p, n), the nucleon. In analogy with ordinary space, where the two spin components of a particle such as an electron (or indeed a proton!) are associated with the eigenvalues $\pm\frac{1}{2}$ of X_3, we associate the charge of the particles with the eigenvalues $\pm\frac{1}{2}$ of an operator I_3, the third of three generators I_1, I_2, I_3 which together form an algebra $SO(3)_I$ with the standard commutation relations

$$[I_i, I_j] = i\varepsilon_{ijk}I_k \tag{7.64}$$

The charge Q is not equal to I_3, but is related by

$$Q = I_3 + \tfrac{1}{2}Y \tag{7.65}$$

with $Y = 1$. For other isotopic multiplets, Y, the *hypercharge*, will take on different values. In fact, since the spectrum of I_3 is symmetric, $\frac{1}{2}Y$ is the average charge of the multiplet.

So far we have introduced a certain amount of additional notation,

†The epithet 'isotopic' is seriously misleading, since isotopes are elements with the same Z but different A. It should strictly be called 'isobaric' spin. The problem can be sidestepped by use of the abbreviation 'isospin'.

but have not really added to the original statement of charge symmetry. That comes by claiming that the strong interaction Hamiltonian commutes not only with I_3, but also with the other generators:

$$[H, I] = 0 \tag{7.66}$$

i.e. that the strong interactions are invariant under the full group $SO(3)_I$ of isotopic rotations. We emphasize again that these rotations, though mathematically identical to the SO(3) rotations of ordinary [3] space, operate in a completely separate internal space. The irreducible representations are labelled by the total isospin quantum number I and have multiplicity $2I + 1$, with I_3 ranging from $+I$ to $-I$ in integer steps.

All strongly interacting particles (hadrons) fall into such isospin multiplets. For example, there are three charge states of the particle known as the pion (π), which fall into an $I = 1$ triplet (π^+, π^0, π^-) with $Y = 0$. The symmetry is manifest at the level of the spectrum in that the masses are approximately equal ($m(\pi^\pm) = 139.6\ \text{MeV}/c^2$, $m(\pi^0) = 135.0\ \text{MeV}/c^2$). In the field of nuclear physics there are many examples of doublets or triplets of nuclei with roughly equal masses.

The invariance claimed in (7.66) can be tested in scattering experiments. The initial and final states can each be decomposed into eigenstates of I, I_3, for which we can invoke the Wigner–Eckart theorem. The invariance of the Hamiltonian under isospin rotations means that the scattering operator T is a scalar operator, so that

$$\langle I', I_3'|T|I, I_3\rangle = T_I \delta_{II'} \delta_{I_3 I_3'} \tag{7.67}$$

which means that the scattering amplitudes for the various charge modes can all be expressed in terms of a small number of eigenamplitudes T_I.

Taking πN scattering as an example, the initial and final states are each made up of a multiplet with $I = 1$ and another with $I = \frac{1}{2}$. The Clebsch–Gordan series for this product contains the irreps $I = 3/2, 1/2$. Thus all possible scattering amplitudes for the different charge configurations are expressible in terms of the two amplitudes $T_{1/2}$, $T_{3/2}$. These can be eliminated to give the relation

$$T(\pi^0\text{p} \to \pi^+\text{n}) = \sqrt{2}[T(\pi^0\text{p} \to \pi^0\text{p}) - T(\pi^+\text{n} \to \pi^+\text{n})] \tag{7.68}$$

between the amplitude for the charge exchange reaction $\pi^0\text{p} \to \pi^+\text{n}$ and those for the elastic reactions $\pi^0\text{p} \to \pi^0\text{p}$ and $\pi^+\text{n} \to \pi^+\text{n}$.

Equation (7.68) is established by explicitly constructing the states of total isospin 1/2, 3/2. The state with $I = 3/2$, $I_3 = 3/2$ is clearly

$$|3/2, 3/2\rangle = |\pi^+\text{p}\rangle \tag{7.69}$$

The state $|3/2, 1/2\rangle$ is obtained by operating on this with the lowering operator I_-:

$$|3/2, 1/2\rangle = (1/\sqrt{3})(\sqrt{2}|\pi^0 p\rangle + |\pi^+ n\rangle) \tag{7.70}$$

while the other state with $I_3 = 1/2$, $|1/2, 1/2\rangle$, is determined by the condition that it is annihilated by I_+:

$$|1/2, 1/2\rangle = (1/\sqrt{3})(|\pi^0 p\rangle - \sqrt{2}|\pi^+ n\rangle) \tag{7.71}$$

Inverting the last two equations we can express $|\pi^0 p\rangle$ and $|\pi^+ n\rangle$ in terms of the isospin eigenstates:

$$\begin{aligned} |\pi^0 p\rangle &= (1/\sqrt{3})(\sqrt{2}|3/2, 1/2\rangle + |1/2, 1/2\rangle) \\ |\pi^+ n\rangle &= (1/\sqrt{3})(|3/2, 1/2\rangle - \sqrt{2}|1/2, 1/2\rangle) \end{aligned} \tag{7.72}$$

By virtue of (7.67) the scattering amplitudes for the different physical processes are given by

$$\begin{aligned} \langle \pi^0 p|T|\pi^+ n\rangle &= (\sqrt{2}/3)(T_{3/2} - T_{1/2}) \\ \langle \pi^0 p|T|\pi^0 p\rangle &= (1/3)(2T_{3/2} + T_{1/2}) \\ \langle \pi^+ n|T|\pi^+ n\rangle &= (1/3)(T_{3/2} + 2T_{1/2}) \end{aligned} \tag{7.73}$$

Subtracting the last two equations we obtain

$$\begin{aligned} \langle \pi^0 p|T|\pi^0 p\rangle - \langle \pi^+ n|T|\pi^+ n\rangle &= (1/3)(T_{3/2} - T_{1/2}) \\ &= (1/\sqrt{2})\langle \pi^0 p|T|\pi^+ n\rangle \end{aligned}$$

which is equation (7.68).

Because the scattering amplitudes are in general complex numbers, equation (7.68) does not give an equality for the cross-sections, which are proportional to their squared moduli. Rather, it means that the square roots $\sqrt{(\frac{1}{2}\sigma_{ce})}$, $\sqrt{\sigma_+}$ and $\sqrt{\sigma_0}$ can be represented as the sides of

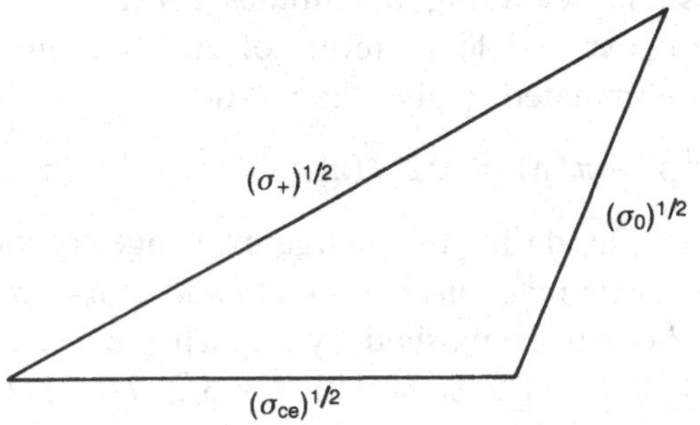

Figure 7.3 Triangle inequalities for πN scattering cross-sections.

a triangle, as illustrated in figure 7.3, and therefore satisfy *triangle inequalities* such as $\sqrt{(\frac{1}{2}\sigma_{ce})} \leq \sqrt{\sigma_+} + \sqrt{\sigma_0}$.

The amplitudes for the other possible scattering channels are

$$\langle \pi^+ p|T|\pi^+ p\rangle = T_{3/2} = \langle \pi^- n|T|\pi^- n\rangle \tag{7.74}$$

the last equation being a result of charge symmetry; similarly

$$\begin{aligned} \langle \pi^- p|T|\pi^+ n\rangle &= \langle \pi^+ n|T|\pi^+ n\rangle \\ \langle \pi^0 n|T|\pi^0 n\rangle &= \langle \pi^0 p|T|\pi^0 p\rangle \\ \langle \pi^0 n|T|\pi^- p\rangle &= \langle \pi^0 p|T|\pi^+ n\rangle \end{aligned} \tag{7.75}$$

This is as far as group theory alone can take us. However, low-energy scattering is dominated by the Δ resonance, which is a particle with $I = 3/2$. Thus in this regime $T_{3/2}$ is much greater than $T_{1/2}$. With the help of this *dynamical* statement we can directly relate the cross-sections for $\pi^- p$ and $\pi^+ p$ elastic scattering:

$$\langle \pi^- p|T|\pi^- p\rangle \simeq (1/3)\langle \pi^+ p|T|\pi^+ p\rangle$$

and hence

$$\sigma(\pi^- p) \simeq (1/9)\sigma(\pi^+ p) \tag{7.76}$$

which is well verified experimentally at the Δ peak.

Problems for Chapter 7

7.1 Taking into account Fermi statistics, what is the degeneracy of the electronic configuration for N: $(1s)^2(2s)^2(2p)^3$?

7.2 Consider the strong decay of an $I = \frac{1}{2}$ nucleonic resonance N* into πN. (i) How many invariant amplitudes govern the decay? (ii) How are the amplitudes for $N^{*+} \to \pi^+ n$ and $\pi^0 p$ related?

7.3 How many $SU(2)_I$-invariant amplitudes are required to describe the processes (i) $\pi N \to \pi\Delta$, (ii) $NN \to N\Delta$?

7.4 How many are required to describe NN → NN? Relate the amplitude for charge exchange, pn → np, to those for pp → pp and pn → pn.

8

The SU(N) Groups and Particle Physics

8.1 The Relation Between SU(2) and SO(3)

We have frequently alluded to the close connection between these two groups: it is now time to clarify the precise nature of their relationship.

The commutation relations of SO(3) summarized in equation (6.28) can be realized by 2×2 matrices X_i. By a choice of phase these matrices can be taken as $X_i = \frac{1}{2}\sigma_i$, where the σ_i are the *Pauli spin matrices*:

$$\sigma_1 = \begin{pmatrix} 0 & 1 \\ 1 & 0 \end{pmatrix} \qquad \sigma_2 = \begin{pmatrix} 0 & -\mathrm{i} \\ \mathrm{i} & 0 \end{pmatrix} \qquad \sigma_3 = \begin{pmatrix} 1 & 0 \\ 0 & -1 \end{pmatrix} \tag{8.1}$$

The reader should indeed check that they satisfy the required commutation relations. So, however, would any cyclic permutation. The precise form of the Pauli matrices arises from the general procedure of Appendix B for solving the eigenvalue problem in quantum mechanics. This gives the matrix elements (6.33) for X_3, X_+ and X_-, where an implicit choice of phase (the Condon–Shortley phase convention) has been made in taking the positive square roots for the normalization factors on the right-hand side. In the case $j = \frac{1}{2}$, with $|\frac{1}{2}\,\frac{1}{2}\rangle$ and $|\frac{1}{2}\,-\frac{1}{2}\rangle$ represented by the column vectors (1, 0) and (0, 1) respectively, X_3 corresponds precisely to the matrix $\frac{1}{2}\sigma_3$, while $X_\pm$ become $\frac{1}{2}\sigma_\pm$, where

$$\sigma_+ = \begin{pmatrix} 0 & 1 \\ 0 & 0 \end{pmatrix} \qquad \sigma_- = \begin{pmatrix} 0 & 0 \\ 1 & 0 \end{pmatrix} \tag{8.2}$$

Accordingly $X_1 \equiv \frac{1}{2}(X_+ + X_-)$ and $X_2 \equiv -\frac{1}{2}\mathrm{i}(X_+ - X_-)$ are represented by $\frac{1}{2}\sigma_1$ and $\frac{1}{2}\sigma_2$ respectively.

These three generators X_i are in fact 2×2 Hermitian, traceless matrices. They actually form a complete basis for such matrices, whose general form must be

$$H = \begin{pmatrix} c & a - ib \\ a + ib & -c \end{pmatrix}$$

with a, b and c real. But this is nothing other than $H = a\sigma_1 + b\sigma_2 + c\sigma_3$. When exponentiated, the three X_i generate the complete group of 2×2 unitary matrices with unit determinant, namely the group SU(2). That is, a general SU(2) matrix can be written as

$$U = \exp(-\tfrac{1}{2}\mathrm{i}\boldsymbol{\sigma}\cdot\boldsymbol{\alpha}) \tag{8.3}$$

To make contact with the geometrical picture of SO(3) we can always write the set of infinitesimal generators $\boldsymbol{\alpha}$ as $\boldsymbol{\alpha} = \boldsymbol{n}\theta$, where $\boldsymbol{n}$ is a unit vector and θ an angle. In this form the exponential series implicit in (8.3) can easily be evaluated since the matrix $\boldsymbol{\sigma}\cdot\boldsymbol{n}$ satisfies $(\boldsymbol{\sigma}\cdot\boldsymbol{n})^2 = \mathbb{1}$. Successive even terms thus give the cosine series for $\frac{1}{2}\theta$ multiplied by the unit matrix, while the odd terms give the sine series multiplied by $-\mathrm{i}\boldsymbol{\sigma}\cdot\boldsymbol{n}$. That is,

$$U = \cos\tfrac{1}{2}\theta - \mathrm{i}\boldsymbol{\sigma}\cdot\boldsymbol{n}\sin\tfrac{1}{2}\theta \tag{8.4}$$

Now any SO(3) matrix O is similarly parametrized by $\boldsymbol{n}$ and θ, as in equation (6.26), so we can consider a homomorphism M: $\mathrm{SU(2)} \to \mathrm{SO(3)}$ between the two groups. The mapping is not 1:1 since the kernel is non-trivial. This is precisely where the 2π periodicity of SO(3) versus the 4π periodicity of SU(2) comes in. For consider the case $\theta = 2\pi$. From (8.4) we see that $U(2\pi) = -\mathbb{1}$. However, the corresponding SO(3) matrix is $R(2\pi) = \mathbb{1} = R(0)$. Thus two distinct elements of SU(2), $U(0)$ and $U(2\pi)$, map onto the identity of SO(3). That is, the kernel K of the homomorphism is non-trivial, consisting of the two elements:

$$K = \left\{\begin{pmatrix} 1 & 0 \\ 0 & 1 \end{pmatrix}, \begin{pmatrix} -1 & 0 \\ 0 & -1 \end{pmatrix}\right\} \tag{8.5}$$

which happens to be the centre of SU(2), forming a Z_2 normal subgroup. Hence the precise relation between SU(2) and SO(3) is

$$\mathrm{SO(3)} \cong \mathrm{SU(2)}/\mathrm{Z}_2 \tag{8.6}$$

The groups differ in their global properties, but share the same algebra, which reflects the structure of the group in the vicinity of the identity. The relationship can be depicted geometrically, as in figure 8.1. Here the vector OP represents $\boldsymbol{\alpha} \equiv \boldsymbol{n}\theta$. The parameter space of SU(2) corresponds to the whole of the interior of a sphere of radius 2π. The upper half-sphere takes care of all angles less than 2π. Angles greater than 2π can be represented by points in the lower half-sphere, since $U_n(\theta) = U_{-n}(2\pi - \theta)$. On the same diagram, however, we can accommodate the entire parameter space of SO(3) in just the upper half of the sphere.

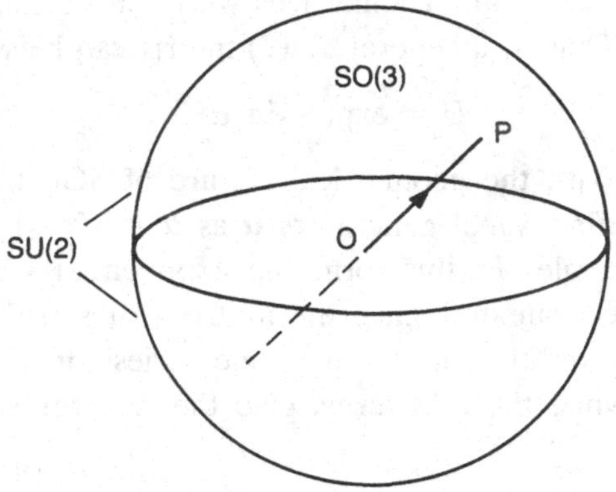

Figure 8.1 Parameter space of SO(3) (upper hemisphere) and SU(2) (entire sphere). The vector OP represents $\boldsymbol{\alpha} = \boldsymbol{n}\theta$.

8.2 SU(2)

(i) Quarks; Isospin as SU(2)

It is now believed that the constituents of nuclei, the proton and the neutron, are themselves made up of still more fundamental entities, the *quarks*. The isospin invariance observed at the hadron level is now understood as a reflection of the same invariance between the two so-called 'flavours' of quarks which make up the nucleons. These are known as the u and d quarks (for 'up' and 'down' respectively: particle physicists are rather unimaginative when it comes to notation!), which transform as a doublet:

$$q = \begin{pmatrix} u \\ d \end{pmatrix} \tag{8.7}$$

under $SU(2)_I$. Thus u is assigned $I_3 = \frac{1}{2}$ and d the value $-\frac{1}{2}$,

As far as quantum numbers are concerned, the nucleons can be considered as being made up of the combinations

$$p = uud$$
$$n = udd \tag{8.8}$$

With three quarks making up the nucleon, whose hypercharge is $Y = 1$, the hypercharge of the up and down quarks must be $\frac{1}{3}$. Equation (7.65) then gives their charges as $Q_u = \frac{1}{2} + \frac{1}{6} = \frac{2}{3}$ and $Q_d = -\frac{1}{2} + \frac{1}{6} = -\frac{1}{3}$. As you should check, this is consistent with the charges of the proton and neutron being the sum of the charges of their constituents.

In group theoretic terms equation (8.8) corresponds to one term in the Clebsch–Gordan series for the triple product of irreps $\frac{1}{2} \otimes \frac{1}{2} \otimes \frac{1}{2} = \frac{3}{2} \oplus \frac{1}{2} \oplus \frac{1}{2}$, as can easily be seen by first forming the direct product $\frac{1}{2} \otimes \frac{1}{2} = 1 \oplus 0$ and then taking the product of the resultant with the remaining factor. The first term in the series is in fact realized by the Δ resonance mentioned in §7.4.

Exactly the same story holds as far as ordinary spin is concerned. The quarks are spin-$\frac{1}{2}$ fermions, so that a 3-quark combination can have total spin $S = \frac{3}{2}$, as in the case of the delta, or $\frac{1}{2}$, as in the case of the nucleon.

The other particle mentioned in §7.4, the pion, is made up from quarks in a rather different way, namely as a quark–antiquark composite $q\bar{q}$. The antiquarks $\bar{q}$ also form an isospin doublet:

$$\bar{q} = \begin{pmatrix} -\bar{d} \\ \bar{u} \end{pmatrix} \tag{8.9}$$

(the minus sign will be explained later). The three charge states are made up by the combinations

$$\pi^+ = -u\bar{d}$$
$$\pi^0 = (1/\sqrt{2})(u\bar{u} - d\bar{d}) \tag{8.10}$$
$$\pi^- = d\bar{u}$$

which in group theoretic terms corresponds to the first possibility in the Clebsch–Gordan decomposition $\frac{1}{2} \otimes \frac{1}{2} = 1 \oplus 0$. The remaining possibility is realized in the particle known as the η. Again, the zero spin of the pion is obtained from the singlet combination of the spins of the quark and antiquark.

In the quark picture isospin invariance has a dynamical origin. That is, the quarks are supposed to interact by the exchange of *gluons*, g, which are spin-1 bosons. From the present point of view their most important property is that they are flavour-neutral, and in particular have $I = 0$. Thus in the fundamental 3-point interaction $q^\dagger qg$ the isospin of the quarks is conserved. In fact the interaction is invariant under a general $SU(2)_I$ which 'mixes' the flavour components of the quark spinor according to $q \to Uq$, and correspondingly $q^\dagger \to q^\dagger U^\dagger$, while leaving the gluon unchanged.

The SU(2) invariance of isospin has a very natural extension to the case when more than two 'flavours' are involved. There are now known to be at least three more types of quark, the 'strange' quark s, the 'charmed' quark c and the 'bottom' quark b (lack of imagination again!), each carrying an additive quantum number conserved by the strong interactions. Including these additional quarks, the fundamental quark–gluon vertex is $\psi^\dagger \psi g$, where ψ is the column vector (u, d, s, c, b). This is now invariant under the larger transformation $\psi \to U\psi$, where U is a matrix of $SU(N_f)$, N_f being the number of flavours, five in this case. However, such symmetries are increasingly broken due to the quark masses, which are produced by the weak and electromagnetic interactions. The case of SU(3), which is broken only at the (10–20)% level, is discussed in more detail in §8.3. With hindsight the generalization of isospin invariance to $SU(N_f)$ seems rather natural, but only when the isospin group is thought of as SU(2) rather than SO(3).

(ii) SU(2) Tensors

Let us now go through this rather more formally, namely to classify the irreducible representations of SU(2), construct the Clebsch–Gordan series, etc. The method used will be different from that used in the study of finite groups, and to some extent for SO(3), where we concentrated on the *matrices* of the representation, and in particular their traces—the *characters*. Instead, for the SU(N) groups it turns out to be easier to work in terms of the *vector spaces* which carry the representations. Moreover, in a mathematical parallel to the quark model, we can build up these vector spaces by multiple products of the N-plets which carry the fundamental [N] representation.

In SU(2) the fundamental [2] irrep acts on 2-component spinors ψ_a, with $a = 1, 2$ corresponding respectively to the upper ($m = \frac{1}{2}$) and lower ($m = -\frac{1}{2}$) components. Since these are *components*, not the vectors themselves (in Dirac notation $\psi_a = \langle a|\psi\rangle$), the transformation is the matrix multiplication

$$\psi'_a = U_{ab}\psi_b \qquad U \in SU(2) \tag{8.11}$$

For a general SU(N) group there are actually two kinds of spinor—*upper* and *lower*. The upper spinors transform according to the conjugate representation, i.e. with the matrices $U^* = (U^{-1})^\dagger$:

$$\psi'^a = U^*_{ab}\psi^b \tag{8.12}$$

That is, ψ^a transforms like $(\psi_a)^*$. Note that the matrices U^* do indeed form a representation, since the mapping $D(U) = U^*$ preserves the group multiplication.

However, in the case of SU(2) the two representations are *equivalent*. For consider the special matrix

$$C = \mathrm{i}\sigma_2 = \begin{pmatrix} 0 & 1 \\ -1 & 0 \end{pmatrix} \tag{8.13}$$

It is an antisymmetric, anti-Hermitian matrix satisfying $C^2 = -1$. Using C to form conjugates of the three Pauli matrices we find

$$\begin{aligned} C\sigma_1 C^{-1} &= (\mathrm{i}\sigma_2)\sigma_1(-\mathrm{i}\sigma_2) = \sigma_2(\mathrm{i}\sigma_3) = -\sigma_1 \\ C\sigma_2 C^{-1} &= \sigma_2 \\ C\sigma_3 C^{-1} &= (\mathrm{i}\sigma_2)\sigma_3(-\mathrm{i}\sigma_2) = \sigma_2(-\mathrm{i}\sigma_1) = -\sigma_3 \end{aligned} \tag{8.14}$$

Now σ_1 and σ_3 are real and σ_2 is pure imaginary. So equations (8.14) can be summarized as

$$C\boldsymbol{\sigma}C^{-1} = -\boldsymbol{\sigma}^* \tag{8.15}$$

If, therefore, we conjugate a general element $U = \exp(-\frac{1}{2}\mathrm{i}\boldsymbol{\sigma}\cdot\boldsymbol{\alpha})$ with C we obtain

$$\boxed{CUC^{-1} = U^*} \qquad (\mathrm{SU}(2)) \tag{8.16}$$

Thus the mapping $U \mapsto U^*$ is an 'inner' automorphism, i.e. one obtained through conjugation. Also recall that the condition for two representations to be equivalent is precisely that their matrices should all be related by the same similarity transformation.

Equation (8.16) can be looked at in quite another way. Using $C^2 = -1$ and the unitarity of U we can write $C^{-1}UC = (U^{\mathrm{T}})^{-1}$. Hence, multiplying on the left by C and on the right by U^{T},

$$UCU^{\mathrm{T}} = C \tag{8.17}$$

and in a similar way the complex conjugate of (8.16) leads to

$$U^{\mathrm{T}}CU = C \tag{8.18}$$

Rather than thinking of C as a matrix it is useful to consider it as a tensor. Its components can be summarized as

$$(C)_{ab} = \varepsilon_{ab} \tag{8.19}$$

where ε_{ab} is the antisymmetric symbol on two indices, with $\varepsilon_{12} = +1$. The content of equation (8.17) is that C, or ε, is an *invariant* tensor. Written in component form it reads

$$\varepsilon'_{ab} := U_{ac}U_{bd}\varepsilon_{cd} = \varepsilon_{ab} \tag{8.20}$$

showing that the transformed tensor $U_{ac}U_{bd}\varepsilon_{cd}$ is identical to the original form (cf. the discussion of the conductivity tensor σ_{ij} in §5.1(ii)). Alternatively, by considering individual elements it is easy to see that

$$\varepsilon'_{ab} = (\det U)\varepsilon_{ab} \tag{8.21}$$

which again is equal to ε_{ab} because U is a *special* unitary matrix, with unit determinant.

With the aid of the matrix C we can then take a lower spinor ψ_a and from it form an upper spinor ψ^a according to

$$\psi^a = C_{ab}\psi_b \tag{8.22}$$

To show that this is a correct equation we must show that ψ^a transforms according to U^*. Now

$$\psi'^a = C_{ab}U_{bc}\psi_c$$

using the invariance of C. But from (8.18), $CU = (U^{\mathrm{T}})^{-1}C = U^*C$. Thus

$$\psi'^a = U^*_{ab}\psi^b \tag{8.23}$$

as required.

The complex conjugate of equation (8.20) shows that C is also invariant under transformation by the U^*. It can therefore be regarded as a tensor endowed with two *upper* rather than two *lower* indices, and we will adopt this convention from now on. Then (8.22) will be rewritten as

$$\psi^a = C^{ab}\psi_b \tag{8.22$'$}$$

So C is to be considered a raising operator, in close analogy to the $g^{\mu\nu}$ of general relativity. In this latter form the index a appears consistently as an upper index on both sides of the equation, while the summation on b is a contraction between an upper and a lower index.

In order to conform to the convention that an upper index can only be contracted with a lower one, the transformation matrices U should now be written as $U_a{}^b$, so that equation (8.11) becomes

$$\psi'_a = U_a{}^b \psi_b \tag{8.11'}$$

for example.

We are now in a position to explain the minus sign in front of the upper component of equation (8.9). In field theory antiparticles are naturally associated with complex conjugate wavefunctions. Thus the natural definition of the antiquark spinor is as an *upper* spinor $\bar{q}^a$ with components $\bar{q}^1 = \bar{u}$, $\bar{q}^2 = \bar{d}$. But if we insist on writing it as a lower spinor, we can do so by the inverse of (8.22′), in this case

$$\bar{q}_a = (C^{-1})_{ab}\bar{q}^b \tag{8.24}$$

with $(C^{-1})_{ab} = -\varepsilon_{ab}$. Taking components we obtain $\bar{q}_1 = -\bar{d}$, $\bar{q}_2 = \bar{u}$.

It has already been remarked that in equation (8.22′) the upper indices a match between the two sides. By the same token the contracted indices b, one an upper and one a lower, effectively 'cancel out'. In more formal language the contraction of an upper spinor with a lower one produces a scalar *invariant*. Thus consider the contraction $\psi^a\varphi_a$, which is the same as $\varepsilon^{ab}\psi_b\varphi_a$. Its transform is

$$\begin{aligned}\psi'^a\varphi'_a \equiv \varepsilon^{ab}\psi'_b\varphi'_a &= \varepsilon^{ab}U_b{}^d\psi_d U_a{}^c\varphi_c \\ &= (U^{\mathrm{T}}CU)^{cd}\psi_d\varphi_c \\ &= \varepsilon^{cd}\psi_d\varphi_c \equiv \psi^c\varphi_c \end{aligned} \tag{8.25}$$

by virtue of equation (8.18). The upper and lower indices of more general tensors transform in exactly the same way and so again form an invariant when contracted.

We can think about equation (8.25) in another way. We know from our discussion of the conductivity tensor in §5.1(ii) that the antisymmetric and symmetric parts of a 2-index tensor transform independently, i.e. form invariant subspaces. The product $\psi_a\varphi_b$ can be decomposed into the three symmetric elements $\psi_1\varphi_1$, $\psi_1\varphi_2 + \psi_2\varphi_1$ and $\psi_2\varphi_2$, and the single antisymmetric element $\psi_1\varphi_2 - \psi_2\varphi_1 \equiv \varepsilon^{ab}\psi_a\varphi_b$. Since there is only one antisymmetric element it must perforce be an invariant! In angular momentum language it corresponds precisely to the spin-zero possibility in the addition of two spin-$\frac{1}{2}$'s.

Let us now consider how the higher-dimensional tensor irreps can be systematically built up from the fundamental [2] representation which acts on the lower spinors ψ_a. The first stage, which we have just been discussing, is to take the Kronecker product $\psi_a\varphi_b$ and decompose it into its symmetric and antisymmetric components:

$$\psi_a\varphi_b = \tfrac{1}{2}(\psi_{(a}\varphi_{b)} + \psi_{[a}\varphi_{b]}) \tag{8.26}$$

where

$$\begin{aligned}\psi_{(a}\varphi_{b)} &:= \psi_a\varphi_b + \psi_b\varphi_a \\ \psi_{[a}\varphi_{b]} &:= \psi_a\varphi_b - \psi_b\varphi_a\end{aligned} \tag{8.27}$$

This symmetrization and antisymmetrization between the indices a and b can be represented pictorially in what are in fact simple examples of the Young tableaux discussed in more detail in §8.4. In this language equation (8.26) is represented as

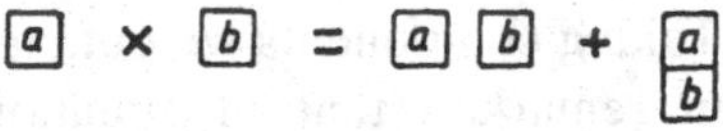

(8.26′)

where the convention is that indices in the same horizontal row are symmetrized, while two indices in the same vertical column are antisymmetrized. In the case of SU(2) no further antisymmetrization is possible because, as we have just seen, a pair of antisymmetric indices transforms like a singlet. Labelling the irreducible representations by their dimensionalities $2j + 1$, equation (8.26′) corresponds to the Clebsch–Gordan series

$$2 \otimes 2 = 3 \oplus 1 \tag{8.26″}$$

Carrying on the process we can now build up a 3-index tensor by multiplying the symmetric tensor $\psi_{(a}\varphi_{b)}$ by another spinor χ_c. The additional index can either be symmetrized with respect to both a and b to form a completely symmetric tensor $\xi_{(abc)}$ or antisymmetrized (contracted) with respect to either, leaving effectively a 1-index object. The latter spans a space of dimension 2, while the number of independent components of $\xi_{(abc)}$ is *four* ($\xi_{(111)}$, $\xi_{(112)}$, $\xi_{(122)}$, $\xi_{(222)}$). Thus the Clebsch–Gordan series is

$$3 \otimes 2 = 4 \oplus 2 \tag{8.28}$$

In pictorial terms, omitting the indices, this is

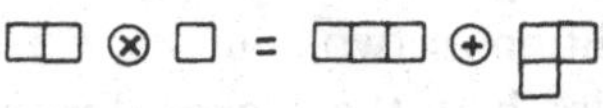

(8.28′)

where the vertical column in the last term can be excised, since it corresponds to an antisymmetrized pair of indices.

It is perhaps worth writing down the explicit decomposition of which

(8.28) is a representation, namely

$$3\psi_{(a}\varphi_{b)}\chi_c = (\psi_{(a}\varphi_{b)}\chi_c + \psi_{(b}\varphi_{c)}\chi_a + \psi_{(c}\varphi_{a)}\chi_b)$$
$$+ (\psi_{(a}\varphi_{b)}\chi_c - \psi_{(b}\varphi_{c)}\chi_a) + (\psi_{(a}\varphi_{b)}\chi_c - \psi_{(c}\varphi_{a)}\chi_b) \qquad (8.28'')$$

The first term is the completely symmetrized form. The second is antisymmetric in a and c, and in fact can be written as $\varepsilon_{ac}\eta_b$, where $\eta_b := \psi_{(d}\varphi_{b)}\chi^d$. The third involves the same spinor η, but in the combination $\varepsilon_{bc}\eta_a$.

It should now be clear that the irreps of SU(2) are carried by tensors with any number of completely symmetrized lower indices $\varphi_{(abc\ldots)}$, or symbolically $\boxed{}\boxed{}\boxed{}\boxed{\cdots}$. If there are n indices, or correspondingly n boxes, the dimensionalilty of the representation is $n + 1$ (since the number of indices equal to 1 can range from 0 to n). When two such tensors, φ of length m and ψ of length n, say, are multiplied together, the resultant tensor is reducible, since it is not completely symmetric with respect to all $m + n$ indices. Invariant subspaces can be formed by antisymmetrizing between an index of φ and another of ψ, which effectively removes both indices and reduces the dimension by 2. Taking $m \geqslant n$, for definiteness, this procedure can be continued until all the n indices of ψ have been used up. Pictorially the Clebsch–Gordan series is

$$\overset{m}{\boxed{}\boxed{}\boxed{}\boxed{\cdots}} \otimes \overset{n}{\boxed{}\boxed{}\boxed{}\boxed{\cdots}} = \sum_{l=|m-n|}^{m+n} \oplus \overset{l}{\boxed{}\boxed{}\boxed{}\boxed{\cdots}} \qquad (8.29)$$

where the number of boxes l on the right-hand side goes from $|m - n|$ to $m + n$ in steps of two. Writing this in terms of the respective dimensionalities $m + 1$, $n + 1$ and $l + 1$, and identifying $m = 2j_1$, $n = 2j_2$ and $l = 2j$, it reads

$$(2j_1 + 1) \otimes (2j_2 + 1) = \sum_{j=|j_1-j_2|}^{j_1+j_2} \oplus (2j + 1) \qquad (8.29')$$

which is just the Clebsch–Gordon series for SO(3), equation (6.42), in another guise.

8.3 SU(3)

(i) Strangeness

As accelerators increased in energy, a plethora of new particles were produced, some of which were relatively long-lived, with lifetimes of the order of 10^{-8}–10^{-10} s. The rate of production of the new particles was consistent with the strong interaction. However, if this same mechanism had been responsible for their decay, one would have expected very much shorter lifetimes. The explanation proposed was that the particles carried a new additive quantum number *strangeness*, S, which was conserved in the strong interactions, but not in the weak interactions. Thus they could be produced strongly in pairs $S\bar{S}$, with no net change in strangeness, but individually they could only decay weakly.

Gradually multiplets began to emerge, of particles with different isospin I and strangeness S, but the same space–time properties, namely the same intrinsic angular momentum (spin) J and parity P, the latter being the sign, $\pm$, produced by spatial reflection $\boldsymbol{x} \to -\boldsymbol{x}$. As mentioned in the previous section, these multiplets have roughly the same mass, but the spread in masses is much greater (~20%) than within isospin multiplets.

The most well-established multiplets are as follows. First we have the pseudoscalar mesons P, with spin-parity $J^P = 0^-$, consisting of π (π^+, π^0, π^-), K (K^+, K^0), $\bar{K}$ ($\bar{K}^0$, $\bar{K}^-$), η and η'—nine particles in all. In figure 8.2 we show their positions on an (I_3, Y) plot. For mesons the hypercharge Y, already introduced in §7.4, is equivalent to strangeness. The general relation between the two is

$$S = Y + B \tag{8.30}$$

where B is the *baryon number*. This is 0 for mesons, which can be made up of $q\bar{q}$ pairs, and 1 for particles such as the nucleon and the delta, which carry the quantum numbers of qqq.

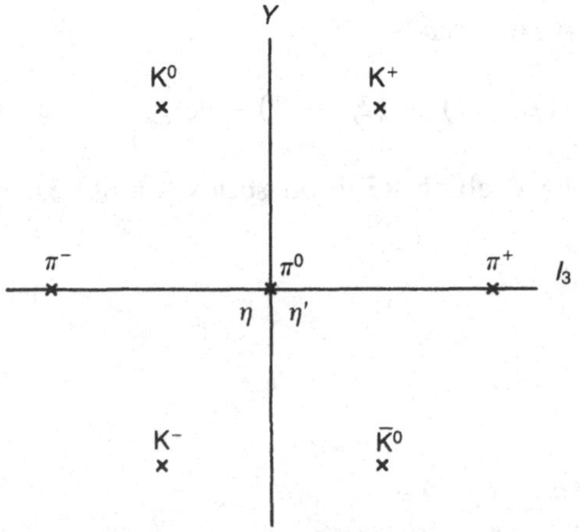

Figure 8.2 The pseudoscalar nonet of SU(3).

Then there is a nonet of vector (spin-1) mesons comprising the ρ in its three charge states (ρ^+, ρ^0, ρ^-), the K* doublet (K^{*+}, K^{*0}) and its antiparticles $\bar{K}^*$ ($\bar{K}^{*0}$, K^{*-}), the ω and the φ. These form an identical hexagonal pattern in (I_3, Y) space.

Turning now to the baryons, the nucleon forms part of an octet whose other members are the Σ (Σ^+, Σ^0, Σ^-), the Ξ (Ξ^0, Ξ^-) and the Λ^0. This multiplet is displayed in figure 8.3. The pattern is the same as for the mesons except that there are only two particles at the centre of the plot, i.e. with $I_3 = Y = 0$.

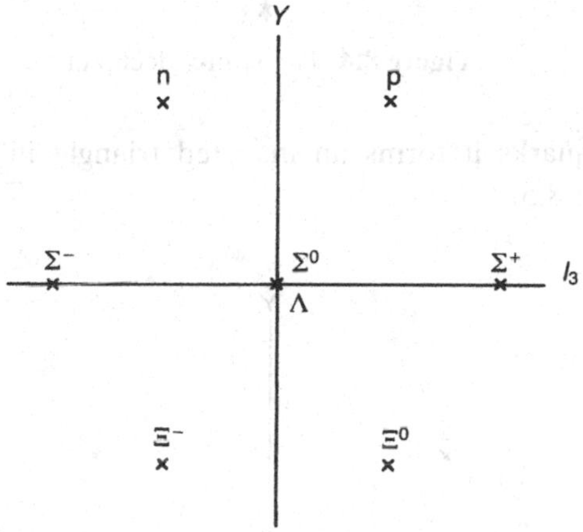

Figure 8.3 The baryon octet.

The delta resonance forms part of a decuplet rather than an octet. The other members are Σ^* and Ξ^*, excited states of the Σ and Ξ respectively, and the famous Ω^-, a particle of hypercharge $Y = -2$ or strangeness $S = -3$. These form a distinctive triangular pattern in (I_3, Y) space, as illustrated in figure 8.4.

These regularities are now understood in terms of an underlying SU(3) symmetry of the strong interactions, which has irreducible representations of just the required dimensionalities. Indeed the Ω^- was predicted on the basis of the incomplete multiplet formed by the Δ, Σ^* and Ξ^*, and its experimental discovery was a triumphant verification of the essential correctness of the SU(3) picture.

At the quark level the picture is that we simply have another quark, s, carrying strangeness, with the strong interactions being invariant under SU(3) transformations of the three quark 'flavours' u, d, s. The strange quark has isospin $I = 0$ and hypercharge $Y = -\frac{2}{3}$. Along with

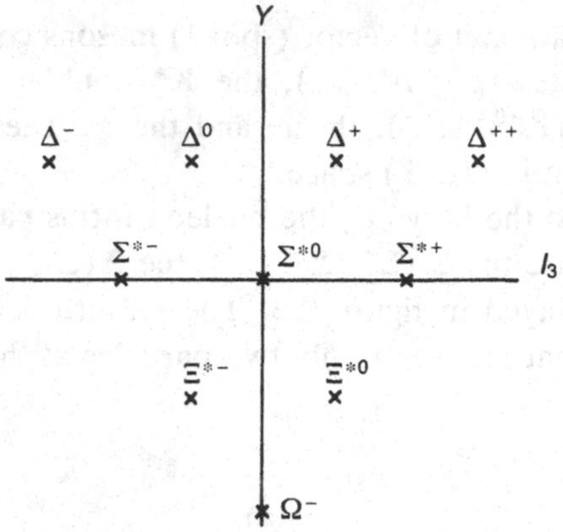

Figure 8.4 The spin-$\frac{3}{2}$ decuplet.

the u and d quarks it forms an inverted triangle in (I_3, Y) space, as shown in figure 8.5.

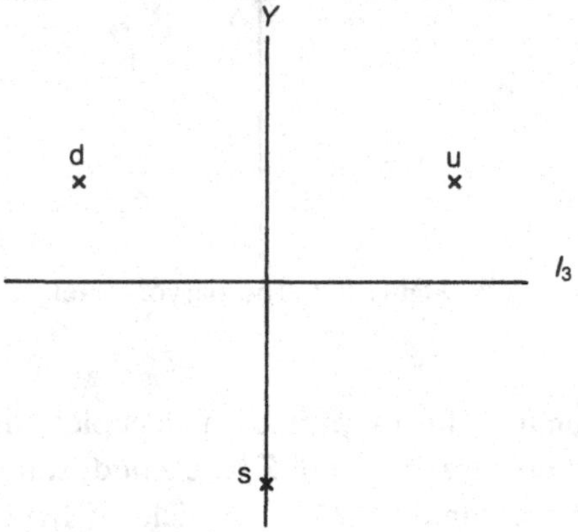

Figure 8.5 The quarks in (I_3, Y) space.

With the inclusion of the strange quark, the correct quantum numbers for all the mesons and baryons listed above can be obtained from the combinations $q\bar{q}$ and qqq respectively. In terms of group theory this just amounts to building up the higher representations of SU(3) from the fundamental 3 and $\overline{3}$ representations.

(ii) SU(3) Tensors

The fundamental [3] irrep is carried by 3-component spinors ψ_a, with

$a = 1, 2, 3$, which transform according to

$$\psi'_a = U_a{}^b \psi_b \qquad U \in \mathrm{SU}(3) \tag{8.31}$$

In contrast to the case of SU(2), complex conjugation can no longer be implemented by an inner automorphism, so that the complex conjugate representation $\overline{3}$, acting on upper spinors ψ^a, is a different, inequivalent irrep.

The invariant tensor of SU(3) is now a 3-index object, the completely antisymmetric symbol ε_{abc}, with $\varepsilon_{123} = +1$. Indeed its transform is

$$\varepsilon'_{abc} = U_a{}^d U_b{}^e U_c{}^f \varepsilon_{def} \tag{8.32}$$

$$= (\det U)\varepsilon_{abc} \tag{8.33}$$

in analogy with equations (8.20) and (8.21). Again, since we are dealing with the *special* unitary matrices, $\det U = 1$.

The tensor ε can be thought of as carrying either upper or lower indices (equation (8.32) is invariant under complex conjugation) and can be used to convert upper to lower indices or vice versa. However, the correspondence now is not between a single lower and a single upper index. Rather, it converts an *antisymmetric pair* of lower indices to an upper index, according to

$$\psi^a = \varepsilon^{abc}\varphi_{[bc]} \tag{8.34}$$

It is only the antisymmetric part of the tensor φ_{ab} which can be so raised: the symmetric part gives no contribution because of the antisymmetry of ε_{abc}.

Scalar invariants can be formed by contraction with ε_{abc}, which is equivalent to contracting an upper with a lower index. For example, from a lower spinor χ_a and an antisymmetric 2-index tensor $\varphi_{[bc]}$ we can form the invariant

$$S := \varepsilon^{abc}\chi_a\varphi_{[bc]}$$

$$= \chi_a\psi^a \tag{8.35}$$

The dimensionality of tensors with more indices can be reduced by similar contractions. That is, we can break a multi-index tensor into its irreducible components by the processes of symmetrization, antisymmetrization and contraction between upper and lower indices.

Let us now consider building up the higher-dimensional irreps of SU(3) by taking Kronecker products of the fundamental 3 and $\overline{3}$ representations and projecting out the invariant subspaces. Taking first the case $3 \otimes \overline{3}$, which corresponds physically to the $q\bar{q}$ picture of mesons, we have

$$\psi_a\bar{\psi}^b = (\psi_a\bar{\psi}^b - \tfrac{1}{3}\delta_a^b\psi_c\bar{\psi}^c) + \tfrac{1}{3}\delta_a^b\psi_c\bar{\psi}^c \tag{8.36}$$

The last term is a singlet under SU(3) transformations. The first part, which has been constructed so as to be traceless, transforms irreducibly, and has dimensionality 8. We have thus constructed the Clebsch–Gordan series for $3 \otimes \bar{3}$:

$$3 \otimes \bar{3} = 8 \oplus 1 \tag{8.36'}$$

The decomposition written explicitly in equation (8.36) can be written symbolically in terms of the Young tableaux, which we have still to explain properly. On the left-hand side the lower spinor ψ_a is represented by a single box □. As we have seen, the upper spinor $\bar{\psi}^b$ is equivalent to a tensor with a pair of antisymmetric lower indices $[de]$, say, which are represented pictorially by two boxes one above the other:

In a Kronecker product the boxes may be combined according to certain rules, which in the present case give

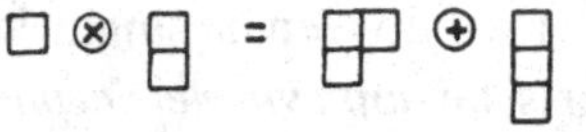

(8.36″)

The last term represents the completely antisymmetric combination of three indices, which is the scalar formed as in equations (8.36). The first is the mixed-symmetry object we first encountered in equation (8.28′). In SU(2) it has dimension 2, but, as we have just seen, there are 8 independent components in SU(3).

This product structure gives precisely the meson multiplets observed in nature. Identifying $\psi = (u, d, s)$, equation (8.36) gives the explicit $q\bar{q}$ combinations which correspond to the individual members. Thus, for the pseudoscalar octet, writing $P_a^b = \psi_a\bar{\psi}^b - \frac{1}{3}\delta_a^b\psi_c\bar{\psi}^c$, we have $P_1^2 = u\bar{d} = \pi^+$, etc. The diagonal members are somewhat more complicated. For example, $P_1^1 = u\bar{u} - \frac{1}{3}(u\bar{u} + d\bar{d} + s\bar{s})$, which can be re-expressed as $\pi^0/\sqrt{2} + \eta_8/\sqrt{6}$, where π^0 is as given in equation (8.10) and $\eta_8 = (1/\sqrt{6})(u\bar{u} + d\bar{d} - 2s\bar{s})$. Written as a matrix, P takes the form

$$P = \begin{pmatrix} \pi^0/\sqrt{2} + \eta_8/\sqrt{6} & \pi^+ & K^+ \\ \pi^- & -\pi^0/\sqrt{2} + \eta_8/\sqrt{6} & K^0 \\ K^- & \bar{K}^0 & -2\eta_8/\sqrt{6} \end{pmatrix} \tag{8.37}$$

The weak and electromagnetic interactions, which break SU(3), cause mixing between the $I = Y = 0$ member of the octet η_8 and the singlet state $\eta_1 := (u\bar{u} + d\bar{d} + s\bar{s})/\sqrt{3}$, to produce the mass eigenstates η, η'.

This mixing turns out to be fairly small in the case of the pseudoscalar mesons, but is large for the vector mesons V.

Turning now to the baryons, let us build up the product $3 \otimes 3 \otimes 3$, which should give the appropriate irreps generated by qqq. To begin with we first form the product $3 \otimes 3$ carried by tensors $\psi_a \varphi_b$. Since both indices are lower, there is no possibility of contraction: all we can do is form the symmetric and antisymmetric parts, which transform irreducibly:

$$\psi_a \varphi_b = \tfrac{1}{2}(\psi_{(a}\varphi_{b)} + \psi_{[a}\varphi_{b]}) \tag{8.38}$$

The last term has a pair of antisymmetric indices, and so is equivalent to an upper spinor, with three independent components. By subtraction, the first symmetric tensor must have six, which is correct $(6 = \tfrac{1}{2} \times n(n + 1),$ with $n = 3)$. Thus the Clebsch–Gordan series corresponding to (8.38) is

$$3 \otimes 3 = 6 \oplus \bar{3} \tag{8.38'}$$

or in terms of Young tableaux

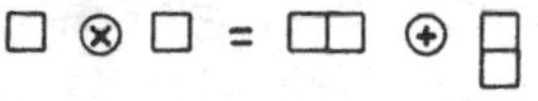

$$\square \otimes \square = \square\square \oplus \begin{array}{c}\square\\\square\end{array} \tag{8.38''}$$

Continuing the process, we multiply the result of equation (8.38) by the lower spinor χ_c. In the product $\psi_{(a}\varphi_{b)}\chi_c$ we can symmetrize and antisymmetrize with respect to the additional label to produce a completely symmetric tensor and a tensor of mixed symmetry, exactly as in equation (8.28''). The only difference from SU(2) is in the dimensionality of the irreps carried by these tensors. In the case of SU(3) the completely symmetric tensor $\xi_{(abc)}$ has ten independent indices (three components where the indices are all the same, like $\xi_{(111)}$, six where two indices are the same, like $\xi_{(112)}$, and one where they are all different, namely $\xi_{(123)}$). By subtraction, the tensor of mixed symmetry must therefore have dimension $6 \times 3 - 10 = 8$. In fact it is just the octet we have already met, since each pair of antisymmetrized lower indices $[ac]$ or $[bc]$ is equivalent to an upper index. Thus we have constructed the Clebsch–Gordan series

$$6 \otimes 3 = 10 \oplus 8 \tag{8.39'}$$

which in terms of Young tableaux reads

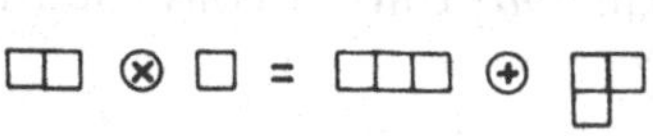

$$\square\square \otimes \square = \square\square\square \oplus \begin{array}{c}\square\square\\\square\end{array} \tag{8.39''}$$

As remarked above, the second term in (8.38) is equivalent to an upper spinor ζ^d carrying the $\bar{3}$ representation. Thus the product

$\psi_{[a}\varphi_{b]}\chi_c$ carries the representation $3 \otimes \bar{3}$, whose decomposition we have already obtained in (8.36′). Altogether, then.

$$3 \otimes 3 \otimes 3 = 10 \oplus 8 \oplus 8 \oplus 1 \tag{8.40′}$$

Again we see irreps of precisely the dimensionalities observed in the physical spectrum.

It is easy to identify the individual members of the completely symmetric decuplet D_{abc}. The deltas, containing no strange quark, have a, b, c ranging over 1 and 2 only. Thus $D_{111} = uuu = \Delta^{++}$, $D_{112} = (uud + udu + duu)/3 = \Delta^+/\sqrt{3}$, etc. The Σ^* particles contain one strange quark, with $D_{113} = (uus + usu + suu)/3 = \Sigma^{*+}/\sqrt{3}$ and so on, down to $D_{333} = sss = \Omega^-$. The baryon octet can be represented either as a 3-index tensor $B_{[ab]c}$ of mixed symmetry, which emphasizes the quark content, or, by use of ε^{abd}, as a matrix B^d_c, analogous to that for the mesons. In the latter form we have

$$B = \begin{pmatrix} \Sigma^0/\sqrt{2} + \Lambda/\sqrt{6} & \Sigma^+ & p \\ \Sigma^- & -\pi^0/\sqrt{2} + \Lambda/\sqrt{6} & n \\ \Xi^- & \Xi^0 & -2\Lambda/\sqrt{6} \end{pmatrix} \tag{8.41}$$

A systematic way of classifying the irreducible tensors of SU(3) has been given by S Coleman (*Aspects of Symmetry* CUP, Chapter 1). The canonical form of such tensors is taken as $\hat{\varphi}^{(abc\ldots)}_{(ijk\ldots)}$, with a certain number m, say, of symmetrized upper indices and n symmetrized lower indices. The $\hat{}$ signifies that the tensor is traceless with respect to contraction between any upper and any lower index, i.e. $\hat{\varphi}^{(abc\ldots)}_{(ajk\ldots)} = 0$, etc.

If these tensors do indeed form a basis, it must be possible, by symmetrization, antisymmetrization and contraction, to reduce a general tensor $\varphi^{abc\ldots}_{ijk\ldots}$ to a linear combination of such forms. We shall not attempt to give a general proof, but instead show how it works for the generic case of a 4-index tensor φ^{ab}_{ij}. After removal of traces, which correspond to octets such as $\delta^a_i\varphi^{cb}_{cj}$, we are left with the traceless tensor $\hat{\varphi}^{ab}_{ij}$. Symmetrizing and antisymmetrizing in the upper indices gives the decomposition

$$\hat{\varphi}^{ab}_{ij} = \tfrac{1}{2}(\hat{\varphi}^{(ab)}_{ij} + \hat{\varphi}^{[ab]}_{ij}) \tag{8.42}$$

The antisymmetric pair $[ab]$ can be converted into a single lower index according to

$$\hat{\varphi}^{[ab]}_{ij} = \varepsilon^{abk}\chi_{ijk} \tag{8.43}$$

The great simplification noted by Coleman is that χ is completely symmetric in its lower indices and is therefore already in canonical form. One way of seeing this is to contract χ_{ijk} with ε^{pjk}, ε^{ipk} and ε^{ijp} in turn. These contractions will pick out the parts of χ_{ijk} antisymmetric in the pairs (jk), (ki) and (ij) respectively. If all three such contractions are zero then χ must be completely symmetric.

So let us consider $\varepsilon^{ijp}\chi_{ijk}$. Now equation (8.43) is invertible; that is, we can express χ_{ijk} in terms of $\hat{\varphi}^{[ab]}_{ij}$ by

$$\chi_{ijk} = \tfrac{1}{2}\varepsilon_{kcd}\hat{\varphi}^{[cd]}_{ij} \tag{8.44}$$

as you should check by inserting (8.44) into (8.43) and using equation (6.29) with appropriate indices. The latter equation is in fact a special case of the more general relation

$$\varepsilon^{ijp}\varepsilon_{kcd} = \begin{vmatrix} \delta^i_k & \delta^i_c & \delta^i_d \\ \delta^j_k & \delta^j_c & \delta^j_d \\ \delta^p_k & \delta^p_c & \delta^p_d \end{vmatrix} \tag{8.45}$$

which is proved by considering different possible values for the indices ((ijp) must be some permutation of (123), as must be (kcd), and the answer is $+1$ if the two permutations are relatively even and -1 if they are relatively odd). We have actually chosen the indices in (8.45) to be just those which occur when we consider

$$\varepsilon^{ijp}\chi_{ijk} = \tfrac{1}{2}\varepsilon^{ijp}\varepsilon_{kcd}\hat{\varphi}^{[cd]}_{ij} \tag{8.46}$$

But because $\hat{\varphi}$ is traceless, all the contractions required by the Kronecker deltas in the top right-hand submatrix, δ^i_c, δ^j_c, etc., are zero. Thus $\varepsilon^{ijp}\chi_{ijk} = 0$. The proof that $\varepsilon^{pjk}\chi_{ijk}$ and $\varepsilon^{ipk}\chi_{ijk}$ are also zero follows essentially the same lines and is left as an exercise for the sceptical reader.

Thus we have shown that the second term in equation (8.42) can be written in canonical form. Turning now to the first term, we can symmetrize and antisymmetrize the two lower indices ij. The symmetric term will already be in the standard form $\hat{\varphi}^{(ab)}_{(ij)}$. The antisymmetric pair $[ij]$ can be converted into an upper index e, say, but by precisely the same kind of manipulations the resulting tensor $\xi^{(ab)e}$ will in fact be totally symmetric in (abe).

The proof in the general case is not really more complicated. It just involves contracting ε tensors with objects like χ and $\hat{\varphi}$, as in (8.46), but with additional, inert, indices. The dimensionality of the irrep carried by the general canonical tensor $\hat{\varphi}^{(abc\ldots)}_{(ijk\ldots)}$, with m upper and n lower indices, is $\frac{1}{2}(m+1)(n+1)(m+n+2)$, as you are invited to show in problem 8.5. In terms of Young tableaux it corresponds to the figure

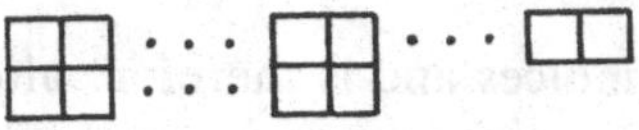

with $m + n$ boxes in the first row and m in the second. Thus each of the first m columns has two boxes, corresponding to two antisymmetrized indices, or one upper one, while the remaining n columns each have just one box, corresponding to a lower index. As we shall see in the next section, the rules for dimensionality of Young tableaux will give the same answer for SU(3), as of course they must.

Now just as was the case for isospin, as discussed in §7.4, SU(3) invariance manifests itself not only in the mass spectrum, but also in scattering experiments. Recall that in πN scattering the amplitudes for all the possible charge channels were expressible in terms of just two complex scattering amplitudes $T_{1/2}$ and $T_{3/2}$ corresponding to the two possible values of the total isospin. In the context of SU(3), the pions and nucleons fall into larger multiplets, namely the pseudoscalar and baryon octets P and B respectively, and the scattering amplitudes for πN are related to those for KN, $\pi\Sigma$, etc. Indeed all the scattering amplitudes for PB scattering are now expressible in terms of *six* complex amplitudes corresponding to the six terms in the Clebsch–Gordan decomposition of $8 \otimes 8$:

$$8 \otimes 8 = 27 \oplus 10 \oplus \overline{10} \oplus 8 \oplus 8 \oplus 1 \qquad (\mathrm{SU}(3)) \qquad (8.47)$$

The representations 27 and $\overline{10}$, which we have not encountered before, correspond to canonical tensors of the form $\hat{\varphi}^{(ab)}_{(ij)}$ and $\varphi^{(abc)}$ respectively. Equation (8.47) can be derived by precisely the procedure outlined above (problem 8.3).

8.4 SU(N); Young Tableaux

For general $N > 3$ the rather simple tensor manipulations we have used so far become rapidly more complicated, and the method of Young tableaux comes into its own. What each tableau represents in fact is a particular process of symmetrization and antisymmetrization of a given tensor to produce a tensor transforming according to some irreducible representation of SU(N). Rules have been developed for giving the dimensionality of the irrep associated with a given tableau and for deriving the Clebsch–Gordan series obtained from the direct product of two such irreps.

(i) *Young Tableaux*

A Young tableau corresponding to a tensor with n indices is an arrangement of n boxes, which must take the general form illustrated in figure 8.6.

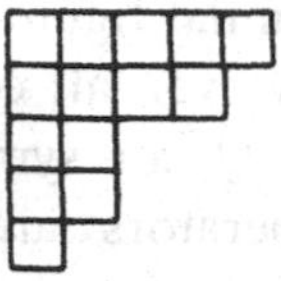

Figure 8.6 A general Young tableau.

The rows are all aligned on the left, and the rules which this figure is meant to exemplify are:

(*a*) Each row must contain no more boxes than the row above. Thus a figure like

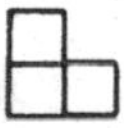

is not a legal Young diagram.

(*b*) The number of rows must not exceed N for SU(N). The reason for this is that columns of boxes represent antisymmetrized indices, and in SU(N) the indices only range from 1 to N.

Each tableau represents a very specific process of symmetrization and antisymmetrization applied to an initial tensor $\psi_{a_1 a_2 \ldots a_n}$, in which each index a_i can take any integer value from 1 to N. The indices are assigned to the boxes of the tableau, working from left to right and top to bottom, as shown in figure 8.7. On this indexed figure are applied the two operators

$$P := \sum p \tag{8.48}$$

where p is a permutation of the indices in a row, and

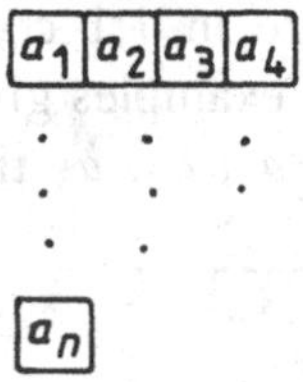

Figure 8.7 Assignment of indices to a Young tableau.

$$Q := \sum \eta_q q \tag{8.49}$$

where q is a permutation of the indices in a column and η_q is its sign (+/− according to whether q is even/odd).

The summation Σ in (8.48) extends over all rows. Thus after application of P, all the rows of the initial figure are symmetrized. Similarly, in (8.49) the summation extends over all columns. Because of the extra factor of η_q, application of Q antisymmetrizes the columns of an indexed figure. The two operators are applied in succession, the combination

$$Y := QP \tag{8.50}$$

being known as the Young operator. Note carefully that since Q is the last operator to act, the resulting tensor is indeed antisymmetric in the sets of indices corresponding to columns. However, the symmetry of the rows produced by P is destroyed by the subsequent action of Q.

Examples

(1) 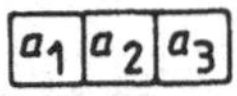

corresponds precisely to $\psi_{(a_1 a_2 a_3)}$, the completely symmetric tensor on three indices.

(2) 

corresponds instead to $\psi_{[a_1 a_2 a_3]}$, the completely antisymmetric tensor on three indices. It can only exist for $N \geqslant 3$.

(3) 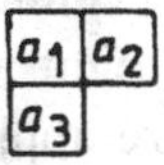

corresponds to
$(e - (13))(e + (12))\psi_{a_1 a_2 a_3}$
$= \psi_{a_1 a_2 a_3} + \psi_{a_2 a_1 a_3} - \psi_{a_3 a_2 a_1} - \psi_{a_2 a_3 a_1}$
$:= \psi_{[a_1 a_3] a_2}$.

As remarked above, the final tensor is no longer symmetric in the indices a_1 and a_2.

The primary significance of Young tableaux is in connection with irreps of the permutation group S_n; their relation to representation of SU(N) is really a secondary one. In fact (Hamermesh, Chapter 7) the different irreps of S_n are in 1:1 correspondence with Young tableaux containing n boxes. The examples given above, with $n = 3$, are a case in point. For *fixed* indices a_1, a_2, a_3 the three tensors represented by the Young tableaux form the bases for the three irreducible representations of S_3. These are respectively the trivial representation, the 1-dimensional representation whereby each permutation is represented by its sign, and the 2-dimensional representation labelled as E in table 4.2 for $D_3 \cong S_3$. Indeed there are only two independent tensors of the form $\psi_{[ac]b}$, since they satisfy the cyclic relation

$$\psi_{[ac]b} + \psi_{[ba]c} + \psi_{[cb]a} = 0 \tag{8.51}$$

as you should verify.

As we have seen above, the processes of symmetrization and antisymmetrization form invariant subspaces under the action of SU(N), allowing us to decompose the representation acting on tensors of a given rank r. In SU(N), as opposed to SO(N), there is no other way of reducing the dimensionality, so tensors of a definite symmetry type are irreducible. (The difference in the case of SO(N) is that lower indices can be contracted together, as in the scalar product $\boldsymbol{a}\cdot\boldsymbol{b}$. Thus, for example, the symmetric part of the conductivity tensor $\sigma_{(ij)}$ of §5.1 can be further decomposed as $\sigma_{(ij)} = (\sigma_{(ij)} - \frac{1}{3}\delta_{ij}\sigma_{(kk)}) + \frac{1}{3}\delta_{ij}\sigma_{(kk)}$, corresponding to the irreps $l = 2$ and 0 of SO(3) respectively.)

In the next two subsections we give without proof the rules for finding the dimensionality of the representation of SU(N) corresponding to a given Young tableau, and for determining the Clebsch–Gordan series resulting from the direct product of two such representations.

(ii) Rules for Dimensionality

(a) Write down the formal quotient of two copies of the tableau, e.g.

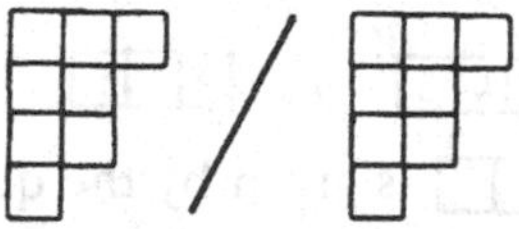

(b) In the numerator fill the boxes, starting with N in the top left-hand corner, increasing by 1 for each successive column and decreasing by 1 for each successive row. Thus in our example the numerator would be

N	$N+1$	$N+2$
$N-1$	N	
$N-2$	$N-1$	
$N-3$		

(*c*) In the denominator, the number to be entered in a given box is the number of boxes to the right of it in the same row plus the number below it in the same column, plus one for itself. Thus in our example the top left-hand box has two boxes to the right and three below; hence the entry is $2 + 3 + 1 = 6$. The remaining entries are filled in the following diagram:

6	4	1
4	2	
3	1	
1		

(*d*) Evaluate the quotient. The numerator and denominator are taken as the products of all their respective entries. In general it will be best to leave them in factored form and cancel as many common factors as possible before evaluating the final result.

Examples

(1) SU(2). The dimensionality of the representation given by , with n boxes, is

as we found before in §8.2.

(2) SU(3). The dimension of 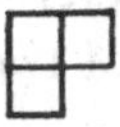 is given by the quotient

$$\begin{array}{|c|c|}\hline 3 & 4 \\ \hline 2 & \\ \cline{1-1}\end{array} \Big/ \begin{array}{|c|c|}\hline 3 & 1 \\ \hline 1 & \\ \cline{1-1}\end{array} = 8$$

That of 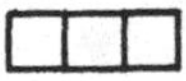 is given by

$$\begin{array}{|c|c|c|}\hline 3 & 4 & 5 \\ \hline\end{array} \Big/ \begin{array}{|c|c|c|}\hline 3 & 2 & 1 \\ \hline\end{array} = 10$$

The dimension of 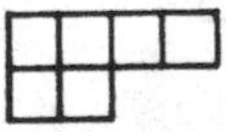 is given by the quotient

$$\begin{array}{|c|c|c|c|}\hline 3 & 4 & 5 & 6 \\ \hline 2 & 3 & & \\ \cline{1-2}\end{array} \Big/ \begin{array}{|c|c|c|c|}\hline 3 & 4 & 2 & 1 \\ \hline 2 & 1 & & \\ \cline{1-2}\end{array} = 27$$

This is the representation we have met before, carried by the traceless tensors $\hat{\varphi}^{(ab)}_{(ij)}$ with two upper and two lower symmetrized indices.

(3) SU(6). The dimensionality of the irrep corresponding to 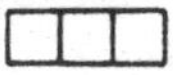 is now

$$\begin{array}{|c|c|c|}\hline 6 & 7 & 8 \\ \hline\end{array} \Big/ \begin{array}{|c|c|c|}\hline 3 & 2 & 1 \\ \hline\end{array} = 56$$

while that of 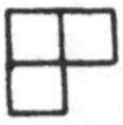 is given by

$$\boxed{\begin{smallmatrix}6&7\\5&\end{smallmatrix}} \Big/ \boxed{\begin{smallmatrix}3&1\\1&\end{smallmatrix}} = 70$$

The group SU(6) has some relevance to particle physics: the low-energy hadronic particle spectrum falls very neatly into irreps of an SU(6) which contains $SU(3)_{I,Y} \times SU(2)_J$. Thus, for example, the <u>56</u> contains precisely the spin-$\frac{1}{2}$ octet, of total multiplicity $8 \times 2 = 16$, and the spin-$\frac{3}{2}$ decuplet, of total multiplicity $10 \times 4 = 40$. The pseudoscalar octet and the vector meson nonet likewise fall into the irrep <u>35</u> generated by tensors φ_β^α, corresponding to the Young tableau

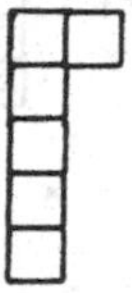

The group $SU(4) \supset SU(2)_I \times SU(2)_J$ is of similar importance in nuclear physics.

(iii) Rules for CG Series

(a) Write down the two tableaux, T_1 and T_2 say, labelling successive rows of T_2 with indices a, b, c, ...:

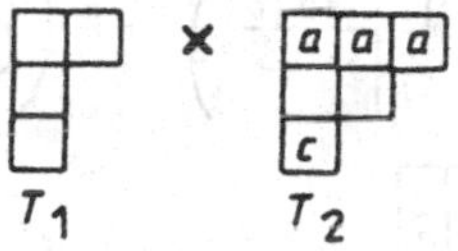

(b) Add boxes $\boxed{a} \cdots \boxed{a}$, $\boxed{b} \cdots \boxed{b}$, $\boxed{c} \cdots$ etc. from T_2 to T_1 one at a time and in that order according to the following rules:

(1) At each stage the augmented T_1 diagram must be a legal Young tableau.

(2) Boxes with the same label, e.g. a, must not appear in the same column (since this would correspond to antisymmetrizing a symmetric object).

(3) At any given box position, define n_a to be the number of a's above and to the right of it. Similarly for n_b, etc. Then we must have $n_a \geqslant n_b \geqslant n_c$ etc.

(c) If two tableaux of the same shape are produced by this process they are counted as different only if the labels are differently distributed.

(d) Cancel columns with N boxes, since they correspond to the trivial representation of SU(N).

Example: $8 \otimes 8$ in SU(3)
We have to multiply

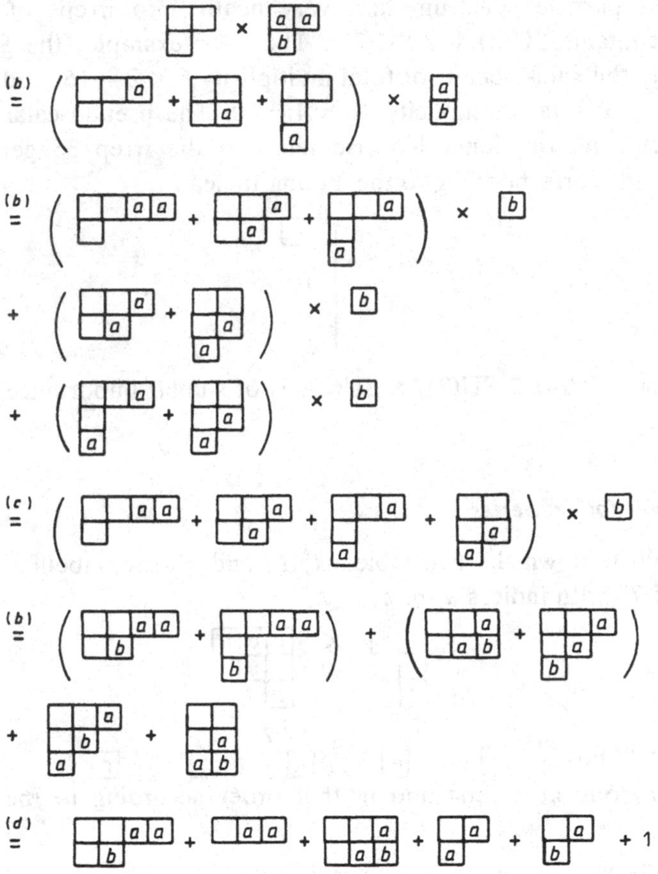

i.e.

$$8 \otimes 8 = 27 + 10 + \overline{10} + 8 + 8 + 1$$

which is just equation (8.47).

Problems for Chapter 8

8.1 Given that the $(2I + 1)_Y$ content of the fundamental representation of SU(3) is $3 = 2_{1/3} \oplus 1_{-2/3}$, use equations (8.36′), (8.40′) to find the $(2I + 1)_Y$ content of the 8- and 10-dimensional representations.

8.2 SU(3) is broken down to $SU(2)_I \times U(1)_Y$ by the weak and electromagnetic interactions, which distinguish the s quark, or the index 3, from the others. One manifestation of this is the differing masses within an octet or decuplet. A common assumption is that the mass-breaking operator behaves like the $s\bar{s}$ component of an octet, which in tensor language means that it can contain one upper index and one lower index that is explicitly given the value 3.

(*a*) In this language show that:

(i) The symmetry-breaking term $dD^{ab3}D_{ab3}$ for the decuplet gives rise to equally spaced masses: $m(\Delta) - m(\Sigma^*) = m(\Sigma^*) - m(\Xi^*) = m(\Xi^*) - m(\Omega^-)$.

(ii) The symmetry-breaking terms $a\bar{B}^a_3 B^3_a + b\bar{B}^3_a B^a_3$ for the baryon octet give rise to the Gell-Mann–Okubo mass relation $2(m(\Xi) + m(\mathrm{N})) = 3m(\Lambda) + m(\Sigma)$.

(*b*) Rederive the above results on the basis of quark content, and on the assumption that the masses are additive. Note that this model is more restrictive, giving the additional (incorrect) prediction $m(\Lambda) = m(\Sigma)$, corresponding to $a = -b$.

8.3 Prove equation (8.47) using the tensor method.

8.4 Show that the dimensionality of the completely symmetric SU(3) tensor $\varphi^{(i_1 i_2 \ldots i_m)}$ is $\frac{1}{2}(m+1)(m+2)$. A tensor $\varphi^{(i_1 i_2 \ldots i_m)}_{(j_1 j_2 \ldots j_n)}$ therefore has dimension $\frac{1}{2}(m+1)(m+2) \times \frac{1}{2}(n+1)(n+2)$.

The *irreducible* tensor $\hat{\varphi}^{(i_1 \ldots i_m)}_{(j_1 \ldots j_n)}$ is traceless with respect to contractions between upper and lower indices. Its dimension is therefore less than φ by the dimension of the space of traces $\varphi^{(i_1 \ldots i_{m-1})}_{(j_1 \ldots j_{n-1})}$. Show that $\dim \hat{\varphi}^{(i_1 \ldots i_m)}_{(j_1 \ldots j_n)} = \frac{1}{2}(m+1)(n+1)(m+n+2)$.

8.5 The Young tableau corresponding to the irreducible tensor $\hat{\varphi}^{(i_1 \ldots i_m)}_{(j_1 \ldots j_n)}$ of problem 8.4 is

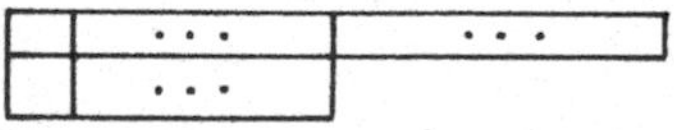

with $m + n$ boxes in the upper row and m in the lower one. Verify the result of problem 8.4 using the rules for dimensionality of Young tableaux.

8.6 Calculate the dimensionality of the following Young tableaux:

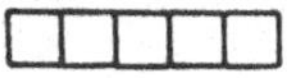

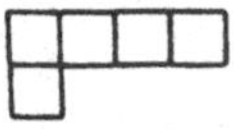

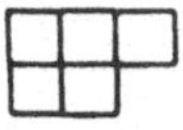

in SU(3)

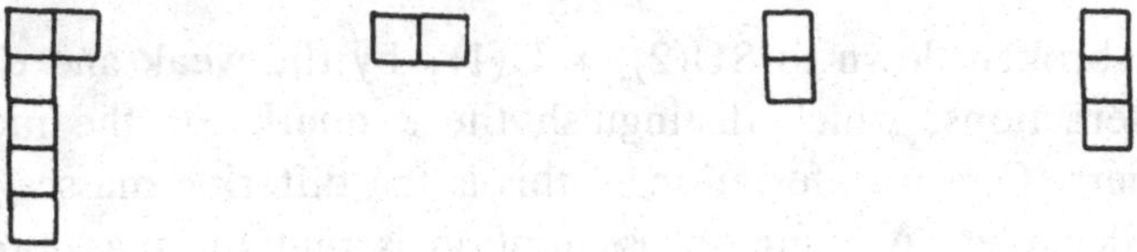 in SU(6)

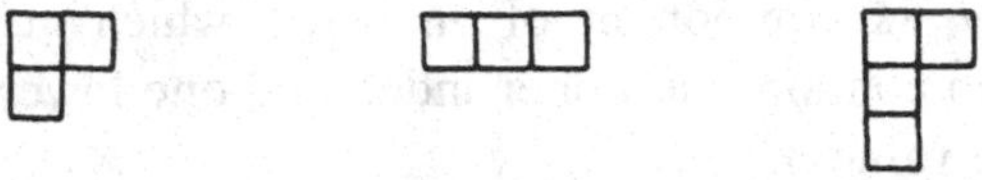 in SU(4)

8.7 Find the Clebsch–Gordan series for

 and 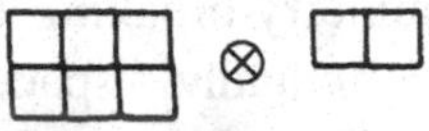 in SU(3)

$6 \otimes \bar{6}$ and $6 \otimes 6 \otimes 6$ in SU(6)

8.8 How many SU(3)-invariant amplitudes are required to describe the process

$$8 \otimes 8 \rightarrow 8 \otimes 10?$$

8.9* How many SU(6)-invariant amplitudes are required to describe the process

$$35 \otimes 56 \rightarrow 35 \otimes 70?$$

9

General Treatment of Simple Lie Groups

Up to now we have treated a number of Lie groups, such as SU(2), SU(3), SO(4) etc, on an individual basis. We have started from the defining representation, from this extracted the algebra and then gone on to enumerate the irreps by a variety of methods—characters, explicit tensor representations, Young tableaux, and in the case of SU(2) and SO(4) by the use of raising and lowering operators $J_{\pm}$.

The general treatment of Lie algebras due to Cartan, Weyl and Dynkin is an extension of the latter method. The generators are split into a set $\{H\}$ which commute with each other, and which accordingly can be assigned simultaneous eigenvalues, and the rest $\{E\}$, which are generalized raising and lowering operators.

The different Lie groups can be classified systematically in terms of this split, and of the nature of the non-vanishing commutation relations between $\{H\}$ and $\{E\}$ and those between the members of $\{E\}$. This classification amounts to specifying a number of (simple) *roots*, whose lengths and scalar products are severely restricted. These properties can be summarized by the Cartan matrix, or equivalently by a Dynkin diagram. It turns out that there are four regular infinite series of Lie algebras, the so-called classical Lie algebras, along with five exceptional Lie algebras which do not belong to such a series.

The representation theory of these Lie algebras is also modelled on the treatment of SU(2) by raising and lowering operators. In particular, recall that the states of a representation $|jm\rangle$ can all be obtained by applying lowering operators to the 'top' state $|jj\rangle$. In the Cartan language this is a state of highest weight, and the concept generalizes to more complicated Lie algebras with more than one commuting generator.

9.1 The Adjoint Representation and the Killing Form

A Lie algebra L of dimension d is specified by a set of d generators T_α closed under commutation:

$$[T_\alpha, T_\beta] = \mathrm{i} f^\gamma_{\alpha\beta} T_\gamma. \tag{9.1}$$

Strictly speaking, the Lie product $[A, B]$ is not necessarily a commutator of operators or matrices, but a skew-symmetric mapping from $L \times L$ onto L satisfying the Jacobi identity

$$[A, [B, C]] + [B, [C, A]] + [C, [A, B]] = 0 \tag{9.2}$$

which is trivially satisfied if $[A, B]$ is indeed a commutator. An important example of a Lie product which is not defined as a commutator is the Poisson bracket of Hamiltonian classical mechanics, defined as

$$[A, B]_{\mathrm{PB}} := \sum_\alpha \left(\frac{\partial A}{\partial q_\alpha} \frac{\partial B}{\partial p_\alpha} - \frac{\partial A}{\partial p_\alpha} \frac{\partial B}{\partial q_\alpha} \right). \tag{9.3}$$

The algebra is a vector space, that is, we can take linear combinations $\lambda_1 A + \lambda_2 B$, with λ_1 and λ_2 in some field, typically real or complex numbers, and the Lie product has to be consistent with the vector space addition, i.e.

$$[\lambda_1 A + \lambda_2 B, C] = \lambda_1 [A, C] + \lambda_2 [B, C]. \tag{9.4}$$

Just as for finite groups, where multiplication by a fixed element permutes the group elements, giving a finite-dimensional representation, the regular representation, so for Lie algebras commutation with a fixed generator maps the generators onto themselves, again giving a representation, of dimension d, called the *adjoint* representation, which we will denote by $\mathcal{A}$. Thus, for fixed T_α in equation (9.1) the second element T_β is mapped onto some linear combination of the generators. In this representation, the structure constants are themselves the matrix representatives of the generators, i.e.

$$(D_{\mathcal{A}}(T_\alpha))^\gamma_\beta = \mathrm{i} f^\gamma_{\alpha\beta}. \tag{9.5}$$

Note that at the moment, while $f^\gamma_{\alpha\beta}$ is clearly antisymmetric in α and β, there is no obvious symmetry involving γ, which is on a different footing from the other two, and written with an upper index to emphasize that fact. However, it is possible to lower this index, to produce a completely antisymmetric $f_{\alpha\beta\gamma}$ by means of the *Killing form.* This is a scalar product defined in terms of the adjoint representation which will prove very useful in the following analysis.

The Killing form is defined as

$$(A, B) := \mathrm{Tr}\,(D_A(A)D_A(B)) \tag{9.6}$$

i.e. the trace of the product of the matrices representing A and B in the adjoint representation. As a shorthand notation we will just write this as $\mathrm{Tr}_A(AB)$. Applying this to the generators themselves we get a metric $g_{\alpha\beta}$, the *Cartan metric*, defined by

$$g_{\alpha\beta} := \mathrm{Tr}_A(T_\alpha T_\beta) = -f^{\delta}_{\alpha\gamma} f^{\gamma}_{\beta\delta}. \tag{9.7}$$

If we use $g_{\delta\gamma}$ to lower the last index on $f^{\delta}_{\alpha\beta}$ we get

$$f_{\alpha\beta\gamma} := f^{\delta}_{\alpha\beta} g_{\delta\gamma} \tag{9.8}$$

which turns out to be totally antisymmetric in α, β and γ. Thus consider $\mathrm{Tr}_A([T_\alpha, T_\beta], T_\gamma])$. Evaluating the commutator and then taking the trace, we get

$$\mathrm{Tr}_A([T_\alpha, T_\beta], T_\gamma]) = \mathrm{i} f^{\delta}_{\alpha\beta} \mathrm{Tr}_A(T_\delta T_\gamma) = \mathrm{i} f_{\alpha\beta\gamma}. \tag{9.9}$$

But because of the cyclic property of the trace we have the identities

$$\mathrm{Tr}\,([A, B]C) = \mathrm{Tr}\,([B, C]A) = \mathrm{Tr}\,([C, A]B). \tag{9.10}$$

Hence $f_{\alpha\beta\gamma} = f_{\gamma\alpha\beta} = -f_{\alpha\gamma\beta}$ etc.

A Lie algebra may contain subalgebras, whose elements are a subset of the original generators which close under commutation and generate a subgroup of the group generated by L. A normal/invariant subgroup is generated by an invariant subalgebra, or *ideal.* This is a subspace I such that *all* commutators involving I lie in that subspace: symbolically $[I, L] \subset I$. The basic building blocks of Lie algebras are *simple* Lie algebras. These are Lie algebras which do not contain any proper ideals (i.e. other than the null vector or L itself). A *semi-simple* Lie algebra is one which does not contain any proper Abelian ideals.

The Cartan metric can tell us directly whether or not a Lie algebra is semi-simple. It turns out that a necessary and sufficient condition for L to be semi-simple is that g be non-singular, i.e. $\det(g) \neq 0$. This is equivalent to the statement that the Killing form is non-degenerate, i.e. $(A, X) = 0\,\forall\, X \in L \Rightarrow A = 0$.

It is now possible to show that a semi-simple Lie algebra can be written as a direct sum of simple Lie algebras. If L is simple, that is the end of the story. Otherwise, suppose L contains an ideal I. Then define P to be its orthogonal complement with respect to the Killing form: $(I, P) = 0$. This is itself a subalgebra, since it is a vector space, and

$$([P, P]I) = ([P, I]P) = (IP) = 0 \tag{9.11}$$

using successively the cyclic property of the trace and the fact that I is an ideal. Thus any element of $[P, P]$ is also orthogonal to I.

Now consider the scalar product of $[I, P]$ with the other members of the algebra.

First, as above,

$$([P, I]P) = (IP) = 0 \tag{9.12}$$

Then

$$([P, I]I) = ([I, I]P) = (IP) = 0 \tag{9.13}$$

using successively the cyclic property of the trace and the closure of I.

Thus $[P, I]$ is orthogonal to all elements of the algebra, and is therefore zero, since the Killing form is non-degenerate. Thus $L = I \oplus P$. If P itself is not simple, we repeat the process, and so on, until L has been expressed as the direct sum of a number of simple algebras.

For real Lie algebras, the Cartan metric also distinguishes between compactness and non-compactness. If g is positive definite, then L is compact. In that case, $g_{\alpha\beta}$ can be diagonalized and rescaled to become just $\delta_{\alpha\beta}$.

9.2 The Cartan Basis of a Lie Algebra

The structure constants determine the Lie algebra, but the converse is clearly not true, because one has the freedom of taking new basis vectors which are linear combinations of the old ones, thus changing the overt form of the commutation relations. As a prototype, consider the SU(2) algebra (scaling away the $\hbar$):

$$[J_i, J_j] = \mathrm{i}\varepsilon_{ijk} J_k. \tag{9.14}$$

By using J_3, and the complex† combinations $J_\pm = J_1 \pm \mathrm{i} J_2$ these can be recast in the form

$$\begin{aligned} [J_3, J_\pm] &= \pm J_\pm \\ [J_+, J_-] &= 2J_3. \end{aligned} \tag{9.15}$$

† Note that by allowing complex linear combinations of the generators, we are going from the real algebra of SU(2) to its complex extension, as a means to the end of determining the irreps of SU(2).

In this form the commutation relations look rather different, but of course they are completely equivalent to the first form. The Cartan way of presenting the commutation relations is a generalization of this second form.

We start by looking for a maximal set of commuting generators. This is exactly what one would like for representations in quantum mechanics, where these generators will usually represent commuting observables which can simultaneously be assigned definite eigenvalues. In SU(2) there is only one such generator, which is arbitrarily chosen to be J_3, but in SU(3) for example there are two such generators, T_3 and T_8, realized in the fundamental representation by the matrices $\frac{1}{2}\mathrm{diag}(1, -1, 0)$ and $\mathrm{diag}(1, 1, -2)/2\sqrt{3}$ respectively.

The maximal set of commuting generators $\{H_i\}$, $i = 1 \ldots r$, forms a basis for the *Cartan subalgebra*, and the number r of such generators is the *rank* of the algebra or group. Thus, in the Lie groups we have encountered so far, SU(2) has rank 1 while SU(3) and SO(4) have rank 2. In general, SU(N) has rank $N - 1$, the number of independent real traceless diagonal $N \times N$ matrices. The simultaneous eigenvalues (weights) of the H_i will be used to label the states of any representation. Thus, part of our commutation relations can be cast as

$$[H_i, H_j] = 0 \tag{9.16}$$

with $i, j = 1 \ldots r$. We wish to take linear combinations $\{E_\alpha\}$ of the remaining $d-r$ generators so that they have the property of step operators with respect to all of the H_i, namely

$$[H_i, E_\alpha] \propto E_\alpha. \tag{9.17}$$

Another way of looking at this is that if we are able to write the commutation relations in this form we have achieved a diagonalization of the adjoint action of H_i. That is, $[H_i, X] \propto X$, with the coefficient on the right-hand side actually being zero for X in the Cartan subalgebra.

Thus we are looking for the eigenvalues, λ, of the equation

$$\det(C_{\alpha\beta} - \lambda g_{\alpha\beta}) = 0 \tag{9.18}$$

where $C_{\alpha\beta} = \mathrm{i} f_{i\alpha\beta}$, for a particular, fixed i, and we have used the Killing form to lower the last index. We are assuming that the algebra has been cast in such a way that the f's are real. In that case C is a pure imaginary, antisymmetric, and therefore Hermitian matrix, with real eigenvalues. Of these, r will be zero, corresponding to equation (9.16), and $d-r$ non-zero, corresponding to equation (9.17). The fundamental theorem of Cartan, which is beyond the scope of the present treatment, is that the non-zero eigenvalues, $(\alpha)_i$, are *non-degenerate*, i.e. there is only one eigenvector, E_α, corresponding to each:

$$[H_i, E_\alpha] = (\alpha)_i E_\alpha. \tag{9.19}$$

We have achieved a nice canonical form for the commutators involving this particular H_i, but what of the other commutators? First consider $[H_j, E_\alpha]$ for $j \neq i$. Looking at its commutator with H_i, we have

$$[H_i, [H_j, E_\alpha]] = [H_j, [H_i, E_\alpha]] - [E_\alpha, [H_i, H_j]] \tag{9.20}$$

using the Jacobi identity. The second commutator on the right-hand side is zero, while the first is given by equation (9.19). So

$$[H_i, [H_j, E_\alpha]] = (\alpha)_i [H_j, E_\alpha]. \tag{9.21}$$

Thus $[H_j, E_\alpha]$ is an eigenvector of the adjoint action of H_i with eigenvalue $(\alpha)_i$. But, given that the eigenvectors are degenerate, this must be proportional to E_α. Calling the constant of proportionality $(\alpha)_j$, we have

$$[H_j, E_\alpha] = (\alpha)_j E_\alpha \tag{9.22}$$

for all j, not just for the particular one from which we started. The r-dimensional vectors $\alpha := ((\alpha)_1 \ldots (\alpha)_r)$, $\beta := ((\beta)_1, \ldots (\beta)_r)$ etc are called the *roots*, α, of the Lie algebra, and the corresponding eigenvectors E_α the *step operators*, or *root vectors*.

9.3 Properties of the Roots and Root Vectors

The next obvious question to ask is: what are the commutation relations between the E_α themselves? To this end, consider the commutator $[H_i, [E_\alpha, E_\beta]]$, which, by the Jacobi identity, is

$$\begin{aligned}[H_i, [E_\alpha, E_\beta]] &= -[E_\alpha, [E_\beta, H_i]] - [E_\beta, [H_i, E_\alpha]] \\ &= (\alpha + \beta)_i [E_\alpha, E_\beta]\end{aligned} \tag{9.23}$$

using equation (9.22). Thus $[E_\alpha, E_\beta]$ is also a root vector, with root $\alpha + \beta$, unless $\alpha + \beta = 0$ or $[E_\alpha, E_\beta] = 0$.

If $\alpha + \beta = 0$, i.e. $\beta = -\alpha$, equation (9.23) shows that $[E_\alpha, E_\beta]$ commutes with all the H_i and hence belongs to the Cartan subalgebra. So

$$\begin{aligned}[E_\alpha, E_{-\alpha}] &= \lambda_i H_i \\ [E_\alpha, E_\beta] &= N_{\alpha\beta} E_{\alpha+\beta} \qquad (\beta \neq -\alpha)\end{aligned} \tag{9.24}$$

where $N_{\alpha\beta} = 0$ if $\alpha + \beta$ is not a root.

In writing equation (9.24) we have assumed that if α is a root, then $-\alpha$ is also a root. This is easily shown by taking the transpose of equation (9.18), in which C is an antisymmetric matrix. Thus, the non-zero roots come in pairs $\pm\alpha$, as was the case in the prototype example of SU(2). Moreover, if the H_i are Hermitian, as is certainly the case for a compact group, equation (9.19) shows that $E_{-\alpha} = E_\alpha^\dagger$.

As a consequence of the particular form of the commutators in the Cartan basis, the scalar products of the $\{H_i\}$ and the $\{E_\alpha\}$ have rather simple properties, which can be cast in a canonical form using the freedom of redefinition which still remains. First consider (H_i, E_α). Using the cyclic property of the trace we have $\mathrm{Tr}_A([H_j, E_\alpha]H_i]) = -\mathrm{Tr}_A([H_j, H_i]E_\alpha) = 0$, since H_j and H_i commute. But the left-hand side is $\alpha_j \mathrm{Tr}_A(E_\alpha H_i)$, and $\alpha_j \neq 0$. Hence

$$(H_i, E_\alpha) = 0. \tag{9.25}$$

In an exactly similar fashion we can show that for $\alpha + \beta \neq 0$ the scalar product

$$(E_\alpha, E_\beta) = 0 \qquad \alpha + \beta \neq 0. \tag{9.26}$$

In contrast, $(E_\alpha, E_{-\alpha})$ must be non-zero for a simple Lie algebra with a non-degenerate metric, and by a suitable normalization it can be chosen to be one:

$$(E_\alpha, E_{-\alpha}) = 1. \tag{9.27}$$

Because the Killing form is non-degenerate, and, in view of equation (9.25), block diagonal in the basis of H_i and E_α, the sub-matrix (H_i, H_j) must also be non-degenerate, and therefore, by an appropriate transformation in the Cartan subalgebra, can be taken as δ_{ij}:

$$(H_i, H_j) = \delta_{ij}. \tag{9.28}$$

With these choices, the quantities λ_i in equation (9.24) turn out to be precisely $(\alpha)_i$. Thus, taking the scalar product of that equation with H_j:

$$([E_\alpha, E_{-\alpha}], H_j) = \lambda_j$$

using the orthogonality of the $\{H_i\}$. On the other hand, using the cyclic property of the trace

$$([E_\alpha, E_{-\alpha}]H_j) = ([H_j, E_\alpha], E_{-\alpha}) = (\alpha)_j(E_\alpha, E_{-\alpha}) = (\alpha)_j$$

showing that $\lambda_j = (\alpha)_j$, as claimed.

To summarize, the commutators and scalar products of the generators in the *Cartan–Weyl basis* are

$$\begin{aligned} [H_i, H_j] &= 0 \\ [H_i, E_\alpha] &= (\alpha)_i E_\alpha \\ [E_\alpha, E_{-\alpha}] &= (\alpha)_i H_i \\ [E_\alpha, E_\beta] &= N_{\alpha\beta} E_{\alpha+\beta} \qquad \alpha+\beta \neq 0 \end{aligned} \tag{9.29}$$

and

$$\begin{aligned} (H_i, H_j) &= \delta_{ij} \\ (H_i, E_\alpha) &= 0 \\ (E_\alpha, E_{-\alpha}) &= 1 \\ (E_\alpha, E_\beta) &= 0 \qquad \alpha+\beta \neq 0. \end{aligned} \tag{9.30}$$

9.4 Quantization of the Roots

Each root, α, is an r-dimensional vector, with components $(\alpha)_i$. We can therefore define the scalar products $\alpha \cdot \beta := (\alpha)_i(\beta)_i$ and $\alpha \cdot H := (\alpha)_i H_i$, with an implicit summation over i. For reasons that will soon become apparent, let us define

$$H_\alpha := \frac{2}{\alpha^2}\alpha \cdot H \tag{9.31}$$

where, in an obvious notation, α^2 stands for $\alpha \cdot \alpha$. Since the H_α are linear combinations of the H_i they commute:

$$[H_\alpha, H_\beta] = 0. \tag{9.32}$$

From the second of equations (9.29) we see that

$$[H_\alpha, E_\beta] = \frac{2\alpha \cdot \beta}{\alpha^2} E_\beta \tag{9.33}$$

while the third is just

$$[E_\alpha, E_{-\alpha}] = \frac{1}{2}\alpha^2 H_\alpha. \tag{9.34}$$

As special cases of equation (9.33) we have

$$[H_\alpha, E_{\pm\alpha}] = \pm 2E_{\pm\alpha}. \tag{9.35}$$

If we compare these last two equations with the commutation relations of the SU(2) generators in equation (9.15), we see that for each α we have an SU(2) algebra, which we will denote by S_α, with the identifications $J_3 = \frac{1}{2}H_\alpha$, $J_\pm = (2/\alpha^2)^{\frac{1}{2}} E_{\pm\alpha}$.

The meaning of the latter equation is that $E_{\pm\alpha}$ are step operators for H_α, raising and lowering its eigenvalues by ± 2, respectively. Since we know that the eigenvalues of J_3 are half-integral, those of H_α must be **integral**.

But now consider the action of one of the other step operators, E_β, on H_α. The step in this case is $2\alpha \cdot \beta/\alpha^2$, which must therefore be an integer. Thus we have the important property

$$\frac{2\alpha \cdot \beta}{\alpha^2} = n = \text{integer}. \tag{9.36}$$

Moreover, these integers are strictly limited since, by the Schwartz inequality, the symmetric product

$$\left(\frac{2\alpha \cdot \beta}{\alpha^2}\right)\left(\frac{2\beta \cdot \alpha}{\beta^2}\right) \leqslant 4. \tag{9.37}$$

Defining the generalized angle between the vectors α and β in the obvious way by

$$\cos\theta = \frac{\alpha \cdot \beta}{|\alpha||\beta|} \tag{9.38}$$

where $|\alpha| := \sqrt{(\alpha^2)}$ and similarly for $|\beta|$, the relation (9.37) simply corresponds to $|\cos\theta| \leq 1$. Thus the pair of integers $n_1 = 2\alpha \cdot \beta/\alpha^2$ and $n_2 = 2\alpha \cdot \beta/\beta^2$ are limited by

$$n_1 n_2 \leqslant 4 \tag{9.39}$$

and then

$$\cos\theta = \tfrac{1}{2}(n_1 n_2)^{\frac{1}{2}}. \tag{9.40}$$

This gives us only a limited number of possibilities. Note that n_1 and n_2 must be relatively positive, so for definiteness we will take $0 \leqslant n_2 \leqslant n_1$. Similarly we will take $0 \leqslant \theta \leqslant \pi/2$. Other possibilities are obtained by reversing the sign of α and/or β or by interchanging the two.

First we note that n_1 cannot in fact be equal to 4. In this case n_2 would have to be 1 and the Schwartz inequality would be saturated, with $\theta = 0$, and in fact we would have $\beta = 2\alpha$. This is, however, not possible. Contrary to what is stated in some books, this does not follow from equation (9.23) with $\beta = \alpha$: the equation then simply reads $0 = 0$, which is consistent, but not very informative!

Instead, let us suppose that 2α is a root, even though it cannot be generated from α by the raising operator E_α. From the above we know that 3α can certainly not be a root, hence $[E_\alpha, E_{2\alpha}] = 0$. Then consider $[E_{-\alpha}, E_{2\alpha}]$. Either (a) $[E_{-\alpha}, E_{2\alpha}] = 0$ or (b) $[E_{-\alpha}, E_{2\alpha}] \propto E_\alpha$. In case (a) we get a contradiction by considering the double commutator $[[E_{-\alpha}, E_\alpha], E_{2\alpha}]$. Using the Jacobi identity, this is zero. On the other hand, using (9.34), it is proportional to $E_{2\alpha}$. In case (b) a similar contradiction is found by considering the double commutator $[[E_{-\alpha}, E_{2\alpha}], E_\alpha]$.

The remaining possibilities (excluding $n_1 = n_2 = 2$, which corresponds to the trivial case $\beta = \alpha$) are:
(a) $n_1 = 0 \Rightarrow n_2 = 0$, which corresponds to $\theta = \pi/2$. The two roots are orthogonal, and there is no restriction on their relative lengths.
(b) $n_1 = n_2 = 1$, corresponding to $\theta = \pi/3$, and $|\alpha| = |\beta|$.
(c) $n_1 = 2, n_2 = 1$, corresponding to $\theta = \pi/4$, and $|\beta| = \sqrt{2}|\alpha|$.
(d) $n_1 = 3, n_2 = 1$, corresponding to $\theta = \pi/6$, and $|\beta| = \sqrt{3}|\alpha|$.

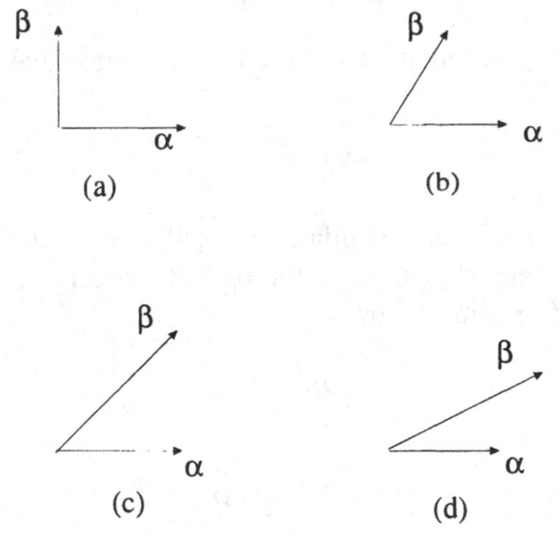

Figure 9.1. Angles and lengths of roots for rank-2 algebras.

These possibilities are illustrated in figure 9.1 for rank 2 groups, where the roots are 2-dimensional vectors. The figure shows only two roots in each case. How do we generate the complete set of roots from these?

We have already touched on one method of generating additional roots: by acting with $E_{\pm\beta}$ on E_α. This generates a *β-root-string* through α, which in general extends from $\alpha + p\beta$ to $\alpha - q\beta$, say, where p and q

are non-negative integers.

The root vectors $E_{\alpha+k\beta}$ corresponding to this string form the basis of an irreducible representation of S_β of dimension $2j+1$, say, where j can be integral or half-integral. At the upper and lower ends of the string the eigenvalues of $\frac{1}{2}H_\beta$ are $\pm j$. Thus

$$jE_{\alpha+p\beta} = [\tfrac{1}{2}H_\beta, E_{\alpha+p\beta}] = \frac{(\alpha+p\beta)\cdot\beta}{\beta^2}E_{\alpha+p\beta}. \tag{9.41}$$

Hence

$$j = \frac{(\alpha+p\beta)\cdot\beta}{\beta^2} \tag{9.42}$$

and similarly

$$-j = \frac{(\alpha-q\beta)\cdot\beta}{\beta^2} \tag{9.43}$$

from which we have

$$\begin{aligned} q+p &= 2j \\ q-p &= 2\alpha\cdot\beta/\beta^2 = n_2. \end{aligned} \tag{9.44}$$

The second equation is equivalent to (9.36), while the first relates the length of the root-string to p and q. In particular, if $p=0$, i.e. if $[E_\beta, E_\alpha] = 0$, the multiplicity is $2j+1 = q+1 = n_2+1$. Similarly, the multiplicity of the α-string through β is n_1+1. Using these root-strings, and including the negative of each root found, we can extend the diagrams of figure 9.1 to obtain the full set of roots, which are shown in figure 9.2.

The first diagram corresponds to the non-simple group SO(4), which, as we have already seen, is locally isomorphic to SU(2) × SU(2). The algebra consists of two commuting sets of SU(2) generators. The second corresponds to SU(3), with which we already have some familiarity. There are six roots and two H's, as required. It is not an accident that the diagram is essentially the same as figure 8.2, which shows the quantum numbers in the adjoint representation. In this case, with $n_1 = n_2 = 1$, each root-string is of length 2. If $[E_\alpha, E_\beta]$ were not equal to zero, we would generate the fourth diagram, which it turns out corresponds to the exceptional group G_2. In the third diagram, where $n_1 = 2$, $n_2 = 1$, the strings are of length 2 and 3, respectively. $[E_\beta, E_\alpha]$ must be zero since $\beta+\alpha$ does not have one of the permitted lengths or angles. This third diagram is the root diagram for SO(5).

It will be noted that these diagrams have a good deal of symmetry. This is a consequence of *Weyl reflections*. These arise from the SU(2)

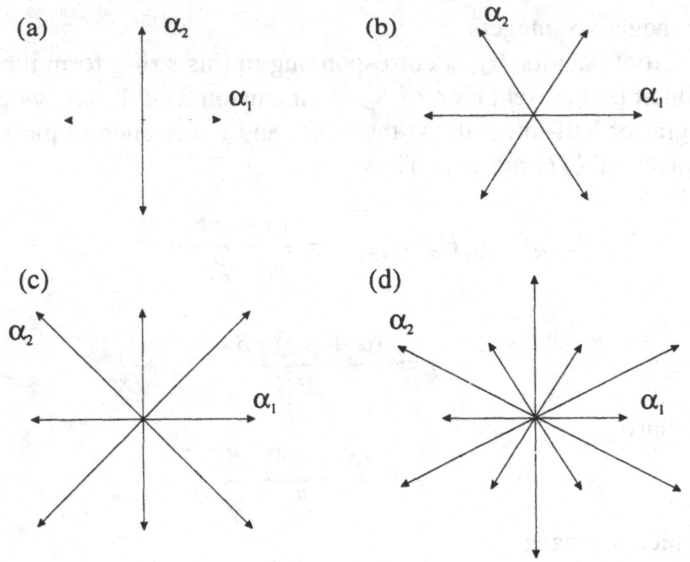

Figure 9.2. The full set of roots for the rank-2 algebras (a) SU(2)×SU(2), (b) SU(3), (c) SO(5) and (d) G_2.

groups associated with each root. The group S_α was obtained by identifying the raising and lowering operators as $J_\pm = E_{\pm\alpha}/|\alpha|$. From these we can define the operator $J_2 := (J_+ - J_-)/2\mathrm{i}$, which has the property that

$$\mathrm{e}^{\mathrm{i}\pi J_2} J_3 \, \mathrm{e}^{-\mathrm{i}\pi J_2} = -J_3 \tag{9.45}$$

or, in view of the identification of J_3,

$$\mathrm{e}^{\mathrm{i}\pi J_2} \alpha \cdot H \, \mathrm{e}^{-\mathrm{i}\pi J_2} = -\alpha \cdot H. \tag{9.46}$$

On the other hand, it has no effect on $v \cdot H$, where v is a vector orthogonal to α, because

$$[v \cdot H, E_{\pm\alpha}] = \pm v \cdot \alpha E_{\pm\alpha} = 0.$$

Thus

$$\mathrm{e}^{\mathrm{i}\pi J_2} v \cdot H \, \mathrm{e}^{-\mathrm{i}\pi J_2} = v \cdot H. \tag{9.47}$$

Then any general linear combination, $x \cdot H$, of the H's is transformed to $\sigma_\alpha(x) \cdot H$, where

$$\begin{aligned} \sigma_\alpha(x) &= \left(x - \frac{x \cdot \alpha}{\alpha^2}\alpha\right) - \frac{x \cdot \alpha}{\alpha^2}\alpha \\ &= x - \frac{2x \cdot \alpha}{\alpha^2}\alpha. \end{aligned} \tag{9.48}$$

In the first line we have shown the decomposition of x into its components perpendicular and parallel to α, and reversed the sign of the latter component. Geometrically, the transformation from x to $\sigma_\alpha(x)$ is a reflection in the plane orthogonal to α.

Now we wish to show that for any root β, its Weyl reflection

$$\sigma_\alpha(\beta) = \beta - \frac{2\beta \cdot \alpha}{\alpha^2}\alpha \tag{9.49}$$

is also a root. Notice that it is a root on the α-string through β. We begin with the defining equation

$$[x \cdot H, E_\beta] = x \cdot \beta E_\beta \tag{9.50}$$

for arbitrary x, and take its transform, which is

$$[\sigma_\alpha(x) \cdot H, \tilde{E}_\beta] = x \cdot \beta \tilde{E}_\beta \tag{9.51}$$

where $\tilde{E}_\beta = \exp(i\pi J_2) E_\beta \exp(-i\pi J_2)$. But the scalar product $x \cdot \beta$ is invariant under Weyl reflections: $x \cdot \beta = \sigma_\alpha(x) \cdot \sigma_\alpha(\beta)$. Moreover, if we define $y := \sigma_\alpha(x)$, which, like x, is a perfectly general vector, equation (9.51) reads

$$[y \cdot H, \tilde{E}_\beta] = y \cdot \sigma_\alpha(\beta) \tilde{E}_\beta. \tag{9.52}$$

This indeed shows that $\sigma_\alpha(\beta)$ is a root, associated with the root vector $\tilde{E}_\beta$.

As a consequence of Weyl reflections and the quantization of the roots, we have the useful property that for two distinct roots α and β

$$\begin{cases} \alpha \cdot \beta < 0 \;\Rightarrow\; \alpha + \beta \text{ is a root} \\ \alpha \cdot \beta > 0 \;\Rightarrow\; \alpha - \beta \text{ is a root.} \end{cases} \tag{9.53}$$

From the possibilities enumerated above for $n_1 \equiv 2\alpha \cdot \beta/\alpha^2$ and $n_1 \equiv 2\alpha \cdot \beta/\alpha^2$ we see that for $\alpha \neq \beta$ either n_1 or n_2 must be ± 1. If $\alpha \cdot \beta > 0$ we must have $n_1 = +1$ or $n_2 = +1$. In the first case, using equation (9.49), $\sigma_\alpha(\beta) = \beta - \alpha$ is a root, as of course is its negative, $\alpha - \beta$. In the second case $\sigma_\beta(\alpha) = \alpha - \beta$ directly. If $\alpha \cdot \beta < 0$ we must have $n_1 = -1$ or $n_2 = -1$, and in each case the appropriate Weyl reflection gives $\alpha + \beta$ as a root.

9.5 Simple Roots—Dynkin Diagrams

Clearly not all the roots are linearly independent; in fact since they are r-dimensional vectors, at most r can be linearly independent.

An extremely useful way of choosing such a basis is by first introducing the concept of positive and negative roots. A *positive* or *negative* root is one whose first non-vanishing component is positive or negative, respectively. By this definition, the set of roots is partitioned into two mutually exclusive sets. Although it depends on our particular choice of axes, the results derived do not depend on that choice, up to symmetries of the root diagram.

Within this concept of positivity one can define a *simple* root. This is a positive root which cannot be expressed as the sum of two other positive roots. It turns out that these simple roots, α_i say†, form a basis for the root diagram, and moreover that any positive root α can be written as

$$\alpha = \sum_i n_i \alpha_i \tag{9.54}$$

where all the non-vanishing n_i are positive integers. Similarly, a negative root can be expressed in the same form, but now all of the non-vanishing n_i are negative integers. This latter statement just follows from the fact that the roots occur in pairs $\pm\alpha$, so that every negative root is the negative of some positive root.

Equation (9.54) is relatively easy to prove. If α is a simple root, then it is trivially true. If not, α can be split into two positive roots: $\alpha = \beta + \gamma$. If either of β or γ is not simple, we can split it further, and so on. The process continues until we indeed have expressed α as a sum of simple roots with positive integer coefficients.

This shows that the simple roots span the space: however, there might be more than r of them, in which case they would be linearly dependent. As a preliminary to proving that this is not the case, and that there are precisely r simple roots, we need first to show that the scalar product of two distinct simple roots α_i and α_j is $\leqslant 0$.

Suppose instead that $\alpha_i \cdot \alpha_j > 0$. Then the second of equations (9.53) shows that $\alpha_i - \alpha_j$ is a root, as of course is its negative, $\alpha_j - \alpha_i$. Now if $\alpha_i - \alpha_j$ is positive we can write $\alpha_i = \alpha_j + (\alpha_i - \alpha_j)$, which expresses α_i as a sum of two positive roots, contrary to the definition of a simple root. Similarly, if $\alpha_j - \alpha_i$ is positive we obtain a contradiction by writing

† Note the distinction between the components $(\alpha)_i$ of a generic root vector α and the simple roots α_i, which are themselves vectors

$\alpha_j = \alpha_i + (\alpha_j - \alpha_i)$.

We are now in a position to prove that the α_i are linearly independent, i.e. that there are no non-trivial coefficients c_i such that $\sum c_i\alpha_i = 0$. Again, we proceed by showing a contradiction if there is some such non-trivial linear combination. By the definition of positivity, this sum can only be zero if some of the coefficients are positive and some negative. Without loss of generality let us assume that the first s of them are positive and the others negative. In that case the equation can be rewritten as

$$\sum_{i=1}^{s} c_i\alpha_i = \sum_{j=s+1} b_j\alpha_j$$

where c_i and $b_j := -c_j$ are both positive. Then

$$\begin{aligned}\left(\sum c_i\alpha_i\right)^2 &= \left(\sum c_i\alpha_i\right)\cdot\left(\sum b_j\alpha_j\right)\\ &= \sum_{i,j} c_i b_j \alpha_\iota . \alpha_j .\end{aligned}$$

But the left-hand side is intrinsically positive, while the right-hand side is a positive linear combination of scalar products $\alpha_i \cdot \alpha_j$ which are each $\leqslant 0$. The only way the two can be reconciled is that all the coefficients c_l, and similarly all the b_j, are zero.

Thus, a simple algebra of rank r has precisely r simple roots α_i, and the other roots can be expressed as integer linear combination of these, in which the non-zero coefficients are either all positive integers or all negative integers.

These simple roots pick out a natural basis for the Cartan subalgebra, namely the H_{α_i} defined as in equation (9.31), which form part of the SU(2) algebra S_{α_i}. In this basis, the *Chevalley* basis, the commutation relations involving the simple roots are

$$\begin{aligned}[H_{\alpha_i}, E_{\alpha_j}] &= \frac{2\alpha_i \cdot \alpha_j}{\alpha_i^2} E_{\alpha_j}\\ \text{and} \qquad\qquad & \qquad\qquad\qquad\qquad (9.55)\\ [E_{\alpha_i}, E_{-\alpha_j}] &= \tfrac{1}{2}\alpha_i^2\,\delta_{ij}\,H_{\alpha_i}.\end{aligned}$$

The latter commutator is zero unless $i = j$ from the definition of simple roots. The $N_{\alpha_i\alpha_j}$ for simple roots are rather easy to calculate, again in terms of scalar products. The only disadvantage with this basis is that the H_{α_i} are not orthogonal, i.e. $(H_{\alpha_i}, H_{\alpha_j}) \propto \alpha_i \cdot \alpha_j$, from equation (9.30).

In figure 9.2 we show one possible choice of simple roots for each of the simple groups of rank 2. Note that the angle between simple roots must be obtuse, since $\alpha_i \cdot \alpha_j < 0$. In fact all the essential information about the group algebra is encoded in these scalar products. One way of presenting this information is by means of the *Cartan matrix* defined by

$$K_{ij} = 2\alpha_i \cdot \alpha_j / \alpha_j^2. \tag{9.56}$$

The diagonal elements are really superfluous, since $K_{ii} = 2 \forall i$ (no summation). If all the roots have the same length, K is symmetric. Note that since the simple roots are linearly independent, K is non-singular, i.e. $\det(K) \neq 0$. The Cartan matrices for SU(3), SO(5) and G_2 are:

$$SU(3): \begin{pmatrix} 2 & -1 \\ -1 & 2 \end{pmatrix} \tag{9.57}$$

$$SO(5): \begin{pmatrix} 2 & -1 \\ -2 & 2 \end{pmatrix} \tag{9.58}$$

$$G_2: \begin{pmatrix} 2 & -1 \\ -3 & 2 \end{pmatrix}. \tag{9.59}$$

The Cartan matrix can be represented in graphical form, as a *Dynkin diagram*. In such a diagram the r simple roots α_i are represented by small circles C_i. Any two circles C_i, C_j are then joined together by n_{ij} lines, where $n_{ij} = K_{ij}K_{ji}$ (no summation). Referring back to equation (9.37) and the subsequent discussion, we see that the only possibilities for n_{ij} are 0, 1, 2 and 3, and indeed we can identify $n_1 = -K_{ij}$, $n_2 = -K_{ji}$.

If $n_{ij} = 0$ the roots are not connected.

If $n_{ij} = 1$ the roots have the same length ($n_1 = n_2 = 1$).

If $n_{ij} = 2$ we have $n_1 = 2$, $n_2 = 1$ or vice-versa.

If $n_{ij} = 3$ we have $n_1 = 3$, $n_2 = 1$ or vice-versa.

When the lengths are unequal we put an arrow pointing towards the root of smaller length.

The Dynkin diagrams corresponding to the rank-2 algebras previously identified are shown in figure 9.3. There are four regular series of Dynkin diagrams of increasing rank r, together with five special diagrams which do not belong to such a series and correspond to the so-called exceptional groups, of which G_2 is the first example. In terms of compact groups the four regular series, labelled A_r, B_r, C_r and D_r correspond respectively to the algebras of SU($r+1$), SO($2r+1$), Sp($2r$) and SO($2r$). The third is the symplectic group, which is important in Hamiltonian mechanics.

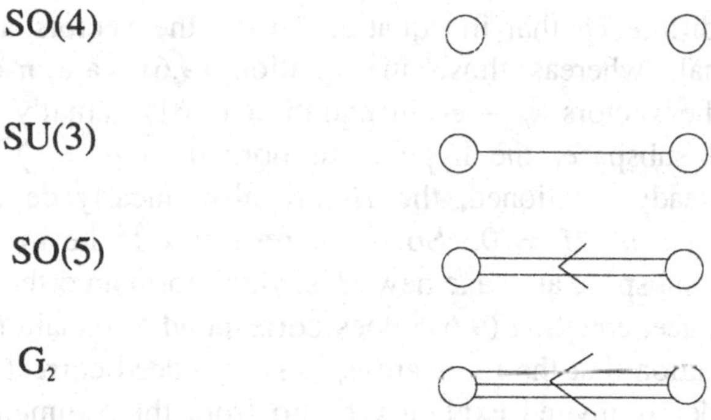

Figure 9.3. Dynkin diagrams for rank-2 algebras.

SU(n)

Let us first look at the algebra of SU(n). The group elements are unitary $n \times n$ matrices of determinant 1. Correspondingly, the elements of the algebra are Hermitian traceless $n \times n$ matrices.

The H_i are clearly the diagonal matrices. There are n of these, but since they are traceless, only $r = n - 1$ are linearly independent. A general element of the Cartan subalgebra can be written as $X = \text{diag}(x_1, x_2, \ldots x_n)$, with $\sum x_i = 0$.

The step operators can be introduced as generalizations of $\sigma_\pm = \frac{1}{2}(\sigma_1 \pm i\sigma_2)$, which are $\begin{pmatrix} 0 & 1 \\ 0 & 0 \end{pmatrix}$ and $\begin{pmatrix} 0 & 0 \\ 1 & 0 \end{pmatrix}$ respectively. These, of course, are not Hermitian, being complex combinations of σ_1 and σ_2. In SU(n) we can define $n^2 - n$ such matrices E_{ab}, whose entries are just a 1 in the (ab) place and zero elsewhere, i.e.

$$(E_{ab})_{kl} = \delta_{ak}\delta_{bl}. \tag{9.60}$$

(In this notation σ_+ would be σ_{12} and σ_- would be σ_{21}.)

Let us examine the commutation relations of the E's with the H's, which for definiteness we can take as $(H_i)_{kl} = \delta_{ik}\delta_{il} - (1/n)\delta_{kl}$. It is easy to see that $[X, E_{ab}] = (x_a - x_b)E_{ab}$, which establishes that E_{ab} is indeed a root vector. In terms of commutation relations with the H_i, this result is equivalent to

$$[x \cdot H, E_{ab}] = x \cdot (e_a - e_b)E_{ab} \tag{9.61}$$

where e_a and e_b are unit n-dimensional vectors pointing in the a and b directions respectively. This is very similar to equation (9.50), and suggests that the root associated with E_{ab} is $e_a - e_b$. There is, however, the subtle difference that in equation (9.50) the vectors are explicitly r-dimensional, whereas those in equation (9.61) are n-dimensional. However, the vectors $e_a - e_b$ in equation (9.61) actually lie in an r-dimensional subspace, the hyperplane normal to $n = \sum e_a$, and, as we have already mentioned, the H_i are also linearly dependent, with $\sum H_i = 0$, i.e. $n \cdot H = 0$. So if we restrict x to lie in this same r-dimensional subspace and take new H's which form an orthonormal basis in this subspace, equation (9.61) does correspond to equation (9.50), and the interpretation that the roots are $e_a - e_b$ is indeed correct. In problem 9.5 the reader is invited explicitly to go from three dimensions to the two-dimensional subspace of the roots in the case of SU(3).

Just as $\sigma_{21} \equiv \sigma_-$ is the lowering operator corresponding to $\sigma_{12} \equiv \sigma_+$, so E_{ba} is the lowering operator corresponding to E_{ab}. This is obvious from equation (9.61), but can also be checked from the commutation relation of E_{ab} with E_{ba}:

$$\begin{aligned}([E_{ab}, E_{ba}])_{kl} &= \sum_p \left[(E_{ab})_{kp}(E_{ba})_{pl} - (E_{ba})_{kp}(E_{ab})_{pl}\right] \\ &= \sum_p \left[\delta_{ak}\delta_{bp}\delta_{bp}\delta_{al} - \delta_{bk}\delta_{ap}\delta_{ap}\delta_{bl}\right] \\ &= \delta_{ak}\delta_{al} - \delta_{bk}\delta_{bl} \\ &= (H_a - H_b)_{kl} \\ &= ((e_a - e_b) \cdot H)_{kl} \, . \end{aligned} \tag{9.62}$$

This is to be compared with the third of equations (9.29), with the same understanding that the dimensionality of the vectors is in fact one less than the nominal n.

In order to make contact with the Dynkin diagram for this group we need the $r = n - 1$ simple roots, which can be chosen as the linearly independent vectors

$$\alpha_i = e_i - e_{i+1} \qquad i = 1 \ldots n-1. \tag{9.63}$$

Consider any particular root $e_i - e_j$ for $i < j$. This can be expressed as

$$e_i - e_j = \sum_{k=i}^{j-1} \alpha_k \tag{9.64}$$

which is a linear combination of the α_k with all positive integer coefficients. Similarly, $e_j - e_i$ is a linear combination with all negative integer coefficients.

Since the e_i are orthogonal unit vectors, each α_i is normalized to $\alpha_i^2 = 2$. The scalar product of two distinct simple vectors $\alpha_i \cdot \alpha_j$ is zero unless they have an e_k in common, i.e. unless $i = j \pm 1$, and then $\alpha_i \cdot \alpha_{i+1} = (e_i - e_{i+1}) \cdot (e_{i+1} - e_{i+2}) = -1$. Thus, for these adjacent simple roots we have $K_{ij} = 2\alpha_i \cdot \alpha_j / \alpha_j^2 = -1$. The Dynkin diagram is thus a simple chain, with each circle connected to its neighbours by a single line, as shown in figure 9.4.

Figure 9.4. Dynkin diagram for SU(r + 1).

The corresponding Cartan matrix is

$$K = \begin{pmatrix} 2 & -1 & 0 & 0 & \dots & \dots & 0 \\ -1 & 2 & -1 & 0 & \dots & \dots & 0 \\ 0 & -1 & 2 & -1 & \dots & \dots & 0 \\ \dots & \dots & \dots & \dots & \dots & \dots & \dots \\ \dots & \dots & \dots & \dots & \dots & \dots & \dots \\ 0 & 0 & 0 & \dots & -1 & 2 & -1 \\ 0 & 0 & 0 & \dots & 0 & -1 & 2 \end{pmatrix}. \tag{9.65}$$

An interesting aspect of this matrix is that it is the matrix governing the motion of the points α_i, regarded as unit masses connected by springs with an exponential tension $e^{x_i - x_{i-1}}$. In terms of the relative displacements $y_i := x_i - x_{i-1}$ these equations take the form

$$\ddot{y}_i = -K_{ij} e^{y_j}.$$

These are called the Toda molecule equations, and, although highly non-linear, turn out to be exactly soluble because of the special algebraic properties of K. The same is true for the Cartan matrices of the other Lie groups.

Although we have performed the analysis for the compact group SU(n), the same Dynkin diagram also describes the non-compact groups SU(p, q) and SL(n, $\mathbb{R}$). The group SU(p, q) differs from SU(n) in that it preserves the form $x^\dagger \mathbb{1}_{p,q} y$, where $\mathbb{1}_{p,q}$ is a diagonal matrix with p entries of $+1$ and q of -1, rather than $x^\dagger y$. The group SL(n, $\mathbb{R}$) is simply the group of $n \times n$ real matrices with determinant 1. One can go from the compact group to its non-compact versions by judiciously multiplying some of the generators by i (or equivalently letting some of the parameters become pure imaginary rather than real).

Let us illustrate the procedure for SU(2). The compact group SU(2) has generators $\mathbf{J} = \frac{1}{2}\boldsymbol{\sigma}$ and by virtue of $J^\dagger = J$ the unitary transformation $x \to e^{i\mathbf{J}\cdot\boldsymbol{\theta}}$ preserves the Hermitian bilinear form $x^\dagger y$, where x and y are both two-component column vectors. The Cartan metric of the algebra is just $2\delta_{\alpha\beta}$.

If instead we wish to preserve the form $x_1^\dagger y_1 - x_2^\dagger y_2 = x^\dagger \sigma_3 y$, the condition on the generators is $J^\dagger \sigma_3 = \sigma_3 J$, which is satisfied by $J_1 = \frac{1}{2}i\sigma_1$, $J_2 = \frac{1}{2}i\sigma_2$, $J_3 = \frac{1}{2}\sigma_3$. The transformations generated by these are respectively a real Lorentz transformation, a complex Lorentz transformation, and a rotation. The Cartan metric now becomes diag$(-2, -2, 2)$.

The generators of the group SL(2, $\mathbb{R}$) must all be pure imaginary, so can be taken as $J_1 = \frac{1}{2}i\sigma_1$, $J_2 = \frac{1}{2}\sigma_2$, $J_3 = \frac{1}{2}i\sigma_3$, with corresponding Cartan metric diag$(-2, 2, -2)$. The latter two algebras are clearly isomorphic.

The general procedure for associating a non-compact algebra with a compact one is first to find the maximal compact subalgebra and then multiply the remaining non-compact generators by a factor of i.

SO(2*r*)

In Appendix D we show that the roots of this algebra are $\pm e_i \pm e_j$ for $i < j$, both in the range $1 \ldots r$, and that the simple roots can be taken as

$$\begin{cases} \alpha_i = e_i - e_{i+1} & i = 1 \ldots r-1 \\ \alpha_r = e_{r-1} + e_r . \end{cases} \tag{9.66}$$

From equation (9.66) we see that all the simple roots have the same length. Each is connected to just two neighbours, except for $\alpha_{r-2} = e_{r-2} - e_{r-1}$, which is connected to α_{r-3}, α_{r-1} and also α_r, with scalar product $\alpha_{r-2} \cdot \alpha_r = -1$, the same as those along the chain. However, α_r and α_{r-1} are not connected, since $\alpha_r \cdot \alpha_{r-1} = 0$. The Dynkin diagram thus has the form shown in figure 9.5, differing from the SU(n) diagram by a fork at the end.

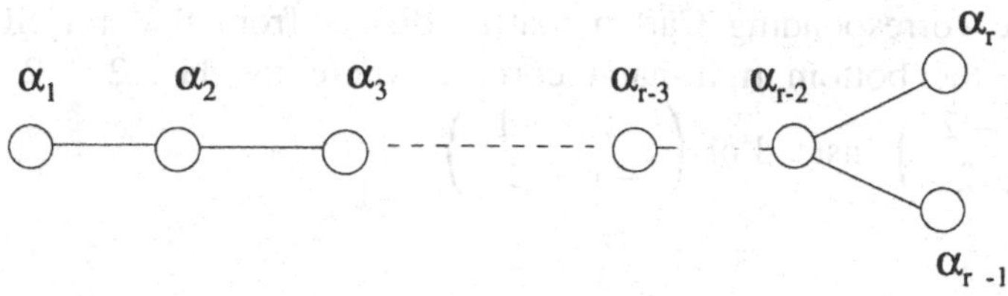

Figure 9.5. Dynkin diagram for SO(2r).

The corresponding Cartan matrix is

$$K = \begin{pmatrix} 2 & -1 & 0 & 0 & \ldots & \ldots & 0 \\ -1 & 2 & -1 & 0 & \ldots & \ldots & 0 \\ 0 & -1 & 2 & -1 & \ldots & \ldots & 0 \\ \ldots & \ldots & \ldots & \ldots & \ldots & \ldots & \ldots \\ \ldots & \ldots & \ldots & \ldots & \ldots & \ldots & \ldots \\ 0 & 0 & 0 & \ldots & 2 & -1 & -1 \\ 0 & 0 & 0 & \ldots & -1 & 2 & 0 \\ 0 & 0 & 0 & \ldots & -1 & 0 & 2 \end{pmatrix}. \tag{9.67}$$

differing from that of equation (9.65) only in the bottom right-hand corner.

Clearly, from (9.66) the simple roots $\alpha_1 \ldots \alpha_{r-1}$ correspond to an SU(r) subalgebra, which in diagrammatic terms is obtained by simply removing the node r from figure 9.5.

SO(2*r* + 1)

In Appendix D we show that SO($2r + 1$) has $2r$ additional roots $\pm e_i$, $i = 1 \ldots r$. In this case the last simple root has to be taken as $\alpha_r = e_r$ rather than $e_{r-1} + e_r$. This is the only simple root which is different in nature from the others. Its scalar product with with α_{r-1} is $\alpha_r \cdot \alpha_{r-1} = -1$, as usual, but its square is $\alpha_r^2 = 1$ rather than 2. Thus $n_{ij} = 4/2 = 2$, which means that the Dynkin diagram ends with a double line, with the arrow pointing to the last simple root α_r, as shown in figure 9.6.

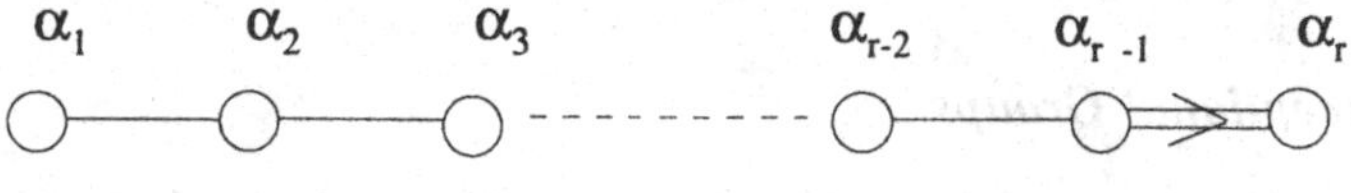

Figure 9.6. Dynkin diagram for SO($2r + 1$).

The corresponding Cartan matrix differs from that for SU($r+1$) only in the bottom right-hand corner, where the last 2×2 block is $\begin{pmatrix} 2 & -2 \\ -1 & 2 \end{pmatrix}$ instead of $\begin{pmatrix} 2 & -1 \\ -1 & 2 \end{pmatrix}$.

Sp(2*r*)

As mentioned above, the symplectic group is important in classical Hamiltonian mechanics, where the dynamics of a system with r degrees of freedom is expressed in terms of a set of r generalized coordinates $\{q_\alpha\}$ and their conjugate momenta $\{p_\alpha\}$. A canonical transformation is a transformation to new coordinates $\{Q_\alpha\}$ and conjugate momenta $\{P_\alpha\}$ such that the Poisson brackets of equation (9.3) remain the same. The condition for this is that the Jacobian of the transformation $J := \partial(Q, P)/\partial(q, p)$ should satisfy $JIJ^T = I$, where I is the $2r \times 2r$ matrix

$$I = \begin{pmatrix} 0 & \mathbb{1} \\ -\mathbb{1} & 0 \end{pmatrix} \tag{9.68}$$

where $\mathbb{1}$ is the unit $r \times r$ matrix. This can be written in direct product form as $(\mathrm{i}\sigma_2) \times \mathbb{1}$. In terms of the infinitesimal generators, defined by $J = \mathbb{1} + \mathrm{i}\varepsilon G + \ldots$, the condition $JIJ^T = I$ translates to $GI = -IG^T$. The general form of G which satisfies this condition is easily seen to be

$$G = \begin{pmatrix} A + S_3 & S_+ \\ S_- & A - S_3 \end{pmatrix} \tag{9.69}$$

where A is an arbitrary antisymmetric $r \times r$ matrix and S_3 and $S_\pm$ arbitrary symmetric $r \times r$ matrices.

The roots of this algebra are shown in Appendix D to be $\pm e_i \pm e_j$ and $\pm 2e_i$. Again, it is only the last simple root which needs to be changed, this time to $2e_r$. Now the $r - 1$ simple roots $e_i - e_{i+1}$ have length $\sqrt{2}$, while the last root $2e_r$ has length 2. So the situation is reversed from SO($2r+1$) and the Dynkin diagram now has the arrow in the opposite direction, as shown in figure 9.7. Correspondingly, the Cartan matrix is the transpose of that for SO($2r+1$).

The Exceptional Groups

We have thus found the Dynkin diagrams for the four classical series of Lie algebras. However, G_2 does not fall into any of these series, and we need to ask if there are any others. The possible types of Dynkin diagram

Figure 9.7. Dynkin diagram for SO($2r+1$).

are severely restricted by the fact that the only possible angles between the simple roots are $\pi/2$, $2\pi/3$, $3\pi/4$ and $5\pi/6$, or in other words, if we denote the normalized vectors $\alpha_i/|\alpha_i|$ by ε_i, the only possible scalar products $\varepsilon_i \cdot \varepsilon_j$ are $0, -1/2, -1/\sqrt{2}$ and $-\sqrt{3}/2$. In fact $(-2\varepsilon_i \cdot \varepsilon_j)^2 = 0, 1, 2$ or 3, the number of links between them.

We will not derive all of these restrictions, but to give a flavour let us show the most important one, namely that the Dynkin diagram cannot contain a closed loop of single links. This comes from the obvious inequality that $(\sum \varepsilon_i)^2 > 0$, where the sum extends over the whole diagram or some subset of it (equality is not possible, since the α_i are linearly independent). But

$$\left(\sum_i \varepsilon_i\right)^2 = V - \sum_{i<j}(-2\varepsilon_i \cdot \varepsilon_j) \qquad (9.70)$$

where V is the number of vertices. Moreover, if we are considering a loop of single links the individual terms in the sum are either 0 or 1, and are therefore equal to their squares. That is,

$$\sum_{i<j}(-2\varepsilon_i \cdot \varepsilon_j) = \sum_{i<j}(-2\varepsilon_i \cdot \varepsilon_j)^2 = L \qquad (9.71)$$

the total number of links. From (9.70) and (9.71) we can deduce that $V > L$. But a closed loop of single links will always have $V = L$. ⊗

Additional rules can be deduced in a similar manner, which show that in fact no loops of any kind are allowed and that the number of lines emanating from any vertex must be less than 4. The net result is that four additional diagrams are possible beyond G_2 and the four classical series. These are the diagram for F_4, shown in figure 9.8, and those for $E_6 \ldots E_8$, shown in figure 9.9. E_5 would be the same as SO(10).
The reason that the E series does not extend beyond $r = 8$ is the rule

$$\frac{1}{p+1} + \frac{1}{q+1} + \frac{1}{r+1} > 1$$

where p, q and r are the lengths of the three branches from the vertex α_{r-3}.

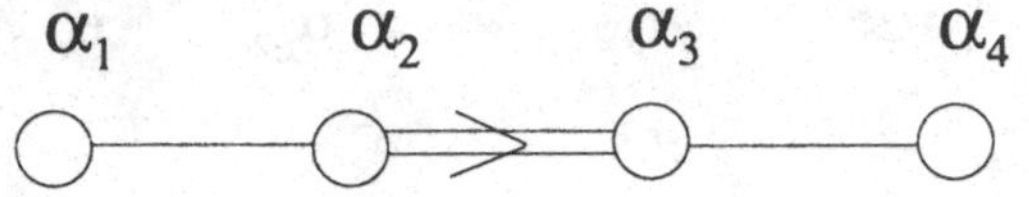

Figure 9.8. Dynkin diagram for F_4.

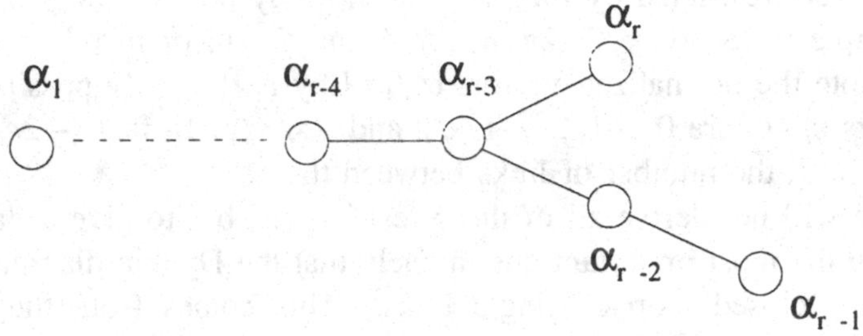

Figure 9.9. Dynkin diagram for E_r. $r = 6, 7, 8$.

The exceptional groups seem to be of more mathematical than physical interest. Thus, in the same way that SU(2) can be regarded as the automorphism group of the quaternions (which can be represented by the Pauli matrices), G_2 turns out to be the group of automorphisms of the octonions. The group E_6 appears in some models of grand unification (unification of the strong, weak and electromagnetic interactions), and E_8 plays a prominent rôle in string theory.

9.6 Representations and Weights

A representation is defined by the action of the elements of the algebra on the states of a vector space. For example, for SU(2) we have (cf. Appendix B)

$$\begin{aligned} J_3|j\,m\rangle &= m\hbar|j\,m\rangle \\ J_\pm|j\,m\rangle &= \hbar\sqrt{(j \mp m)(j \pm m + 1)}|j\,m \pm 1\rangle. \end{aligned} \tag{9.72}$$

This gives the representation matrices of equation (6.33) for J_3, J_+ and J_-, from which we can derive those for J_1 and J_2, and then by exponentiation the matrix for any element of the group. This representation of SU(2) is a finite-dimensional representation, with $|m| \leqslant j$, and the state with the highest value, j, of m satisfies

$$J_+|j\,j\rangle = 0. \tag{9.73}$$

The representation theory of compact Lie groups is based on this approach, with the difference, however, that in general we will have several elements H_i of the Cartan subalgebra analogous to J_3 and several step operators E_α analogous to $J_\pm$.

As far as the H's are concerned, a state of the representation can be labelled by their eigenvalues μ_i:

$$H_i|\mu\rangle = \mu_i|\mu\rangle. \tag{9.74}$$

μ, known as the *weight*, is to be thought of as an r-component vector with components μ_i. Multiplying this equation on the left by a step operator E_α gives

$$E_\alpha H_i|\mu\rangle = E_\alpha \mu_i|\mu\rangle. \tag{9.75}$$

Then, writing $E_\alpha H_i$ as $[E_\alpha, H_i] + H_i E_\alpha$, with the commutator given by equation (9.29), we obtain

$$H_i(E_\alpha|\mu\rangle) = (\mu+\alpha)_i(E_\alpha|\mu\rangle) \tag{9.76}$$

in exact analogy with equation (B4). This equation establishes that E_α shifts the weight μ to $\mu+\alpha$, i.e. that

$$E_\alpha|\mu\rangle \propto |\mu+\alpha\rangle. \tag{9.77}$$

In general, we have more than one step operator, and can shift the weight in different directions. For example, operating on this latter equation with E_β gives

$$E_\beta E_\alpha|\mu\rangle \propto |\mu+\alpha+\beta\rangle$$

etc. The weights μ therefore lie on a lattice, the *weight lattice*, i.e. they are displaced by α's in the same way as the roots themselves. The root lattice is a subset of the weight lattice, obtained by displacements from the origin. The roots are in fact the weights of the adjoint representation. Other representations will be generated by displacements from a different starting point.

Let us investigate what these starting points might be. Recall that $\frac{1}{2}H_\alpha := 2\alpha\cdot H/\alpha^2$ forms the third member of the SU(2) algebra S_α, so that H_α must have integral eigenvalues. But, from (9.74) the eigenvalue of H_α on the state $|\mu\rangle$ is $2\alpha\cdot\mu/\alpha^2$. Hence

$$\frac{2\alpha\cdot\mu}{\alpha^2} = \text{integer}. \tag{9.78}$$

If we shift μ to $\mu + \beta$ by acting with a different step operator E_β, we also require $2\alpha \cdot (\mu + \beta)/\alpha^2$ to be an integer. This condition is indeed satisfied by virtue of equation (9.36).

In the previous section we introduced the idea of the simple roots α_i, which formed a basis in terms of which any general root α could be expanded as $\sum n_i \alpha_i$. A similar construction can be made for the weights, where the basis consists of the *fundamental weights*, λ_i, defined by

$$\frac{2\lambda_i \cdot \alpha_j}{\alpha_j^2} = \delta_{ij}. \tag{9.79}$$

The reader familiar with solid state physics will recognize this construction as essentially the reciprocal lattice. A general weight μ can be expanded as

$$\mu = \sum_{i=1}^{r} m_i \lambda_i \tag{9.80}$$

where we have anticipated the fact that the coefficients m_i must be integers. Thus, if we take the scalar product of this equation with $2\alpha_j/\alpha_j^2$, the left-hand side is an integer and the right-hand side is just m_j.

From now on, let us use as our step operators those corresponding to the simple roots α_i. For a finite-dimensional representation, repeated action of E_{α_i} on the states of the representation must terminate for some μ, for which $E_{\alpha_i}|\mu\rangle = 0$, just as for the top state $|j\, j\rangle$ of an SU(2) representation $J_+|j\, j\rangle = 0$. The analogue of this state for a general Lie group is the *highest weight* state, which is annihilated by all the step operators of the simple roots, i.e.

$$E_{\alpha_i}|\mu\rangle = 0 \qquad i = 1 \ldots r. \tag{9.81}$$

The labels of this state label the whole representation, and are equivalent to the single label j of SU(2). The remaining states of the representation are obtained by acting with the step operators $E_{-\alpha_i}$, which are the analogues of J_-.

The notation 'highest weight' arises from equation (9.80), where the weights are expressed as integer multiples of the fundamental weights. The weights can therefore be ordered by their height, the first non-vanishing coefficient m_i. Dynkin has shown that a necessary and sufficient condition for a weight μ to be the highest weight of an irreducible representation is that all the integers m_i are $\geqslant 0$.

As an example of this more general procedure let us consider the fundamental representation of SU(3), for which the highest weight is just λ_1, or alternatively $(m_1, m_2) = (1, 0)$.

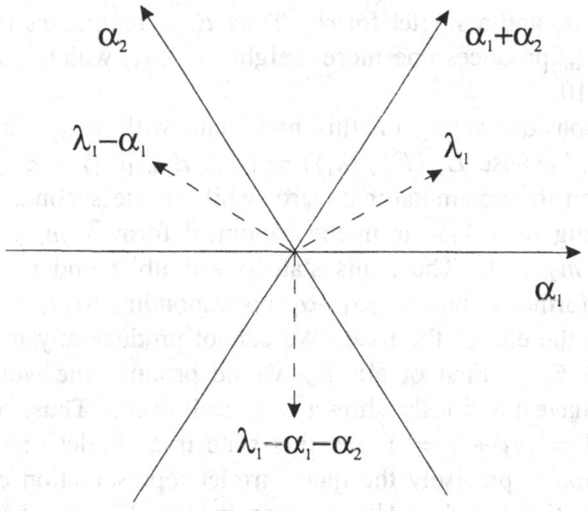

Figure 9.10. The triplet representation of SU(3) with highest weight λ_1.

Taking the frame of reference indicated in figure 9.2, so that $\alpha_1 = (1, 0)$, $\alpha_2 = (-1, \sqrt{3})/2$ (normalized to 1), the solution of equation (9.79) is

$$\begin{aligned} \lambda_1 &= \frac{1}{2}\left(1, \frac{1}{\sqrt{3}}\right) \\ \lambda_2 &= \left(0, \frac{1}{\sqrt{3}}\right). \end{aligned} \tag{9.82}$$

The weight λ_1 is shown in figure 9.10. Its end is actually the centre of the triangle contained by α_1 and $\alpha_1 + \alpha_2$. To find the other states of the representation we need to shift λ_1 by $-\alpha_1$ and $-\alpha_2$, if indeed this is possible.

Acting on a weight $|\mu\rangle$ with a step operator $E_{\pm\beta}$ will in general produce a series of weights from $|\mu + p\beta\rangle$ down to $|\mu - q\beta\rangle$, entirely analogous to the β-string of roots through α. By similar arguments to those which led to equation (9.44) we can establish that

$$\begin{aligned} q + p &= 2j \\ q - p &= 2\mu \cdot \beta/\beta^2. \end{aligned} \tag{9.83}$$

For the fundamental weight λ_1 we know that $p = 0$ for both α_1 and α_2. In that case, the length of the series of weights is $2j + 1 = q + 1 = 2\lambda_1 \cdot \alpha_i/\alpha_i^2 + 1 = m_i + 1$, which is 2 for α_1 and 1 for α_2. In other words, the representation of S_{α_1} generated by acting with $E_{-\alpha_1}$ on λ_1 is a doublet for α_1 and a singlet for α_2. Thus $E_{-\alpha_2}$ annihilates the state $|\lambda_1\rangle$, whereas $E_{-\alpha_1}$ produces one more weight $\lambda_1 - \alpha_1$, which again is shown in figure 9.10.

Now consider acting on this new state with $E_{-\alpha_2}$. It is still true that $p = 0$, because $E_{\alpha_2}(E_{-\alpha_1}|\lambda_1\rangle) = [E_{\alpha_2}, E_{-\alpha_1}]|\lambda_1\rangle + E_{-\alpha_1}E_{\alpha_2}|\lambda_1\rangle$. In the first term the commutator is zero, while in the second, $E_{\alpha_2}|\lambda_1\rangle = 0$. Re-expressing $\mu = \lambda_1 - \alpha$ in the canonical form $\sum m_i\lambda_i$ we find that $m_1 = -1$, $m_2 = 1$. Thus, this state is a doublet under S_{α_2}, and $E_{-\alpha_2}$ produces a further weight $\lambda_1 - \alpha_1 - \alpha_2$, corresponding to $m_1 = 0, m_2 = -1$.

This is the end of the road. We cannot produce any more states by acting with $E_{\pm\alpha_1}$. First of all, E_{α_1} would produce the weight $\lambda_1 - \alpha_2$, which we have previously shown is not allowed. Thus, again $p = 0$ and $2j + 1 = m_1 + 1 = 1$, i.e. this state is a singlet under S_{α_1}. This representation is precisely the quark triplet representation of figure 8.5, with S_{α_1} identified as the SU(2) of isospin. The SU(2) subgroup S_{α_2} has been called U-spin, and can be used effectively in the derivation of many of the properties of SU(3) representations. The hypercharge Y is the combination $(m_1 + 2m_2)/3$. Note that λ_1 has height 1, while the other two weights have height -1.

An exactly parallel procedure will establish that the representation with highest weight λ_2 is the antiquark triplet representation $\bar{3}$. A way of identifying these, and indeed any other irreducible representation, is to annotate the Dynkin diagram by placing m_i above the vertex corresponding to the simple root α_i. Thus for SU(3) the two fundamental representations just discussed would be associated with the diagrams figure 9.11(a) and (b) respectively.

Figure 9.11. Dynkin diagram notation for the **3** and $\bar{\mathbf{3}}$ representations of SU(3).

The representation characterized by a highest weight $\lambda_1 + \lambda_2$, i.e. $(m_1, m_2) = (1, 1)$, is in fact the adjoint representation, since from equation (9.79) we see that $\lambda_1 + \lambda_2$ is precisely the root $\alpha_1 + \alpha_2$. The weights of this representation consist of the six roots, together with two degenerate weights $\mu = 0$. In terms of figure 8.2 these correspond to the states π^0 and the octet combination η_8.

We do not have the space here to establish the correspondence between this approach and the tensorial approach of Chapter 8, but it can in fact be shown (see Georgi, *Lie Algebras in Particle Physics* (Benjamin 1982)) that the representation of SU(3) characterized by the highest weight (m_1, m_2) is precisely that described by the irreducible tensors of problem 8.4, with $m = m_1$ and $n = m_2$ and dimensionality $\frac{1}{2}(m_1 + 1)(m_2 + 1)(m_1 + m_2 + 2)$.

The above example has illustrated the technique for finding all the weights of an irrep, starting from the highest weight, but to calculate the representation matrices themselves we need to calculate the constants of proportionality in equation (9.77). These are the analogues of the factors $((j \mp m)(j \pm m + 1))^{\frac{1}{2}}$ in equation (9.72) which gave us the matrices $D^{(j)}$ for SU(2). These constants can be calculated by establishing a recursion relation.

Writing

$$E_{-\alpha}|\mu\rangle = N_{-\alpha,\mu}|\mu - \alpha\rangle \tag{9.84}$$

so that

$$N_{-\alpha,\mu} = \langle \mu - \alpha|E_{-\alpha}|\mu\rangle \tag{9.85}$$

consider the expectation value $\langle\mu|[E_\alpha, E_{-\alpha}]|\mu\rangle$. This is

$$\langle\mu|[E_\alpha, E_{-\alpha}]|\mu\rangle = \langle\mu|\alpha \cdot H|\mu\rangle = \alpha \cdot \mu.$$

On the other hand, using $E_{-\alpha} = E_\alpha^\dagger$, the left-hand side is $|N_{-\alpha,\mu}|^2 - |N_{\alpha,\mu}|^2$. Moreover, by taking the complex conjugate of (9.85) we see that $N^*_{-\alpha,\mu} = N_{\alpha,\mu-\alpha}$. Hence

$$|N_{\alpha,\mu-\alpha}|^2 - |N_{\alpha,\mu}|^2 = \alpha \cdot \mu. \tag{9.86}$$

So if $E_{-\alpha}$ is acting on a state $|\mu\rangle$ for which $E_\alpha|\mu\rangle = 0$, the series of states it produces is $|\mu\rangle, \ldots |\mu - q\alpha\rangle$ with $q = 2\mu \cdot \alpha/\alpha^2$ and we have the set of difference equations

$$\begin{array}{rcrcl}
|N_{\alpha,\mu-\alpha}|^2 & - & 0 & = & \alpha \cdot \mu \\
|N_{\alpha,\mu-2\alpha}|^2 & - & |N_{\alpha,\mu-\alpha}|^2 & = & \alpha \cdot (\mu - \alpha) \\
\ldots & & \ldots & & \ldots \\
|N_{\alpha,\mu-q\alpha}|^2 & - & |N_{\alpha,\mu-(q-1)\alpha}|^2 & = & \alpha \cdot (\mu - (q-1)\alpha) \\
0 & - & |N_{\alpha,\mu-q\alpha}|^2 & = & \alpha \cdot (\mu - q\alpha)
\end{array} \tag{9.87}$$

whose solution is

$$\begin{aligned}
|N_{\alpha,\mu-r\alpha}|^2 &= -\sum_{s=r}^{q} \alpha \cdot (\mu - s\alpha) \\
&= -\tfrac{1}{2}(q - r + 1)(2\alpha \cdot \mu - (q + r)\alpha^2)
\end{aligned}$$

$$= \tfrac{1}{2}\alpha^2 r(q - r + 1). \qquad (9.88)$$

Notice that with the identifications $q = 2j$, $r = j - m$ this correctly reproduces the factor $(j - m)(j + m + 1)$ of equation (9.72). The phases are not fixed by equation (9.88), and clearly one has to be rather careful for a general Lie algebra to ensure that the phases chosen for the series produced by lowering by different step operators are consistent where they overlap.

It is worth remarking that essentially the same analysis can be used to find the constants $N_{\alpha\beta}$ occurring in equation (9.24). They are just the constants $N_{\alpha,\mu}$ for the adjoint representation, in which E_α acts on the other step operators, E_β, by commutation.

Problems for Chapter 9

9.1 Show that in the usual Cartesian basis the Cartan matrix for SU(2) is $g_{ij} = 2\delta_{ij}$.

9.2 Write down the commutation relations for SU(3), taking $\alpha = \alpha_1$, $\beta = \alpha_2$, $\gamma = -(\alpha_1 + \alpha_2)$ in figure 9.2(b). For the non-vanishing $N_{\alpha\beta}$ you will need to use equation (9.86).

9.3 The *Weyl group* of a root diagram is the group of transformations generated by the Weyl reflections. Show that the Weyl group of SU(3) is S_3. Are there any symmetry transformations which are not contained in the Weyl group?

9.4 Derive the complete root diagram (figure 9.2(d)) for G_2 from the two primitive roots α_1 and α_2.

9.5 Show that the unit vectors $(e_a - e_b)/\sqrt{2}$, $(e_b - e_c)/\sqrt{2}$ and $(e_c - e_a)/\sqrt{2}$ can be transformed respectively into α, β and γ of figure 9.2(b) by a $\pi/4$ rotation about the z-axis, followed by an appropriate rotation about the x-axis. What are the corresponding transforms of the matrices H_i?

9.6 From the fact that the sum of the angles between any three vectors emanating from a point is less than or equal to 360°, and that the simple roots are linearly independent, verify that the only allowable Dynkin diagrams of rank 3 are those for SU(4) (○—○—○) and SO(7) (○—○==○).

9.7 Show that in any Dynkin diagram in which two roots are connected by a single link, another valid Dynkin diagram can be obtained by shrinking that link to nothing, i.e. replacing the two roots and their connecting link by a single root. [Hint: if the two roots are α and β, show that $(\alpha+\beta)^2 = \alpha^2 = \beta^2$ and that the scalar products of all other roots with $\alpha+\beta$ in the reduced diagram are the same as those with either α or β in the original diagram.]

9.8 In SU(3) the *triality* of an irrep (m_1, m_2) is defined as $t = m_1 - m_2 \pmod 3$. Thus, the fundamental triplet representation 3 has $t = 1$, the fundamental $\overline{3}$ representation $t = -1$ ($\equiv 2$), and the trivial representation $t = 0$. Show that triality labels the equivalence classes Λ/R, where Λ is the weight lattice and R the root lattice (the set of integral multiples of α_1 and α_2). What are the trialities of the irreps 6, 8, 10, $\overline{10}$ and 27?

9.9 Find the dimensionalities of the two fundamental representations of SO(5).

10

Representations of the Poincaré Group

10.1 Lorentz Transformations

So far we have been dealing primarily with non-relativistic systems. In particular we have frequently used the non-relativistic expression for the kinetic energy T of a particle of mass m:

$$T = p^2/2m \tag{10.1}$$

which in its quantum mechanical guise gives rise to the Schrödinger equation. The scattering processes discussed in the last chapter are not necessarily non-relativistic, but we did not need to specify the dynamics explicitly: all that was used was its invariance under an internal SU(N) symmetry group.

Non-relativistic physics is characterized by the assumption of a universal time t, which is independent of the observer, and the statement that the laws of physics are the same for all inertial frames, i.e. for all frames moving with a constant velocity with respect to each other.

The relation between the coordinates of an *event* (something happening at a certain time and place) in two frames S and S′ which coincide at time $t = 0$ and are moving with relative velocity $\boldsymbol{v}$ is

$$\begin{aligned} t' &= t \\ \boldsymbol{x}' &= \boldsymbol{x} - \boldsymbol{v}t \end{aligned} \tag{10.2}$$

as illustrated in figure 10.1, where P represents the point at which the event takes place. Under such a transformation the distance between two points remains invariant:

$$(\boldsymbol{x}_1' - \boldsymbol{x}_2')^2 = (\boldsymbol{x}_1 - \boldsymbol{x}_2)^2 \tag{10.3}$$

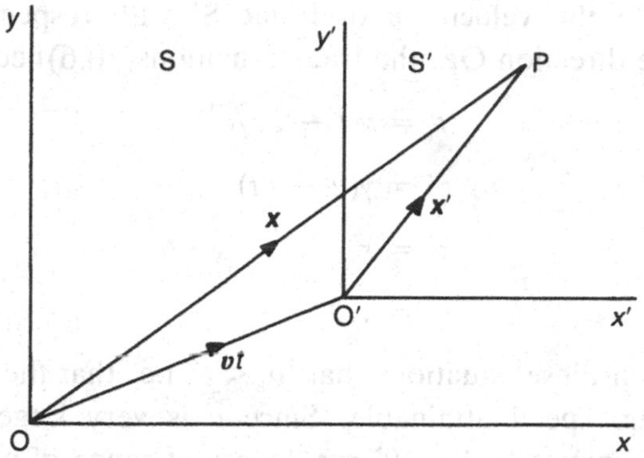

Figure 10.1 Coordinates of an event in the inertial frames S and S′.

as indeed it does under rotations of the coordinate system and a shift of the origin:

$$\boldsymbol{x}' = \boldsymbol{x} + \boldsymbol{a} \tag{10.4}$$

Equation (10.2) leads naturally to the 'commonsense' transformation of velocities. Thus if $\boldsymbol{x}(t)$ represents the position of a particle in the frame S, its velocity $\boldsymbol{u}$ is $\boldsymbol{u} = \mathrm{d}\boldsymbol{x}/\mathrm{d}t$, and similarly $\boldsymbol{u}' = \mathrm{d}\boldsymbol{x}'/\mathrm{d}t$, with

$$\boldsymbol{u}' = \boldsymbol{u} - \boldsymbol{v} \tag{10.5}$$

which corresponds to the usual notion of relative velocity.

However, this apparently reasonable equation is in fact wrong. It founders on the experimentally observed fact that the speed of light, c, is the same in all inertial frames, whereas according to (10.5) the observed speed of a light signal would be less in a frame travelling in the same direction. In order to account for the constancy of c, Einstein proposed a modification of equations (10.2) a modification which is small in non-relativistic situations, but which involves a revolution in our concepts of space and time. No longer is there a universal time: this also changes from frame to frame! The combined equations which make up the *Lorentz transformation* replacing equations (10.2) are

$$\begin{aligned} t' &= \gamma(t - \boldsymbol{x}\cdot\boldsymbol{v}/c^2) \\ \boldsymbol{x}'_\parallel &= \gamma(\boldsymbol{x}_\parallel - \boldsymbol{v}t) \\ \boldsymbol{x}'_\perp &= \boldsymbol{x}_\perp \end{aligned} \tag{10.6}$$

where

$$\gamma = 1/\surd(1 - v^2/c^2) \tag{10.7}$$

Here $\boldsymbol{x}_\parallel$ and $\boldsymbol{x}_\perp$ are respectively the components of $\boldsymbol{x}$ parallel and perpendicular to the velocity $\boldsymbol{v}$ of frame S′ with respect to S. If we choose $\boldsymbol{v}$ in the direction Oz, the transformations (10.6) become

$$\begin{aligned} t' &= \gamma(t - zv/c^2) \\ z' &= \gamma(z - vt) \\ x' &= x \\ y' &= y \end{aligned} \tag{10.6'}$$

It is implicit in these equations that $|\boldsymbol{v}| < c$, i.e. that the speed of light is the maximum speed attainable. Since c is very large by everyday standards (approximately $3 \times 10^8\ \mathrm{m\,s^{-1}}$), a vast range of phenomena are adequately represented by the non-relativistic approximation, in which effectively $c \to \infty$ so that $\gamma \to 1$ in equation (10.7) and equations (10.6) reduce to (10.2). However, small corrections are measurable in atomic physics, while in nuclear and particle physics the difference becomes very important.

To check that equations (10.6) achieve the desired goal of the universality of the speed of light, let us derive the analogue of equation (10.5), namely the transformation of velocities from one frame to another. The velocity in the frame S′ is

$$\boldsymbol{u}' = \frac{\mathrm{d}\boldsymbol{x}'}{\mathrm{d}t'} = \frac{\mathrm{d}\boldsymbol{x}'}{\mathrm{d}t}\bigg/\frac{\mathrm{d}t'}{\mathrm{d}t} \tag{10.8}$$

This breaks up into two different expressions for the parallel and transverse velocities,

$$\begin{aligned} \boldsymbol{u}'_{\parallel} &= \frac{\boldsymbol{u}_{\parallel} - \boldsymbol{v}}{1 - \boldsymbol{u}\cdot\boldsymbol{v}/c^2} \\ \boldsymbol{u}'_{\perp} &= \frac{\boldsymbol{u}_{\perp}}{\gamma(1 - \boldsymbol{u}\cdot\boldsymbol{v}/c^2)} \end{aligned} \tag{10.9}$$

which differ crucially from (10.5) by the presence of the factors in the denominator.

Then

$$\begin{aligned} \boldsymbol{u}'^2 &= (\boldsymbol{u}'_{\parallel})^2 + (\boldsymbol{u}'_{\perp})^2 \\ &= [\boldsymbol{u}_{\parallel}^2 - 2\boldsymbol{u}\cdot\boldsymbol{v} + \boldsymbol{v}^2 + (1 - v^2/c^2)(\boldsymbol{u}^2 - \boldsymbol{u}_{\parallel}^2)]/(1 - \boldsymbol{u}\cdot\boldsymbol{v}/c^2)^2 \\ &= [\boldsymbol{u}_{\parallel}^2 v^2/c^2 - 2\boldsymbol{u}\cdot\boldsymbol{v} + \boldsymbol{u}^2 + v^2(1 - \boldsymbol{u}^2/c^2)]/(1 - \boldsymbol{u}\cdot\boldsymbol{v}/c^2)^2 \\ &= c^2 - c^2(1 - v^2/c^2)(1 - \boldsymbol{u}^2/c^2)/(1 - \boldsymbol{u}\cdot\boldsymbol{v}/c^2)^2 \end{aligned} \tag{10.10}$$

Thus consistently $|\boldsymbol{u}'| \leqslant c$, and if $\boldsymbol{u}$ is the velocity in the original frame of a pulse of light with $|\boldsymbol{u}| = c$, its speed $|\boldsymbol{u}'|$ measured in any other frame is also c.

Under a Lorentz transformation the invariant is no longer the distance between two points as in equation (10.3); instead it is the combination

$$c^2\tau_{12}^2 := c^2(t_1 - t_2)^2 - (\boldsymbol{x}_1 - \boldsymbol{x}_2)^2 \tag{10.11}$$

Thus, taking $t_2 = \boldsymbol{x}_2 = 0$, dropping the '1' subscript and using the Lorentz transformation (10.6′), we get

$$\begin{aligned} c^2(t')^2 - (\boldsymbol{x}')^2 &= \gamma^2[c^2(t - zv/c^2)^2 - (z - vt)^2] - x^2 - y^2 \\ &= \gamma^2(c^2t^2 - z^2)(1 - v^2/c^2) - x^2 - y^2 \\ &= c^2t^2 - \boldsymbol{x}^2 \end{aligned} \tag{10.12}$$

by virtue of equation (10.7).

10.2 4-vector Notation

A rotation about a fixed centre leaves invariant the squared length $\boldsymbol{x}^2$ of a vector, or more generally the scalar product of two vectors $\boldsymbol{x}\cdot\boldsymbol{y}$:

$$\boldsymbol{x}\cdot\boldsymbol{y} = x_i y_i = \text{const} \tag{10.13}$$

As we have just seen, a Lorentz transformation leaves invariant the 'interval' $c^2\tau^2 = c^2t^2 - \boldsymbol{x}^2$. Examination of equation (10.11) shows that it also leaves invariant the combination $c^2t_1t_2 - \boldsymbol{x}_1\cdot\boldsymbol{x}_2$. It is tempting to regard this as some kind of generalized scalar product, with, however, the important sign difference between the first and second terms. To this end we think of t, or rather ct, which has the same dimensions as $\boldsymbol{x}$, as an additional coordinate x_0, and define

$$x_\mu := (x_0; \boldsymbol{x}) = (ct; \boldsymbol{x}) \tag{10.14}$$

This is called a *4-vector*, with the Greek index μ running from 0 to 3. In order to obtain the relative minus sign in τ^2 between the temporal and spatial contributions, we define a related object with indices raised:

$$x^\mu := (x_0; -\boldsymbol{x}) = (ct; -\boldsymbol{x}) \tag{10.15}$$

The generalized (length)^2 of the 4-vector x_μ is then defined as

$$x^2 := x_\mu x^\mu \tag{10.16}$$

with an implied summation over the repeated index μ. Writing out the summation explicitly we obtain

$$x^2 = x_0^2 - \boldsymbol{x}^2 = c^2t^2 - \boldsymbol{x}^2 = c^2\tau^2 \tag{10.17}$$

More generally the 'scalar product' of two position 4-vectors x_1, x_2 is defined as

$$x_1 \cdot x_2 := (x_1)_\mu (x_2)^\mu = c^2 t_1 t_2 - \boldsymbol{x}_1 \cdot \boldsymbol{x}_2 \tag{10.18}$$

precisely the quantity which is left invariant by a Lorentz transformation.

The Lorentz transformation (10.6′) can be written in a very elegant form, which brings out the similarity with rotations, and the crucial difference, by introducing an 'angle' ζ whose hyperbolic cosine is γ. Recall (equation (3.4)) that a rotation through θ about the z axis gives the transformation of coordinates

$$\begin{aligned} x_1' &= x_1 \cos\theta - x_2 \sin\theta \\ x_2' &= x_1 \sin\theta + x_2 \cos\theta \\ x_3' &= x_3 \\ x_0' &= x_0 \end{aligned} \tag{10.19}$$

with $(\boldsymbol{x}')^2 = \boldsymbol{x}^2$ as a consequence of the trigonometric identity $\cos^2\theta + \sin^2\theta = 1$. Rewriting equations (10.6′) in terms of x_0 and defining ζ by

$$\gamma = \cosh\zeta \tag{10.20}$$

so that

$$\tanh\zeta = v/c$$

we obtain

$$\begin{aligned} x_1' &= x_1 \\ x_2' &= x_2 \\ x_3' &= x_3 \cosh\zeta - x_0 \sinh\zeta \\ x_0' &= -x_3 \sinh\zeta + x_0 \cosh\zeta \end{aligned} \tag{10.21}$$

The equations look rather similar, but in fact there is a world of difference between ordinary and hyperbolic trigonometric functions. Note that the sinh term has the same sign in the equations for x_3' and x_0', as opposed to the change of sign of the sine term between x_1' and x_2'. This pattern of signs is necessary to give the invariance of $x_0^2 - \boldsymbol{x}^2$ as a consequence of the hyperbolic trigonometric identity $\cosh^2\zeta - \sinh^2\zeta = 1$.

Just as in ordinary 3-space a general vector $\boldsymbol{a}$ is an object which transforms the same way under rotations as the position vector $\boldsymbol{x}$, so a general 4-vector a_μ is an object which transforms in the same way as x_μ under Lorentz transformations. What other 4-vectors are there? Well, one way of constructing additional 4-vectors is to divide, or differentiate, by Lorentz-invariant quantities. In particular we can define a velocity 4-vector by differentiating x_μ not with respect to t but rather with respect to the proper time τ, which we know to be an invariant. Thus

$$U_\mu := \frac{\mathrm{d}x_\mu}{\mathrm{d}\tau} = (c;\, \boldsymbol{u}) \bigg/ \frac{\mathrm{d}\tau}{\mathrm{d}t} \tag{10.22}$$

in which

$$\mathrm{d}\tau/\mathrm{d}t = \surd(1 - \boldsymbol{u}^2/c^2) = 1/\gamma(u) \tag{10.23}$$

As a 4-vector, the squared length of U_μ should be an invariant. From equations (10.22) and (10.23) it is

$$U^2 = \gamma^2(u)(c^2 - \boldsymbol{u}^2) = c^2 \tag{10.24}$$

The *4-momentum* vector p_μ of a particle of mass m is just its 4-velocity times the mass:

$$p_\mu = mU_\mu = \gamma(u)(mc;\, m\boldsymbol{u}) \tag{10.25}$$

The spatial part is just the non-relativistic momentum, except for the additional factor $\gamma(u)$. The zeroth component is to be identified up to a factor of c with the energy E. That is,

$$\begin{aligned} \boldsymbol{p} &= \gamma(u)m\boldsymbol{u} \\ E/c &= \gamma(u)mc \end{aligned} \tag{10.26}$$

The invariant associated with p_μ is, from (10.24), $p^2 = m^2c^2$. That is,

$$p^2 = m^2c^2 = (E/c)^2 - \boldsymbol{p}^2 \tag{10.27}$$

Solving for E we obtain the relativistic relation between energy and momentum,

$$E/c = (m^2c^2 + \boldsymbol{p}^2)^{1/2} \tag{10.28}$$

which replaces equation (10.1).

For a particle at rest, with $\boldsymbol{p} = 0$, equation (10.28) reduces to the famous $E = mc^2$, the energy a particle possesses merely by virtue of its mass. In the non-relativistic regime, where $|\boldsymbol{p}| \ll mc$, we may expand the square root by the binomial theorem to obtain

$$E = mc^2 + p^2/2m + O(p^4) \tag{10.29}$$

reproducing the non-relativistic expression for the kinetic energy $T = E - mc^2$. Note that equation (10.26) implies that the correct relativistic expression for the momentum is not $\boldsymbol{p} = m\boldsymbol{u}$, but rather $\boldsymbol{p} = (E/c^2)\boldsymbol{u}$, which reduces to the former in the non-relativistic limit.

10.3 The Lorentz Group SO(3, 1)

The Lorentz transformation of equation (10.21) is a 'pure' Lorentz transformation, or *boost*, along the z direction without any rotation. There are obviously similar boosts along the x and y directions. These pure Lorentz transformations do not themselves form a group, since their products will in general involve rotations. The Lorentz group comprises all linear transformations, boosts, rotations and inversions which leave $c^2\tau^2 = x_0^2 - \boldsymbol{x}^2$ invariant. The *proper* Lorentz group excludes the inversions, e.g. $x_0 \to x_0$, $\boldsymbol{x} \to -\boldsymbol{x}$, which are not continuously connected to the identity.

If we think of x_μ as a column vector, the transformations are realized as real 4×4 matrices Λ, say, so that

$$x' = \Lambda x \tag{10.30}$$

(As in the scalar product $\boldsymbol{x}\cdot\boldsymbol{y} = x^\mu y_\mu$, summation always occurs between upper and lower indices, so the matrix elements of Λ must be written as Λ_μ^ν, with one upper and one lower index.)

The invariant interval can be written in matrix form as

$$c^2\tau^2 = x^{\mathrm{T}}\eta x \tag{10.31}$$

where η is the matrix

$$\eta = \mathrm{diag}(1, -1, -1, -1) \tag{10.32}$$

It is this matrix, which for consistency we must consider as having two upper indices $\eta^{\mu\nu}$, which is responsible for raising and lowering indices. That is,

$$x^\mu = \eta^{\mu\nu}x_\nu \tag{10.33}$$

Then after the transformation (10.30) we have

$$x'^{\mathrm{T}}\eta x' = x^{\mathrm{T}}(\Lambda^{\mathrm{T}}\eta\Lambda)x \tag{10.34}$$

In order to ensure the invariance of the interval we therefore need

$$\Lambda^{\mathrm{T}}\eta\Lambda = \eta \tag{10.35}$$

A real orthogonal matrix O belonging to O(4) would satisfy the condition $O^{\mathrm{T}}O = \mathbb{1}$, the unit 4×4 matrix. Instead the condition(10.35) makes Λ a pseudo-orthogonal matrix of type O(3, 1), the '3' and the '1' reflecting the occurrence of three minus signs and one plus sign in the metric tensor $\eta_{\mu\nu}$. Taking the determinant of equation (10.35) we see that $(\det\Lambda)^2 = 1$. The choice $\det\Lambda = +1$ gives the proper subgroup SO(3, 1), which is isomorphic to the proper Lorentz group.

The proper Lorentz group is also closely connected with the group SL(2, $\mathbb{C}$) of complex 2×2 matrices with unit determinant. The relationship is the same kind of 2:1 relationship which holds between SO(3) and SU(2). From a Lorentz vector v_μ one can form the 2×2 matrix

$$V := v_\mu \sigma^\mu \tag{10.36}$$

where σ_μ is made up of the unit matrix and the three Pauli matrices:

$$\sigma_\mu := (1; \boldsymbol{\sigma}) \tag{10.37}$$

Equation (10.36) can be inverted by tracing with $\tilde{\sigma}_\mu := (1; -\boldsymbol{\sigma})$:

$$v_\mu = \tfrac{1}{2}\,\mathrm{Tr}\,(\tilde{\sigma}_\mu V) \tag{10.38}$$

by virtue of the tracelessness of the Pauli matrices and the property $\sigma_i\sigma_j = \mathrm{i}\varepsilon_{ijk}\sigma_k$.

The explicit form of V is

$$V = \begin{pmatrix} v_0 - v_3 & -v_1 + \mathrm{i}v_2 \\ -v_1 - \mathrm{i}v_2 & v_0 + v_3 \end{pmatrix} \tag{10.36}$$

from which one readily sees that $v^2 = \det V$. Thus a transformation of the form

$$V' = AVA^\dagger \tag{10.39}$$

with $\det A = 1$ leaves $\det V$, and hence v^2, invariant. The SL(2, $\mathbb{C}$) matrix A therefore induces a (proper) Lorentz transformation on v_μ. The reader is invited to explore this connection further in problem 10.1.

Generators of SO(3, 1)

By differentiation of the matrices of finite rotations we have already obtained in §6.2 the matrix generators of SO(3). Pure rotations do not affect the time coordinate, so we can trivially extend the matrix generators to 4×4 matrices by including an extra null row and column for x_0:

$$X_1 = \begin{pmatrix} 0 & 0 & 0 & 0 \\ 0 & 0 & 0 & 0 \\ 0 & 0 & 0 & -\mathrm{i} \\ 0 & 0 & \mathrm{i} & 0 \end{pmatrix}$$

$$X_2 = \begin{pmatrix} 0 & 0 & 0 & 0 \\ 0 & 0 & 0 & \mathrm{i} \\ 0 & 0 & 0 & 0 \\ 0 & -\mathrm{i} & 0 & 0 \end{pmatrix} \tag{10.40}$$

$$X_3 = \begin{pmatrix} 0 & 0 & 0 & 0 \\ 0 & 0 & -\mathrm{i} & 0 \\ 0 & \mathrm{i} & 0 & 0 \\ 0 & 0 & 0 & 0 \end{pmatrix}$$

As a 4×4 matrix the boost of equation (10.21) is

$$B_3 = \begin{pmatrix} \cosh \zeta & 0 & 0 & -\sinh \zeta \\ 0 & 1 & 0 & 0 \\ 0 & 0 & 1 & 0 \\ -\sinh \zeta & 0 & 0 & \cosh \zeta \end{pmatrix} \tag{10.41}$$

The corresponding generator, Y_3 say, is given by the derivative of B_3 at $\zeta = 0$:

$$Y_3 := \mathrm{i}\frac{\mathrm{d}B_3(\zeta)}{\mathrm{d}\zeta}\bigg|_{\zeta=0} = \begin{pmatrix} 0 & 0 & 0 & -\mathrm{i} \\ 0 & 0 & 0 & 0 \\ 0 & 0 & 0 & 0 \\ -\mathrm{i} & 0 & 0 & 0 \end{pmatrix} \tag{10.42}$$

which can be characterized as

$$(Y_3)_\mu^{\ \nu} = -\mathrm{i}(\eta_{\mu 0}\eta_3^{\nu} - \eta_{\mu 3}\eta_0^{\nu}) \tag{10.43}$$

The matrices for the generators Y_1 and Y_2 are similarly obtained. In fact the general formula corresponding to equation (10.43) is $(Y_k)_\mu^{\ \nu} = -\mathrm{i}(\eta_{\mu 0}\eta_k^{\nu} - \eta_{\mu k}\eta_0^{\nu})$.

By explicit matrix multiplication (reader please check!) the commutator $[Y_3, X_1]$ is found to be

$$[Y_3, X_1] = \mathrm{i}Y_2 \tag{10.44}$$

Similarly the commutator of Y_1 and Y_2 turns out to be

$$[Y_1, Y_2] = -\mathrm{i}X_3 \tag{10.45}$$

The full algebra of commutation relations can be summarized by

$$
\begin{aligned}
[X_i, X_j] &= \mathrm{i}\varepsilon_{ijk}X_k \\
[X_i, Y_j] &= \mathrm{i}\varepsilon_{ijk}Y_k \\
[Y_i, Y_j] &= -\mathrm{i}\varepsilon_{ijk}X_k
\end{aligned}
\tag{10.46}
$$

From these we note that the generators of rotations close on themselves, forming the subalgebra SO(3). This is not true of the boosts, as the last equation shows. The middle equation can be regarded as saying that $\boldsymbol{Y}$ transforms as a vector under rotations.

The structure of the algebra can be simplified by defining the linear combinations

$$
\boldsymbol{X}^{(\pm)} := \tfrac{1}{2}(\boldsymbol{X} \pm \mathrm{i}\boldsymbol{Y}) \tag{10.47}
$$

In terms of these combinations the commutation relations become

$$
\begin{aligned}
[X_i^{(+)}, X_j^{(+)}] &= \mathrm{i}\varepsilon_{ijk}X_k^{(+)} \\
[X_i^{(-)}, X_j^{(-)}] &= \mathrm{i}\varepsilon_{ijk}X_k^{(-)} \\
[X_i^{(+)}, X_j^{(-)}] &= 0
\end{aligned}
\tag{10.48}
$$

That is, they break up into two independent SU(2) algebras in a very similar way to the decomposition of the SO(4) algebra, equation (7.18). The difference lies in the crucial 'i' in equations (10.47), which means that SO(3, 1) is locally isomorphic to SL(2, ℂ) rather than to SU(2) × SU(2).

The representation theory of SO(3, 1) is now easy. The irreducible representations are just enumerated as (j_1, j_2), with $(2j_1 + 1)(2j_2 + 1)$ degrees of freedom, where j_1 labels an irrep of the first SU(2) and j_2 the second. Apart from the trivial representation (0, 0), the lowest-dimensional irreps, with two degrees of freedom each, are the *Weyl* representations $(0, \frac{1}{2})$ and $(\frac{1}{2}, 0)$. In particle physics they are the representations used for the quantum field describing neutrinos. However, the *Dirac* representation used to describe the electron is actually the reducible representation $(\frac{1}{2}, 0) \oplus (0, \frac{1}{2})$, which acts on 4-component spinors. The reason for this is the need to enlarge the algebra (10.46′) to include the parity operator Π which implements spatial inversion. Indeed, under parity $\boldsymbol{X} \to \boldsymbol{X}$ and $\boldsymbol{Y} \to -\boldsymbol{Y}$, so $\boldsymbol{X}^{(+)}$ and $\boldsymbol{X}^{(-)}$ are interchanged: $\boldsymbol{X}^{(+)} \leftrightarrow \boldsymbol{X}^{(-)}$. Thus a representation including parity must be symmetric in j_1, j_2.

The representation $(\frac{1}{2}, \frac{1}{2})$, with 4 degrees of freedom, corresponds to the defining representation of SO(3, 1), giving the transformations of a

4-vector. The Lorentz transformation is induced by the double transformation of equation (10.39) (see problem (10.1). The representation $(1, 0) \oplus (0, 1)$, with 6 degrees of freedom, is carried by antisymmetric tensors $F_{\mu\nu}$. In electromagnetism the six independent components F_{0i}, F_{ij} can be identified with the $\boldsymbol{E}$ and $\boldsymbol{B}$ fields according to $E_i = -F_{0i}$, $B_i = -\frac{1}{2}\varepsilon_{ijk}F_{jk}$

An important point which is worth making at this stage is that these finite-dimensional representations are *non-unitary*. Thus the inverse of equation (10.47) is

$$\begin{aligned} \boldsymbol{X} &= \boldsymbol{X}^{(+)} + \boldsymbol{X}^{(-)} \\ \boldsymbol{Y} &= \mathrm{i}(\boldsymbol{X}^{(-)} - \boldsymbol{X}^{(+)}) \end{aligned} \tag{10.48}$$

This means that with Hermitian generators $\boldsymbol{X}^{(\pm)}$, the generators of rotations $\boldsymbol{X}$ are also Hermitian, but the generators of boosts $\boldsymbol{Y}$ are *anti*-Hermitian. Hence when exponentiated they produce anti-unitary matrices. The difference can be traced right back to the minus sign in the interval $c^2\tau^2$. SO(3, 1) is a non-compact group: the hyperbolic parameter ζ of pure Lorentz transformations ranges over the whole real axis, in contrast to the angles which parametrize rotations.

10.4 The Poincaré Group

The Poincaré group is another name for the inhomogeneous Lorentz group; that is, the Lorentz group we have been considering so far with the addition of space–time translations. The most general transformation of this group is

$$x'_\mu = \Lambda_\mu^\nu x_\nu + a_\mu \tag{10.49}$$

where a_μ is a constant 4-vector, independent of x. The invariant under such transformations is the finite interval $c^2\tau_{ab}^2$ or the infinitesimal $c^2\,\mathrm{d}\tau^2$. Since the translations $x_\mu \to x_\mu + a_\mu$ cannot be represented as a 4×4 matrix acting on x, we have to proceed in a different way in order to obtain the algebra of the Poincaré group.

To this end we introduce the differential operator

$$\partial_\mu := (\partial/\partial x_0; -\boldsymbol{\nabla}) \tag{10.50}$$

The minus sign here is not a mistake: it is necessary if ∂_μ is to transform in the same way as x_μ. Thus, for example,

$\partial_\mu(x^2) = \partial_\mu(x_0^2 - \mathbf{x}^2) = 2(x_0; \mathbf{x}) = 2x_\mu$. The upper index operator $\partial^\mu = (\partial/\partial x_0; \boldsymbol{\nabla})$ transforms in the opposite way to x_μ, and is the more natural way of introducing the derivative, particularly if one is considering general relativity, where the metric tensor $g_{\mu\nu}$ does not have the rather trivial form of $\eta_{\mu\nu}$.

Consider first an infinitesimal Lorentz transformation, so that

$$\Lambda_{\mu\nu} = \eta_{\mu\nu} - \varepsilon_{\mu\nu} \tag{10.51}$$

In index notation equation (10.35) reads

$$\Lambda_{\mu\rho}\Lambda^{\mu}{}_{\nu} = \eta_{\rho\nu} \tag{10.35$'$}$$

and on substituting (10.51) we find that ε must be antisymmetric:

$$\varepsilon_{\mu\nu} = -\varepsilon_{\nu\mu} \tag{10.52}$$

There are thus six independent components, i.e. six infinitesimal parameters. Of these the three ε_{ij} correspond to rotations and the three ε_{0k} to pure boosts.

The transformation induced by $\Lambda_{\mu\nu}$, namely $x'_\mu = x_\mu - \varepsilon_\mu{}^\nu x_\nu$, can alternatively be obtained by the action of the infinitesimal form of $\exp(-\frac{1}{2}\mathrm{i}\varepsilon_{\rho\sigma}L^{\rho\sigma})$ on x_μ, where the $L_{\rho\sigma}$ are the generalized angular momentum operators:

$$L_{\rho\sigma} := \mathrm{i}(x_\rho\partial_\sigma - x_\sigma\partial_\rho) \tag{10.53}$$

(From here on we will work in natural units, in which $\hbar$ and c are both numerically equal to unity.) Thus

$$\begin{aligned}(1 - \tfrac{1}{2}\mathrm{i}\varepsilon_{\rho\sigma}L^{\rho\sigma})x_\mu &= x_\mu + \tfrac{1}{2}\varepsilon_{\rho\sigma}(x^\rho\partial^\sigma - x^\sigma\partial^\rho)x_\mu \\ &= x_\mu + \varepsilon_{\rho\sigma}x^\rho\eta^\sigma_\mu \\ &= x_\mu - \varepsilon_\mu{}^\rho x_\rho\end{aligned} \tag{10.54}$$

as required. In deriving (10.54) we have used the antisymmetry of $\varepsilon_{\mu\nu}$ and the identity $\partial^\sigma(x_\mu) = \eta^\sigma_\mu$.

This last identity in its operator form $[\partial_\sigma, x_\mu] = \eta_{\mu\sigma}\mathbb{1}$ allows us to calculate the commutation relations of the differential operators $L_{\rho\sigma}$. The result, after a certain amount of algebra, is

$$[L_{\mu\nu}, L_{\rho\sigma}] = -\mathrm{i}(\eta_{\mu\rho}L_{\nu\sigma} - \eta_{\mu\sigma}L_{\nu\rho} + \eta_{\nu\sigma}L_{\mu\rho} - \eta_{\nu\rho}L_{\mu\sigma}) \tag{10.55}$$

The commutation relations are more transparent when we split the generators into those for rotations:

$$J_i = \tfrac{1}{2}\varepsilon_{ijk}L_{jk} \qquad \text{(i.e. } J_3 = L_{12}, \text{ etc.)} \tag{10.56a}$$

and boosts:

$$K_i = L_{0i} \tag{10.56b}$$

Given that we are now working in units where $\hbar = 1$, the J_i are identical with the orbital angular momentum generators L_i. Thus

$$J_i = -\tfrac{1}{2}\mathrm{i}\varepsilon_{ijk}(x_j\nabla_k - x_k\nabla_j)$$

i.e.

$$\boldsymbol{J} = -\mathrm{i}\boldsymbol{x} \times \boldsymbol{\nabla} = \boldsymbol{x} \times \hat{\boldsymbol{p}}$$

We know that the J_i should have the standard commutation relations of angular momentum, but let us just check this from the definition of (10.56a) and the commutation relations (10.55). Exploiting the antisymmetry of the epsilon symbols, and noting that $\eta_{lp} = -\delta_{lp}$, we obtain

$$\begin{aligned}[J_i, J_j] &= \tfrac{1}{4}\varepsilon_{ilm}\varepsilon_{jpq}[L_{lm}, L_{pq}] \\ &= \mathrm{i}\varepsilon_{ipm}\varepsilon_{jpq}L_{mq}\end{aligned}$$

which, upon using the epsilon identity (6.29), correctly gives

$$[J_i, J_j] = -\mathrm{i}L_{ji} = \mathrm{i}\varepsilon_{ijk}J_k$$

The evaluation of the other commutators proceeds in a similar fashion and reproduces the commutation relations of equations (10.46), with $\boldsymbol{X}$ and $\boldsymbol{Y}$ replaced by $\boldsymbol{J}$ and $\boldsymbol{K}$ respectively.

Let us now go on to consider an infinitesimal translation

$$x'_\mu = x_\mu - \varepsilon_\mu \tag{10.57}$$

In a similar way, this transformation can be obtained by the infinitesimal version of the operator $\exp(\mathrm{i}\varepsilon_\rho P^\rho)$ acting on x_μ, where P_μ is the Lorentz 4-momentum operator

$$P_\mu = \mathrm{i}\partial_\mu = (\mathrm{i}\partial_0; -\mathrm{i}\boldsymbol{\nabla}) \tag{10.58}$$

Thus

$$(1 + \mathrm{i}\varepsilon_\rho P^\rho)x_\mu = x_\mu - \varepsilon_\rho\eta^\rho_\mu = x_\mu - \varepsilon_\mu$$

as required.

The momentum operators P_μ clearly commute among themselves:

$$[P_\mu, P_\nu] = 0 \tag{10.59}$$

and the commutator of P_μ with $L_{\rho\sigma}$ is straightforward to calculate:

$$[P_\mu, L_{\rho\sigma}] = \mathrm{i}(\eta_{\mu\rho}P_\sigma - \eta_{\mu\sigma}P_\rho) \tag{10.60}$$

So in 4-vector form the full algebra of the Poincaré group is given by equations (10.55), (10.59) and (10.60) In terms of $\boldsymbol{J}$, $\boldsymbol{K}$, the commutation relations (10.60) read

$$[P_0, J_i] = 0$$
$$[P_i, J_j] = i\varepsilon_{ijk}P_k \tag{10.61}$$

and

$$[P_0, K_i] = iP_i$$
$$[P_i, K_j] = iP_0\delta_{ij} \tag{10.62}$$

Equations (10.61) merely state that P_0 is a scalar and P_i a vector under rotations. Equations (10.62) reflect the fact that under a boost along the ith axis both P_0 and P_i are affected, but the other two spatial components are unaffected.

Massive Representations; Helicity

Particles, such as the electron, nucleon or pion, are characterized by a definite mass and spin. The mass is the invariant associated with the 4-momentum according to $m^2 = p^2$ (equation (10.27) in natural units). The spin refers to the transformation property of the particle state in its rest-frame, where $p_\mu = (m; \mathbf{0})$. In terms of the operators of the Poincaré algebra, the mass and spin are the eigenvalues of two quadratic Casimir operators, analogues of $\boldsymbol{J}^2$ for SO(3), which commute with all of the generators.

The first of these is just $P^2 = P_\mu P^\mu$. It is Lorentz-invariant and translationally invariant, which means that

$$[P^2, P_\mu] = 0$$
$$[P^2, L_{\rho\sigma}] = 0 \tag{10.63}$$

as can be checked by direct calculation from (10.60) with the help of the identity $[A, BC] = [A, B]C + B[A, C]$.

A covariant spin operator, the *Pauli–Lubanski vector*, can be defined as follows:

$$W_\mu := -\tfrac{1}{2}\varepsilon_{\mu\nu\rho\sigma}L^{\nu\rho}P^\sigma \tag{10.64}$$

Here $\varepsilon_{\mu\nu\rho\sigma}$ is the completely antisymmetric tensor on four indices, with $\varepsilon_{0123} = +1$. W^2 is again Lorentz-invariant and translationally invariant, so that it also commutes with all the generators:

$$[W^2, P_\mu] = 0$$
$$[W^2, L_{\rho\sigma}] = 0 \tag{10.65}$$

as you are invited to check explicitly.

Acting on a rest-frame eigenstate of P_μ, when $P_\mu \to \bar{p}_\mu := (m; \mathbf{0})$, the operator W_μ effectively reduces to w_μ, where

$$w_0 = 0 \qquad w_i = \tfrac{1}{2} m\varepsilon_{ijk} L_{jk} = mJ_i \tag{10.66a}$$

and so

$$W^2 \to w^2 = -m^2 \boldsymbol{J}^2 \tag{10.66b}$$

So states can be taken as simultaneous eigenstates of P^2 and W^2 and accordingly labelled as $|m, s \ldots\rangle$, with

$$P^2|m, s \ldots\rangle = m^2|m, s \ldots\rangle$$
$$W^2|m, s \ldots\rangle = -m^2 s(s+1)|m, s \ldots\rangle \tag{10.67}$$

What are the additional quantum numbers we can use to label the states? They must be the eigenvalues of operators which commute with each other. In view of equation (10.59) we are free to consider states of definite 4-momentum p_μ. Since the mass is already fixed, it is only necessary to specify in addition the 3-momentum $\boldsymbol{p}$, the energy being determined by equation (10.28). We cannot simultaneously give a definite value for the third component of the angular momentum operator J_3 because $\boldsymbol{J}$ and $\boldsymbol{P}$ do not commute. However, there *is* an angular momentum operator which commutes with $\boldsymbol{P}$, namely the *helicity*. This operator is the component of spin along the direction of the momentum, $\boldsymbol{J}\cdot\hat{\boldsymbol{p}}$, and its eigenvalues are conventionally labelled as λ. Thus the complete specification of the momentum eigenstate of a massive particle is $|m, s; \boldsymbol{p}, \lambda\rangle$, with

$$P_\mu|m, s; \boldsymbol{p}, \lambda\rangle = p_\mu|m, s; \boldsymbol{p}, \lambda\rangle$$
$$\boldsymbol{J}\cdot\boldsymbol{P}|m, s; \boldsymbol{p}, \lambda\rangle = \lambda|\boldsymbol{p}||m, s; \boldsymbol{p}, \lambda\rangle \tag{10.68}$$

This is the (infinite-dimensional) vector space on which the Poincaré group acts. We now need to find how the states transform under a given Poincaré transformation. Translations are trivial because the states are eigenstates of the generator of translations. Thus a translation $x \to x - a$ is simply represented by the phase factor $\exp(ip\cdot a)$.

As far as homogeneous Lorentz transformations are concerned, we first note that the state $|m, s; \boldsymbol{p}, \lambda\rangle$ can be obtained by a suitable Lorentz transformation Λ_p from a rest-frame state $|m, s; \mathbf{0}, s_3\rangle$, with $s_3 = \lambda$. Note that in the rest-frame the helicity operator becomes indeterminate, so we just have to choose an arbitrary component of spin, conventionally taken as the z component, s_3. The required Lorentz transformation Λ_p consists of a rotation $R(\boldsymbol{p})$ which rotates the z axis into the direction of $\boldsymbol{p}$ followed by a pure boost B_p in the $\boldsymbol{p}$ direction: $B_p = \exp(-\mathrm{i}\boldsymbol{\zeta}\cdot\boldsymbol{K})$, where $\sinh\zeta = |\boldsymbol{p}|/m$. Then

$$|m, s; \boldsymbol{p}, \lambda\rangle = \Lambda_p|m, s; \mathbf{0}, s_3 = \lambda\rangle \qquad (10.69)$$

Now consider the effect on $|m, s; \boldsymbol{p}, \lambda\rangle$ of the transformation $\Lambda_{p'\leftarrow p}$ which takes the momentum p into p':

$$\begin{aligned}\Lambda_{p'\leftarrow p}|m, s; \boldsymbol{p}, \lambda\rangle &= \Lambda_{p'\leftarrow p}\Lambda_p|m, s; \mathbf{0}, s_3 = \lambda\rangle \\ &= \Lambda_{p'}(\Lambda_{p'}^{-1}\Lambda_{p'\leftarrow p}\Lambda_p)|m, s; \mathbf{0}, s_3 = \lambda\rangle \qquad (10.70)\end{aligned}$$

The reason for writing it this way is that the bracketed combination of Lorentz transformations $\Lambda_{p'}^{-1}\Lambda_{p'\leftarrow p}\Lambda_p$ takes the rest-frame momentum $\bar{p}$ to p, then from p to p', and finally from p' back to $\bar{p}$ again. Thus it is a transformation leaving $\bar{p}_\mu = (m; \mathbf{0})$ invariant, and as such can only be a *rotation*, known as the Wigner rotation, which we will denote by R_W. But the rest-frame state $|m, s; \mathbf{0}, s_3 = \lambda\rangle$ transforms under rotations according to the irrep $D^{(s)}$, so we can write

$$\begin{aligned}\Lambda_{p'\leftarrow p}|m, s; \boldsymbol{p}, \lambda\rangle &= \Lambda_{p'}R_\mathrm{W}|m, s; \mathbf{0}, s_3 = \lambda\rangle \\ &= \Lambda_{p'}\sum_{\lambda'} D^s_{\lambda'\lambda}(R_\mathrm{W})|m, s; \mathbf{0}, s_3 = \lambda'\rangle \\ &= \sum_{\lambda'} D^s_{\lambda\lambda'}(R_\mathrm{W})|m, s; \boldsymbol{p}', \lambda'\rangle \qquad (10.71)\end{aligned}$$

giving the required representation of the Lorentz group elements.

Equation (10.71) is an example of what is known as an *induced representation*, whereby the representation of a larger group is obtained from that of a subgroup. The subgroup in question is the *stability group*, or *little group* of a standard momentum $\bar{p}_\mu$, i.e. the subset of Lorentz transformations which leave $\bar{p}_\mu$ invariant. In the massive case we can take $\bar{p}_\mu = (m; \mathbf{0})$, whose stability group is the 3-dimensional rotation group SO(3), or SU(2). It is as well to realize, however, that the Wigner rotation R_W is quite a complicated object. It depends not only on the Lorentz transformation $\Lambda_{p'\leftarrow p}$, but also on Λ_p and $\Lambda_{p'}$ in rather an intricate way.

Massless Representations

The group theory is quite different in the case of massless particles. For such particles, with $p^2 = 0$, a rest-frame does not exist, so the little group, the subgroup of the Lorentz group leaving a canonical momentum invariant, is no longer the rotation group SU(2).

Let us take the standard momentum as $\omega_\mu := \omega(1; 0, 0, 1)$. That is, the momentum is along the z direction, and in natural units has the same magnitude as the energy ω. The stability group of ω_μ turns out to be isomorphic to the 2-dimensional Euclidean group E_2, generated by J_3, $L_1 := J_1 + K_2$ and $L_2 := J_2 - K_1$. Clearly ω_μ is invariant under rotations $\exp(-iJ_3\varphi)$ about the z axis, but it is not so obvious that it is unchanged by the combined rotations and boosts generated by L_1 and L_2. However, from equations (10.61) and (10.62) we see that

$$[L_1, P_0] = [K_2, P_0] = -iP_2 \tag{10.72}$$

which is effectively zero when acting on the state $|\omega_\mu\rangle$, which is an eigenstate of P_2 with $P_2|\omega_\mu\rangle = 0$. As far as the 3-momentum is concerned,

$$\begin{aligned}[L_1, P_i] &= [J_1, P_i] + [K_2, P_i] \\ &= -i\varepsilon_{i1k}P_k - iP_0\delta_{i2}\end{aligned} \tag{10.73}$$

Hence

$$[L_1, P_1] = 0$$

while

$$\begin{aligned}[L_1, P_2] &= i(P_3 - P_0) \\ [L_1, P_3] &= -iP_2\end{aligned}$$

both of which give zero when applied to $|\omega_\mu\rangle$. Similarly $[L_2, P_\mu]|\omega_\mu\rangle = 0$.

The commutation relations of J_3, L_1 and L_2 are

$$\begin{aligned}[J_3, L_1] &= iJ_2 - iK_1 = iL_2 \\ [J_3, L_2] &= -iJ_1 - iK_2 = -iL_1 \\ [L_1, L_2] &= iJ_3 - iJ_3 = 0\end{aligned} \tag{10.74}$$

As already mentioned, this is isomorphic to the Euclidean group of rotations, generated by J_3, and translations, generated by L_1 and L_2, in 2 dimensions. The quadratic Casimir operator of E_2 is the combination

$L_1^2 + L_2^2$. It obviously commutes with L_1 and L_2, and $[J_3, L_1^2 + L_2^2] = \{L_1, iL_2\} - \{L_2, iL_1\} = 0$. So the states may be labelled by the eigenvalues of this Casimir operator and one of the three generators, conventionally taken as J_3. The combinations $L_\pm := L_1 \pm iL_2$ are respectively raising and lowering operators for J_3, since

$$[J_3, L_\pm] = iL_2 \pm L_1 = \pm L_\pm \tag{10.75}$$

to be compared with equation (B3). There is, however, a big difference from the rotation group. There, for a given value of the Casimir operator $\boldsymbol{J}^2$, the spectrum of J_3 was limited because of the inequality $\langle J_3^2 \rangle \leqslant \langle \boldsymbol{J}^2 \rangle$. Now there is no relation between J_3 and the Casimir operator $L_1^2 + L_2^2$. So application of the raising operator will generate an infinite tower of eigenvalues of J_3, and although spaced apart by integers, their actual values will not in general be either integers or half-integers, *unless* the states are taken to be null states of both L_1 and L_2, i.e. $L_1| \ \rangle = L_2| \ \rangle = 0$. The value of the quadratic Casimir operator is then also zero. Thus the representations of physical interest, which correspond to zero-mass particles, are characterized by a zero eigenvalue of $L_1^2 + L_2^2$ and are specified by a single value of J_3, which is just the helicity.

From this analysis it is quite clear that what one means by the 'spin' s of a massless particle is quite different from that of a massive particle. In the massive case one has a representation of SU(2), with the Casimir operator $\boldsymbol{J}^2$ having the eigenvalue $s(s+1)$ and the z component J_3 taking values integrally spaced between $+s$ and $-s$. In the massless case the relevant group is not SU(2) at all, and the only angular momentum operator which is specified is the helicity, which is assigned the value s in a given irrep. The simplest example of this is the neutrino (ν), a massless fermion produced in beta decay, which has $\lambda = -\frac{1}{2}$. Its antiparticle, the antineutrino ($\bar{\nu}$), has $\lambda = +\frac{1}{2}$.

Helicity is, by its definition, rotationally invariant. For massless particles it is also Lorentz-invariant, since no boost is possible which would reverse the sign of the 3-momentum. However, it does reverse sign under parity, the operator which implements the space inversion $\boldsymbol{x} \to -\boldsymbol{x}$. Under parity the momentum $\boldsymbol{p}$ is a true (polar) vector, changing sign just like $\boldsymbol{x}$, whereas $\boldsymbol{J}$ is a pseudo (axial) vector, remaining unchanged. Hence helicity, $\boldsymbol{J}\cdot\hat{\boldsymbol{p}}$, is a pseudoscalar, changing sign under parity. If, therefore, the massless particle in question takes part in reactions which conserve parity, it must possess both helicities $\pm s$. In group theory terms this means using a reducible representation of the proper Poincaré group. The photon is just such a case, occurring with $\lambda = \pm 1$, corresponding to left and right circular polarization respectively. There is, however, no state with $\lambda = 0$, as there would be for a massive particle of spin 1.

For a massless state with momentum ω_μ, the Pauli–Lubanski vector W_μ becomes effectively

$$W_\mu = -\tfrac{1}{2}\varepsilon_{\mu\nu\rho\sigma}L^{\nu\rho}\omega^\sigma$$

with components

$$W_\mu = \omega(J_3;\ L_1,\ L_2,\ J_3)$$

Thus $W^2 = -\omega^2(L_1^2 + L_2^2)$, which, as we have seen, has eigenvalue 0. In fact

$$W_\mu|\omega_\mu\rangle = \lambda\omega(1;\ 0,\ 0,\ 1)|\omega_\mu\rangle = \lambda P_\mu|\omega_\mu\rangle \tag{10.76}$$

which is another way of characterizing the helicity for massless particles.

Writing the states as $|p_\mu;\ \lambda\rangle$, the analogue of equation (10.71) is now just

$$\Lambda_{p'\leftarrow p}|p_\mu;\ \lambda\rangle = \exp(-\mathrm{i}\Phi_\mathrm{W}\lambda)|p'_\mu;\ \lambda\rangle \tag{10.77}$$

where Φ_W is the angular parameter multiplying J_3 in the Wigner E_2 transformation of $|\omega_\mu;\ J_3 = \lambda\rangle$. The other parameters do not enter because $L_1|\omega_\mu\rangle = L_2|\omega_\mu\rangle = 0$, and there is no change of helicity because J_3 is diagonal. Equation (10.77) is in agreement with our previous remark that for massless particles helicity is invariant under both boosts and rotations. In general, however, these transformations do produce a phase factor.

10.5* Angular Momentum States

(i) Single-particle States; Potential Scattering

So far we have discussed states of definite 4-momentum, eigenstates of the operator P_μ and of the helicity $\boldsymbol{J}\cdot\hat{\boldsymbol{p}}$. In a given Lorentz frame the labels giving the orientation of the 3-momentum may be exchanged for angular momentum labels, as was done in the spinless case in §7.3(ii). There the appropriate angular momentum was the orbital angular momentum l and its z component m. In the case of particles with spin, however, the transformation properties of the spin labels also play a role, and it is more useful to construct eigenstates of the total angular momentum J and its z component M. As we have just seen, in the case of massless particles one is bound to use the helicity to describe the internal 'spin' state. For massive particles one *could* use the z component of the intrinsic spin, but the rotational invariance of helicity results in a much simpler angular momentum decomposition.

Eigenstates of total angular momentum $|pJM\lambda\rangle$ are defined as follows:

$$|pJM\lambda\rangle = N_J \int \mathrm{d}R\, (D^J_{M\lambda}(R))^* |\boldsymbol{p}, \lambda\rangle \tag{10.78}$$

Here $N_J = [(2J+1)/4\pi]^{1/2}$ and R is a rotation which takes the z axis into the direction $\hat{\boldsymbol{p}}$, so that $\boldsymbol{p} = R\boldsymbol{p}_3$, where $\boldsymbol{p}_3 = (0, 0, p)$. Correspondingly the state $|\boldsymbol{p}, \lambda\rangle$ is defined by a rotation from the standard helicity state $|\boldsymbol{p}_3, \lambda\rangle$:

$$|\boldsymbol{p}, \lambda\rangle = U(R)|\boldsymbol{p}_3, \lambda\rangle \tag{10.79}$$

There is some arbitrariness in R, since any initial rotation about the z axis will not change the helicity or the final momentum $\boldsymbol{p}$. A convention widely adopted is that of Jacob and Wick (1954 *Annals of Physics (NY)* **1** 404), whereby if $\boldsymbol{p}$ has polar angles (θ, φ), R is taken as the rotation with Euler angles $(\varphi, \theta, -\varphi)$, i.e.

$$R = R(\varphi, \theta, -\varphi) = R_3(\varphi)R_2(\theta)R_3(-\varphi) \tag{10.80}$$

This particular rotation can be characterized as a rotation through θ about the normal to the plane containing $\boldsymbol{p}$ and $\boldsymbol{p}_3$. (It is a conjugate rotation to $R_2(\theta)$, with its axis rotated through φ—see figure 10.2.) Adopting the Jacob–Wick convention, the integration measure $\mathrm{d}R$ in (10.78) can just be taken as the solid angle $\mathrm{d}(\cos\theta)\,\mathrm{d}\varphi$.

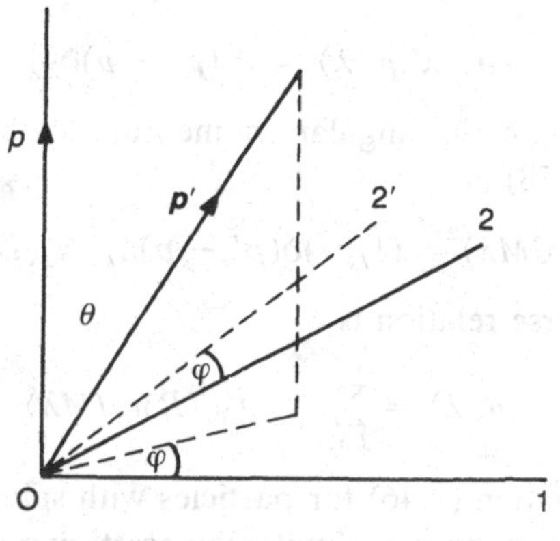

Figure 10.2 The Jacob–Wick rotation $R(\varphi, \theta, -\varphi)$ is a rotation through θ about the axis 02′.

The definition given in equation(10.78) effectively uses the continuum version of the projection operator defined in equation (5.22), and the proof that $|pJM\lambda\rangle$ indeed transforms in the correct way follows very similar lines. Thus if S is a general rotation,

$$\begin{aligned} U(S)|pJM\lambda\rangle &= N_J \int \mathrm{d}R\,(D^J_{M\lambda}(R))^* U(S)U(R)|\boldsymbol{p}_3, \lambda\rangle \\ &= N_J \int \mathrm{d}R'(D^J_{M\lambda}(R))^* U(R')|\boldsymbol{p}_3, \lambda\rangle \end{aligned}$$

where $R' = SR$, so that $R = S^{-1}R'$. Then $D^J_{M\lambda}(R) = \Sigma_{M'} D^J_{MM'}(S^{-1}) \times D^J_{M'\lambda}(R')$ and

$$U(S)|pJM\lambda\rangle = \sum_{M'} D^J_{M'M}(S) N_J \int \mathrm{d}R'(D^J_{M'\lambda}(R'))^* U(R')|\boldsymbol{p}_3, \lambda\rangle$$

There is a subtlety here: we cannot immediately identify $N_J \int \mathrm{d}R' \ldots$ with the state $|pJM'\lambda\rangle$ because R' is not necessarily of the form $R_3(\varphi')R_2(\theta')R_3(-\varphi')$. However, the construction (10.78) is actually independent of the third Euler angle. Thus

$$\begin{aligned} &(D^J_{M\lambda}(RR_3(\gamma)))^* U(RR_3(\gamma))|\boldsymbol{p}_3, \lambda\rangle \\ &\qquad = (D^J_{M\lambda}(R))^* \mathrm{e}^{\mathrm{i}\gamma\lambda} U(R) \exp(-\mathrm{i}J_3\gamma)|\boldsymbol{p}_3, \lambda\rangle \\ &\qquad = (D^J_{M\lambda}(R))^* U(R)|\boldsymbol{p}_3, \lambda\rangle \end{aligned}$$

because $|\boldsymbol{p}_3, \lambda\rangle$ is an eigenstate of J_3 with eigenvalue λ. So indeed we have

$$U(S)|pJM\lambda\rangle = \sum_{M'} D^J_{M'M}(S)|pJM'\lambda\rangle \tag{10.81}$$

as required.

If the plane wave states $|\boldsymbol{p}, \lambda\rangle$ are normalized in the non-covariant fashion

$$\langle \boldsymbol{p}', \lambda'|\boldsymbol{p}, \lambda\rangle = \delta^3(\boldsymbol{p}' - \boldsymbol{p})\delta_{\lambda'\lambda} \tag{10.82}$$

then the overlap with the angular momentum states is easily obtained from equation (10.78) as

$$\langle \boldsymbol{p}', \lambda'|pJM\lambda\rangle = (1/p^2)\delta(p' - p)\delta_{\lambda'\lambda} N_J (D^J_{M\lambda}(R'))^* \tag{10.83}$$

and hence the inverse relation is

$$|\boldsymbol{p}, \lambda\rangle = \sum_{J,M} N_J D^J_{M\lambda}(R)|pJM\lambda\rangle \tag{10.84}$$

which replaces equation (7.46) for particles with spin.

In §7.3(ii) we were concerned with the scattering of a spinless particle off a central potential. We exploited the rotational invariance of the unperturbed Hamiltonian and the potential V to simplify the unitarity relation for the matrix elements of the scattering operator between orbital angular momentum states and thereby introduce the phase shifts δ_l. A similar analysis can be carried out for particles with spin, provided we use the helicity basis. The rotational invariance of S now tells us that

$$^{\text{out}}\langle p'J'M'\lambda'|S|pJM\lambda\rangle^{\text{out}} = (1/p^2)\delta(p'-p)\delta_{J'J}\delta_{M'M}R^J_{\lambda'\lambda}(p) \quad (10.85)$$

The analogue of equation (7.54) is now

$$^{\text{out}}\langle \boldsymbol{p}', \lambda'|S|\boldsymbol{p}, \lambda\rangle^{\text{out}}$$
$$= (1/p^2)\delta(p'-p)[\delta(\Omega'-\Omega)\delta_{\lambda'\lambda} + (\mathrm{i}p/2\pi)f_{\lambda'\lambda}(\theta', \varphi')] \quad (10.86)$$

where it is convenient to take the initial momentum $\boldsymbol{p}$ along the z axis. In natural units we no longer distinguish between p and $k := p/\hbar$. Inserting the angular momentum expansion (10.84) into equation (10.86) and writing

$$R^J_{\lambda'\lambda} = 1 + 2\mathrm{i}pf^J_{\lambda'\lambda} \quad (10.87)$$

in analogy to (7.55), gives

$$(\mathrm{i}p/2\pi)f_{\lambda'\lambda}(\theta', \varphi') = \sum_J N_J^2(D^J_{\lambda\lambda'}(R'))^*(2\mathrm{i}pf^J_{\lambda'\lambda})$$

so that

$$f_{\lambda'\lambda}(\theta', \varphi') = \sum_J (2J+1)\mathrm{e}^{\mathrm{i}(\lambda-\lambda')\varphi'}d^J_{\lambda\lambda'}(\theta')f^J_{\lambda'\lambda} \quad (10.88)$$

It is this equation which replaces the partial wave expansion (7.56), to which it reduces for spinless particles. Unitarity is now not quite so simple as in that case: unitarity of S reduces to the statement that $R^J_{\lambda'\lambda}(p)$ is a unitary *matrix*, of dimension $(2s+1) \times (2s+1)$ for massive particles. If this matrix can be diagonalized by exploiting some additional invariance, we can again introduce phase shifts, which afford a useful parametrization of the low-energy scattering amplitude.

If, for example, we consider electrons scattering off a spinless atom, the interaction respects parity invariance. Now, under the parity operation angular momentum is invariant, whereas linear momentum changes spin, $\boldsymbol{p} \to -\boldsymbol{p}$. Thus the combination $\boldsymbol{J}\cdot\hat{\boldsymbol{p}}$, namely the helicity, also changes sign, i.e. $\lambda \to -\lambda$. There are phases associated with the quantum mechanical states, but these cancel out in elastic scattering, and the invariance is expressed in the equality

$$f^J_{\lambda'\lambda} = f^J_{-\lambda', -\lambda} \quad (10.89)$$

Thus R^J or f^J is diagonalized by going over to the basis vectors $(1/\sqrt{2})(|pJM\frac{1}{2}\rangle \pm |pJM, -\frac{1}{2}\rangle)$. The corresponding eigenvalues of R^J are unitary numbers expressible in terms of phase shifts as $\exp(2i\delta^J_\pm)$ respectively. Hence we can write

$$\begin{aligned} f^J_{1/2,1/2} &= f^J_{-1/2,-1/2} = \tfrac{1}{2}(f^J_+ + f^J_-) \\ f^J_{1/2,-1/2} &= f^J_{-1/2,1/2} = \tfrac{1}{2}(f^J_+ - f^J_-) \end{aligned} \tag{10.90}$$

where $f^J_\pm = (1/p)\exp(i\delta^J_\pm)\sin(\delta^J_\pm)$. At low energies only a few phase shifts will be appreciable.

In the helicity formalism the expression for the unpolarized cross-section is remarkably simple. Since helicity is a measurable quantity, the amplitudes for different helicity configurations do not interfere. The generalization of equation (7.30) is just

$$\frac{d\sigma}{d\Omega} = \sum_{\lambda'} \overline{\sum_{\lambda}} |f_{\lambda'\lambda}(\theta, \varphi)|^2 \tag{10.91}$$

where $\bar{\Sigma}$ represents an average over the initial helicities. The azimuthal phase in equation (10.88) is then irrelevant, in accordance with the fact that there is no preferred z axis. However, interference effects do occur, with a corresponding φ dependence, in the case of polarization in either the initial or final state.

(ii) Two-particle States; Relativistic Scattering

The angular momentum decomposition for the scattering of two particles into two others follows very similar lines to that for scattering of a single particle off a potential. The main technicality is taking care of normalization.

Since we are considering relativistic scattering we will adopt the covariant normalization explained in Appendix E. The initial two-particle state $|\boldsymbol{p}_1\lambda_1; \boldsymbol{p}_2\lambda_2\rangle$ is therefore normalized as

$$\langle \boldsymbol{p}'_1\lambda_1; \boldsymbol{p}'_2\lambda_2|\boldsymbol{p}_1\lambda_1; \boldsymbol{p}_2\lambda_2\rangle = (2E_1)(2E_2)\delta^3(\boldsymbol{p}'_1 - \boldsymbol{p}_1)\delta^3(\boldsymbol{p}'_2 - \boldsymbol{p}_2)\delta_{\lambda'_1\lambda_1}\delta_{\lambda'_2\lambda_2} \tag{10.92}$$

Because of the conservation of the total 4-momentum $P := p_1 + p_2$ it is useful to relabel the state as $|P\hat{\boldsymbol{q}}; \lambda_1\lambda_2\rangle$, where $\boldsymbol{q}$ is the relative momentum in the centre-of-mass frame. By essentially the same manipulations used in Appendix E to evaluate the density of states factor $d\rho$, this can be recast as

$$\langle P'\hat{\boldsymbol{q}}'; \lambda_1'\lambda_2'|P\hat{\boldsymbol{q}}; \lambda_1\lambda_2\rangle = \frac{1}{\mathcal{N}^2}\,\delta^4(P' - P)\delta(\Omega' - \Omega)\delta_{\lambda_1'\lambda_1}\delta_{\lambda_2'\lambda_2} \quad (10.92')$$

where

$$\mathcal{N}^2 = \frac{\mathrm{d}\rho}{\mathrm{d}q} = \frac{q}{16\pi^2 W} \quad (10.93)$$

In defining the matrix element T_{fi} in equation (E5) we explicitly extracted the factor of $\delta^4(P' - P)$ which occurs again in (10.92'). It can therefore be considered as the matrix element of an operator $T(W)$ between states $\langle \hat{\boldsymbol{q}}'; \lambda_3\lambda_4|$ and $|\hat{\boldsymbol{q}}; \lambda_1\lambda_2\rangle$ normalized as

$$\langle \hat{\boldsymbol{q}}'; \lambda_1'\lambda_2'|\hat{\boldsymbol{q}}; \lambda_1\lambda_2\rangle = \frac{1}{\mathcal{N}^2}\,\delta(\Omega' - \Omega)\delta_{\lambda_1'\lambda_1}\delta_{\lambda_2'\lambda_2} \quad (10.94)$$

It is to these centre-of-mass states that an angular momentum decomposition can be applied, in close analogy to the procedure previously used for single-particle states. As far as phases are concerned, the state $|\hat{\boldsymbol{q}}; \lambda_1\lambda_2\rangle$ is again defined by a rotation $R(\varphi, \theta, -\varphi)$ from a standard state $|\boldsymbol{k}; \lambda_1\lambda_2\rangle$ in which the relative momentum is along the z axis:

$$|\hat{\boldsymbol{q}}; \lambda_1\lambda_2\rangle = U(R)|\boldsymbol{k}; \lambda_1\lambda_2\rangle \quad (10.95)$$

However, $|\boldsymbol{k}; \lambda_1\lambda_2\rangle$ is itself ambiguous: particle 1 is in the standard configuration with its momentum along the z axis, but particle 2 has its momentum in the negative z direction, and at $\theta = \pi$ the azimuthal angle φ is not defined. The Jacob–Wick convention is to use the rotation $R_2(\pi)$ and to multiply by the factor $(-1)^{s-\lambda}$:

$$|\boldsymbol{k}; \lambda_1\lambda_2\rangle := |\boldsymbol{k}; \lambda_1\rangle(-1)^{s_2-\lambda_2}\mathrm{e}^{-\mathrm{i}\pi J_2}|\boldsymbol{k}; \lambda_2\rangle \quad (10.96)$$

The reason for this factor is that for a particle at rest

$$(-1)^{s-\lambda}R_2(\pi)|s, s_3 = \lambda\rangle = |s, s_3 = -\lambda\rangle.$$

The appropriate definition of angular momentum states, analogous to equation (10.78), is

$$|JM; \lambda_1\lambda_2\rangle = N_J\mathcal{N}\int \mathrm{d}R\,(D^J_{M\lambda}(R))^*|\hat{\boldsymbol{q}}; \lambda_1\lambda_2\rangle \quad (10.97)$$

where $\lambda = \lambda_1 - \lambda_2$. As before, we can verify that this transforms in the correct way. Crucial to the argument is the fact that the second index λ is the eigenvalue of J_3 for the standard state $|\boldsymbol{k}; \lambda_1\lambda_2\rangle$. The factor of $\mathcal{N}$ has been included so that the normalization of the angular momentum states is just

$$\langle J'M'; \lambda_1'\lambda_2'|JM; \lambda_1\lambda_2\rangle = \delta_{J'J}\delta_{M'M}\delta_{\lambda_1'\lambda_1}\delta_{\lambda_2'\lambda_2} \quad (10.98)$$

while the overlap with the plane wave states is

$$\langle \hat{q}'; \lambda_1'\lambda_2'|JM; \lambda_1\lambda_2\rangle = N_J\mathcal{N}(D^J_{M\lambda}(R'))^*\delta_{\lambda_1'\lambda_1}\delta_{\lambda_2'\lambda_2} \tag{10.99}$$

Now consider the scattering amplitude $\langle \hat{q}'; \lambda_3\lambda_4|T(W)|\hat{q}; \lambda_1\lambda_2\rangle$, where it is convenient to take the initial direction $\hat{q}$ along the z axis. Inserting two complete sets of angular momentum states and exploiting the rotational invariance of T, which implies that

$$\langle J'M'; \lambda_3\lambda_4|T(W)|JM; \lambda_1\lambda_2\rangle = \delta_{J'J}\delta_{M'M}\langle \lambda_3\lambda_4|T^J(W)|\lambda_1\lambda_2\rangle \tag{10.100}$$

it is straightforward to obtain

$$\langle \hat{q}'; \lambda_3\lambda_4|T(W)|\hat{q}; \lambda_1\lambda_2\rangle = \sum_J N_J^2\mathcal{N}'\mathcal{N}(D^J_{\lambda\mu}(R'))^*\langle \lambda_3\lambda_4|T^J(W)|\lambda_1\lambda_2\rangle$$

where $\mu := \lambda_3 - \lambda_4$. Writing out the rotation matrix and the factors of $\mathcal{N}, \mathcal{N}'$ explicitly, this becomes

$$\langle \hat{q}'; \lambda_3\lambda_4|T(W)|\hat{q}; \lambda_1\lambda_2\rangle = \frac{4\pi W}{\sqrt{(qq')}}\sum_J (2J+1)\mathrm{e}^{\mathrm{i}(\lambda-\mu)\varphi'}\,\mathrm{d}^J_{\lambda\mu}(\theta')\langle \lambda_3\lambda_4|T^J(W)|\lambda_1\lambda_2\rangle \tag{10.101}$$

which is the relativistic analogue of equation (10.88)

The simplicity of this equation is very striking if one compares it with analogous expansions using the canonical J_3 labelling instead of helicity, and we hope that the reader who has persevered this far will forgive the length of the derivation. The crucial point was the invariance of the helicity operator under rotations. As far as unitarity is concerned one will usually need to invoke some extra conservation laws to diagonalize the matrix $\langle \lambda_3\lambda_4|T^J(W)|\lambda_1\lambda_2\rangle$ and introduce phase shifts. In pion–nucleon scattering, where the pion is spinless, the analysis is essentially identical to the discussion leading to equation (10.90).

Problems for Chapter 10

10.1 Show that the Lorentz transformation $v'_\mu = \Lambda_{\mu\nu}v^\nu$ induced by $V' = AVA^\dagger$ has

$$\Lambda_{\mu\nu}(A) = \tfrac{1}{2}\,\mathrm{Tr}\,(\tilde{\sigma}_\mu A\sigma_\nu A^\dagger)$$

thus defining a mapping $A \to \Lambda(A)$ from SL(2, $\mathbb{C}$) into SO(3, 1).

By considering a further transformation $V'' = BV'B^{\dagger}$ show that $\Lambda(BA) = \Lambda(B)\Lambda(A)$, so that the mapping is a homomorphism. Identify the kernel of the homomorphism as the centre of SU(2), i.e. $A = \pm\mathbb{1}$, thus showing that the mapping is 2:1.

10.2 Verify explicitly that $A = \exp(-\frac{1}{2}\mathrm{i}\sigma_3\theta)$ generates the rotation $v_1' = v_1 \cos\theta - v_2 \sin\theta$, $v_2' = v_2 \cos\theta + v_1 \sin\theta$.

10.3 Show that the element A corresponding to the boost taking the rest-frame momentum $\bar{p}_\mu := (m; \mathbf{0})$ to a general momentum p_μ is $A = (p\cdot\sigma/m)^{1/2}U$, where U is an arbitrary U(2) matrix.

10.4 In the Weyl representation $D^{(1/2,0)}$ of the Lorentz group $D(X) = \frac{1}{2}\boldsymbol{\sigma}$, $D(Y) = -\frac{1}{2}\mathrm{i}\boldsymbol{\sigma}$, evaluate the boost matrix $D(B) = \exp(-\frac{1}{2}\boldsymbol{\sigma}\cdot\boldsymbol{\zeta})$ and show that for $\sinh\zeta = |\boldsymbol{p}|/m$ it can be written as

$$D(B) = (p\cdot\sigma + m)/[2m(E + m)]^{1/2}$$

Verify that $D(B^2) = p\cdot\sigma/m$, in agreement with problem 10.3.

10.5 The Dirac representation is defined by $L_{\mu\nu} = \frac{1}{2}\sigma_{\mu\nu}$, where $\sigma_{\mu\nu} = \frac{1}{2}\mathrm{i}[\gamma_\mu, \gamma_\nu]$ and the γ matrices are defined by the *anticommutation* relations

$$\{\gamma_\mu, \gamma_\nu\} = 2\eta_{\mu\nu}$$

Verify that the commutation relations (10.55) are indeed satisfied.

10.6 A possible set of γ matrices (Weyl) is

$$\gamma_0 = \begin{pmatrix} 0 & \mathbb{1} \\ \mathbb{1} & 0 \end{pmatrix} \qquad \gamma_i = \begin{pmatrix} 0 & -\sigma_i \\ \sigma_i & 0 \end{pmatrix}$$

where each block is a 2×2 matrix. Show that

$$J_i = \begin{pmatrix} \frac{1}{2}\sigma_i & 0 \\ 0 & \frac{1}{2}\sigma_i \end{pmatrix} \qquad K_i = \begin{pmatrix} \frac{1}{2}\mathrm{i}\sigma_i & 0 \\ 0 & -\frac{1}{2}\mathrm{i}\sigma_i \end{pmatrix}$$

which explicitly exhibits the direct sum form $(0, \frac{1}{2}) \oplus (\frac{1}{2}, 0)$. Note that $\gamma_0 \boldsymbol{J} \gamma_0 = \boldsymbol{J}$, $\gamma_0 \boldsymbol{K} \gamma_0 = -\boldsymbol{K}$, so that γ_0 plays the role of the parity operator Π.

10.7 Show that in the Dirac representation, $D(B) = (p\cdot\gamma\gamma_0 + m)/[2m(E + m)]^{1/2}$ and $D(B^2) = p\cdot\gamma\gamma_0/m$ where B is again the boost from $\bar{p} = (m; \mathbf{0})$ to p.

10.8 Check that the definitions (Pauli)

$$\gamma_0 = \begin{pmatrix} \mathbb{1} & 0 \\ 0 & -\mathbb{1} \end{pmatrix} \qquad \gamma_i = \begin{pmatrix} 0 & \sigma_i \\ -\sigma_i & 0 \end{pmatrix}$$

give another possible choice for the γ matrices.

10.9 Using the helicity formalism show that the rest-frame decay amplitude of a particle of spin s and z component $s_3 = \lambda$ to a two-particle state $|\boldsymbol{p}, \lambda_3 : -\boldsymbol{p}, \lambda_4\rangle$ has the angular dependence $d^s_{\lambda\mu}(\theta)\exp[\mathrm{i}(\lambda - \mu)\varphi]$, where $\mu = \lambda_3 - \lambda_4$.

11

Gauge Groups

11.1 The Electromagnetic Potentials; Gauge Transformations

In MKS units, Maxwell's equations for the electric and magnetic fields take the form

$$\nabla \cdot \boldsymbol{B} = 0 \tag{11.1}$$

$$\nabla \times \boldsymbol{E} = -\frac{\partial \boldsymbol{B}}{\partial t} \tag{11.2}$$

$$\nabla \cdot \boldsymbol{D} = \rho \tag{11.3}$$

$$\nabla \times \boldsymbol{H} = \boldsymbol{j} + \frac{\partial \boldsymbol{D}}{\partial t} \tag{11.4}$$

The first two equations do not involve the sources ρ and $\boldsymbol{j}$, and for that reason are called the homogeneous equations. Their structure is such that they can be automatically satisfied by defining $\boldsymbol{B}$ and $\boldsymbol{E}$ in terms of subsidiary quantities, the electromagnetic potentials $\boldsymbol{A}$ and φ, leaving only the two homogeneous equations to worry about. Thus equation (11.1) is automatically satisfied by the *ansatz*

$$\boldsymbol{B} = \nabla \times \boldsymbol{A} \tag{11.5}$$

since $\nabla \cdot (\nabla \times \boldsymbol{A}) = 0$ identically. Substituting in equation (11.2) we obtain

$$\nabla \times \left(\boldsymbol{E} + \frac{\partial \boldsymbol{A}}{\partial t} \right) = 0$$

which in turn is automatically satisfied if we set

$$\boldsymbol{E} = -\frac{\partial \boldsymbol{A}}{\partial t} - \nabla \varphi \tag{11.6}$$

since again $\nabla \times (\nabla \varphi) = 0$ for any φ.

It is of the utmost importance to note that the electromagnetic potentials $\boldsymbol{A}$ and φ are not determined uniquely by equations (11.5) and (11.6). Thus the magnetic field $\boldsymbol{B}$ is unaltered by the addition to $\boldsymbol{A}$ of the gradient of a scalar, according to

$$\boldsymbol{A} \rightarrow \boldsymbol{A}' = \boldsymbol{A} + \boldsymbol{\nabla}\Omega \tag{11.7a}$$

The electric field $\boldsymbol{E}$ is also unchanged if *at the same time* we transform φ according to

$$\varphi \rightarrow \varphi' = \varphi - \frac{\partial \Omega}{\partial t} \tag{11.7b}$$

Equations (11.7a) and (11.7b) together constitute a *gauge transformation*.

Turning now to the inhomogeneous equations, we substitute for $\boldsymbol{H} = \boldsymbol{B}/\mu_0$ and $\boldsymbol{D} = \varepsilon_0 \boldsymbol{E}$ in equation (11.4) to obtain

$$\boldsymbol{\nabla} \times (\boldsymbol{\nabla} \times \boldsymbol{A}) = \mu_0 \boldsymbol{j} - \varepsilon_0 \mu_0 \left(\frac{\partial^2 \boldsymbol{A}}{\partial t^2} + \boldsymbol{\nabla} \frac{\partial \varphi}{\partial t} \right)$$

Using the identity $\boldsymbol{\nabla} \times (\boldsymbol{\nabla} \times \boldsymbol{A}) = \boldsymbol{\nabla}(\boldsymbol{\nabla}\cdot\boldsymbol{A}) - \boldsymbol{\nabla}^2 \boldsymbol{A}$ and identifying $\varepsilon_0 \mu_0 = 1/c^2$, this becomes

$$\left(\frac{1}{c^2} \frac{\partial^2}{\partial t^2} - \boldsymbol{\nabla}^2 \right) \boldsymbol{A} = \mu_0 \boldsymbol{j} - \boldsymbol{\nabla} \left(\boldsymbol{\nabla}\cdot\boldsymbol{A} + \frac{1}{c^2} \frac{\partial \varphi}{\partial t} \right) \tag{11.8}$$

Now we can take advantage of the gauge degree of freedom to set

$$\boldsymbol{\nabla}\cdot\boldsymbol{A} + \frac{1}{c^2} \frac{\partial \varphi}{\partial t} = 0 \tag{11.9}$$

Equation (11.9) is known as the *Lorentz gauge* condition. If it is not satisfied by a given $\boldsymbol{A}$, φ, the gauge transformation (11.7) gives an $\boldsymbol{A}'$, φ' which do satisfy it, provided that Ω is a solution of

$$\left(\frac{1}{c^2} \frac{\partial^2}{\partial t^2} - \boldsymbol{\nabla}^2 \right) \Omega = \boldsymbol{\nabla}\cdot\boldsymbol{A} + \frac{1}{c^2} \frac{\partial \varphi}{\partial t} \tag{11.10}$$

When the Lorentz condition *is* satisfied, (11.8) takes the simple form

$$\left(\frac{1}{c^2} \frac{\partial^2}{\partial t^2} - \boldsymbol{\nabla}^2 \right) \boldsymbol{A} = \mu_0 \boldsymbol{j} \tag{11.11}$$

Moreover, equation (11.3),

$$\left(-\boldsymbol{\nabla}^2 \varphi - \frac{\partial}{\partial t} (\boldsymbol{\nabla}\cdot\boldsymbol{A}) \right) = \frac{\rho}{\varepsilon_0}$$

then becomes

$$\left(\frac{1}{c^2} \frac{\partial^2}{\partial t^2} - \boldsymbol{\nabla}^2 \right) \varphi = \frac{\rho}{\varepsilon_0} \tag{11.12}$$

That is, $\boldsymbol{A}$ and φ are solutions of decoupled equations, where they are related by the wave operator $(1/c^2)\partial^2/\partial t^2 - \boldsymbol{\nabla}^2$ to their respective sources $\mu_0 \boldsymbol{j}$ and ρ/ε_0. In order for these equations to be consistent with the Lorentz condition, we need

$$\frac{\partial \rho}{\partial t} + \boldsymbol{\nabla}\cdot\boldsymbol{j} = 0 \tag{11.13}$$

which just expresses the conservation of electric charge.

11.2 Interaction with Non-relativistic Electrons

In the non-relativistic regime an electron is described quantum mechanically by a wavefunction ψ whose time development is governed by the Schrödinger equation

$$\mathrm{i}\hbar\frac{\partial \psi}{\partial t} = \hat{H}\psi \tag{11.14}$$

where we have suppressed the 2-component spinor index on ψ needed to describe the spin state of the electron.

For a purely electric field, the potential V is just $e\varphi$, so that $\hat{H} = \hat{\boldsymbol{p}}^2/2m + e\varphi$. In that case the Schrödinger equation reads

$$\mathrm{i}\hbar\frac{\partial \psi}{\partial t} = \left(-\frac{\hbar^2}{2m}\boldsymbol{\nabla}^2 + e\varphi\right)\psi \tag{11.15}$$

Now the wavefunction itself is not a physical quantity: such quantities always involve ψ in the combination $\psi^* \ldots \psi$. Thus the physics should not be changed if we go over to another wavefunction $\psi' = \mathrm{e}^{\mathrm{i}\alpha}\psi$. In group theoretic terms the transformation $\psi \rightarrow \psi'$ is just a (global) U(1) transformation. The epithet 'global' means that at the moment we are considering α to be just a constant, independent of x or t. Indeed the Schrödinger equation is invariant under such a transformation, since the resulting factor of $U(\alpha) := \mathrm{e}^{\mathrm{i}\alpha}$ can simply be cancelled from both sides.

Suppose, however, that we wish to let α be time-dependent. Then on substituting $\psi = \mathrm{e}^{-\mathrm{i}\alpha}\psi'$ we pick up an extra term proportional to $\partial\alpha/\partial t$ on the left-hand side:

$$\mathrm{i}\hbar\frac{\partial \psi'}{\partial t} + \psi'\hbar\frac{\partial \alpha}{\partial t} = \left(-\frac{h^2}{2m}\boldsymbol{\nabla}^2 + e\varphi\right)\psi' \tag{11.15'}$$

But this term can be taken over to the other side and absorbed into a redefinition of φ, namely

$$\varphi' = \varphi - (\hbar/e)\partial\alpha/\partial t \tag{11.16}$$

which is just a gauge transformation of the form of equation (11.7) with Ω identified as $\Omega = (\hbar/e)\alpha$, or equivalently $\alpha = e\Omega/\hbar$. Thus the coupled Schrödinger equation (11.15) is invariant under a t-dependent U(1) transformation, provided that we *simultaneously* perform the appropriate gauge transformation to the electromagnetic potentials, a transformation which does not affect the physical $\boldsymbol{E}$ and $\boldsymbol{B}$ fields.

What happens if we make α truly local, i.e. dependent on x as well as t? Certainly equation (11.15) is not form-invariant as it stands, because now the gradient operator acts non-trivially on α, producing extra terms. Thus

$$\nabla\psi = \nabla(e^{-i\alpha}\psi') = e^{-i\alpha}(\nabla\psi' - i\psi'\nabla\alpha) \tag{11.17}$$

Now the corresponding gauge transformation on $\boldsymbol{A}$ (equation (11.7a) with $\Omega = \hbar e/\alpha$) is

$$\boldsymbol{A} = \boldsymbol{A}' - (\hbar/e)\nabla\alpha \tag{11.18}$$

Thus we can identify the extra term $-i\psi'\nabla\alpha$ in equation (11.17) as $(-ie/\hbar)(\boldsymbol{A}' - \boldsymbol{A})\psi'$ and rearrange the equation as

$$(-i\hbar\nabla - e\boldsymbol{A})\psi = e^{-i\alpha}(-i\hbar\nabla - e\boldsymbol{A}')\psi'$$

Thus the combination $(\hat{\boldsymbol{p}} - e\boldsymbol{A})\psi$ transforms with a simple phase factor, just like ψ itself.

By repeating the argument it is not difficult to show that $(\hat{\boldsymbol{p}} - e\boldsymbol{A})^2\psi$ has the same property. Thus the Schrödinger equation would be invariant under a local U(1) transformation combined with the gauge transformations (11.7) if in the presence of electromagnetic fields it took the form

$$i\hbar\frac{\partial\psi'}{\partial t} = \left[-\frac{\hbar^2}{2m}\left(\nabla - \frac{ie\boldsymbol{A}}{\hbar}\right)^2 + e\varphi\right]\psi' \tag{11.19}$$

rather than equation (11.15) Remarkably this indeed turns out to be the correct form of the non-relativistic Hamiltonian. In terms of the momentum operator we have replaced $\hat{\boldsymbol{p}}$ by $\hat{\boldsymbol{p}} - e\boldsymbol{A}$, which is known as *minimal substitution*. As is shown in Appendix F, the same substitution is necessary in classical mechanics in order to obtain the correct equations of motion governed by the Lorentz force $\boldsymbol{F} = e(\boldsymbol{E} + \boldsymbol{v} \times \boldsymbol{B})$.

From the present approach we have almost been able to predict the necessity for the gauge potentials φ and $\boldsymbol{A}$ with their particular transformation properties by insisting that the global U(1) invariance of the free Schrödinger equation should be promoted to a local invariance. This procedure is known as 'gauging a symmetry'. More interesting features emerge when we apply the same procedure to non-Abelian symmetries such as SU(2) or SU(3). Since these symmetries have their principal relevance in high-energy physics, we will return to the subject in a later section after an excursion into relativistic field equations and quantum field theory.

At this stage we should just mention for completeness that the Dirac relativistic equation for the electron predicts its spin and gives a gyromagnetic ratio of 2, so that there is an additional term

$$H_{\text{spin}} = -(e\hbar/m)\boldsymbol{\sigma}\cdot\boldsymbol{B} \tag{11.20}$$

to be added to the non-relativistic Hamiltonian. However, since this contains the physical field $\boldsymbol{B}$ and no derivatives, it does not affect the invariance of the Schrödinger equation under local U(1) transformations.

11.3 Relativistic Formulation of Electromagnetism

Although they do not appear so, Maxwell's equations are in fact relativistically covariant; that is, they take the same form in all frames of reference connected by the Lorentz transformations of special relativity.

Let us start by looking at the equation expressing charge conservation, equation (11.13), which should indeed be true in any frame. Defining

$$j_\mu := (c\rho; \boldsymbol{j}) \tag{11.21}$$

the equation becomes simply

$$\partial\cdot j = 0 \tag{11.13'}$$

Since ∂ is a 4-vector operator we see that $j_\mu(x)$ as defined in equation (11.21) must also be a 4-vector field, i.e. $j_\mu(\Lambda x) = \Lambda_\mu{}^\nu j_\nu(x)$. Again, if we define

$$A_\mu := (\varphi/c; \boldsymbol{A}) \tag{11.22}$$

the Lorentz condition (11.9) takes the form

$$\partial \cdot A = 0 \tag{11.9'}$$

This seems to imply that $A_\mu(x)$ is a 4-vector field: however, the gauge degree of freedom means that a Lorentz transformation may be accompanied by a gauge transformation, even within the Lorentz gauge. In 4-vector language the gauge transformations of equations (11.7) become

$$A'_\mu = A_\mu - \partial_\mu \Omega \tag{11.7'}$$

and A'_μ still remains in the Lorentz gauge provided that Ω satisfies the homogeneous wave equation $\partial^2\Omega = 0$. With this proviso all the equations of §11.1 can be cast in 4-vector form. Thus the two equations of motion (11.11) and (11.12), derived in the Lorentz gauge, are just

$$\partial^2 A_\mu = \mu_0 j_\mu \tag{11.11'}$$

The right-hand side transforms as a 4-vector, as does the left-hand side, precisely because any accompanying gauge transformation must satisfy $\partial^2\Omega = 0$.

The electromagnetic fields $\boldsymbol{E}$ and $\boldsymbol{B}$ do not fit into 4-vectors, but rather occur as elements of the antisymmetric tensor

$$F_{\mu\nu} := \partial_\mu A_\nu - \partial_\nu A_\mu \tag{11.23}$$

Since $\boldsymbol{E}$ and $\boldsymbol{B}$ are physical quantities, $F_{\mu\nu}$ must be invariant under the gauge transformation (11.7′), as is indeed the case for arbitrary Ω. There are six independent elements of $F_{\mu\nu}$, which can be grouped as F_{0i} and F_{ij}. Evaluating each in turn:

$$\begin{aligned} F_{0i} &= \partial_0 A_i - \partial_i A_0 \\ &= \frac{1}{c}\frac{\partial}{\partial t} A_i + \nabla_i\left(\frac{1}{c}\varphi\right) \\ &= -\frac{1}{c} E_i \end{aligned} \tag{11.24}$$

by virtue of equation (11.6). Again

$$\begin{aligned} F_{12} &= \partial_1 A_2 - \partial_2 A_1 \\ &= -(\boldsymbol{\nabla} \times \boldsymbol{A})_3 = -B_3 \end{aligned}$$

or in general

$$B_i = -\tfrac{1}{2}\varepsilon_{ijk} F_{jk} \tag{11.25}$$

as remarked in 10.3.

11.4 Relativistic Equation of Motion for the Electron

The Schrödinger equation $(-\nabla^2/2m + V)\psi = E\psi$ is derived by the correspondence principle from the non-relativistic expression $T = \boldsymbol{p}^2/2m$ for the kinetic energy of a particle of mass m. By the correspondence principle the total energy $E = T + V$ gives rise to the Hamiltonian operator $\hat{H} = \hat{\boldsymbol{p}}^2/2m + V$, where $\hat{\boldsymbol{p}} = -i\hbar\nabla$ when acting on the wave-function $\psi(x)$. In the time-dependent formalism the eigenvalue E is replaced by the operator $i\hbar\partial\psi/\partial t$.

It is natural when considering relativistic particles to try to apply the correspondence principle to the correct relativistic expression of equation(10.28), namely $E/c = (\boldsymbol{p}^2 + m^2c^2)^{1/2}$, rather than its non-relativistic limit. However, one is immediately confronted with the problem of dealing with the square root and its associated sign ambiguity. One obvious way of attempting to circumvent this difficulty is to apply the correspondence principle, in the form $E \to i\hbar\partial/\partial t$, $\boldsymbol{p} \to -i\hbar\nabla$, to the *squared* relation $E^2/c^2 = \boldsymbol{p}^2 + m^2c^2$, to obtain the relativistic free particle equation

$$(\partial^2 + m^2)\varphi(x) = 0 \tag{11.26}$$

in natural units. This equation is known as the Klein–Gordon equation.

However, we have not really resolved the problem of negative energy solutions, and in fact there are insurmountable difficulties in treating equation (11.26) as a single-particle wave equation. The fundamental problem is that at relativistic energies the equivalence between energy and mass means that particles can be created and destroyed, subject to any selection rules that may be operative. Thus a single-particle equation is actually inconsistent with special relativity. The way out of this dilemma is known as second quantization, whereby $\varphi(x)$ is promoted to the status of an operator capable of creating or annihilating particles. In that context the Klein–Gordon equation does make sense. In the form (11.26) it just describes a free field with no interactions, but these can be incorporated by adding a term $-\partial V(\varphi)/\partial\varphi$ on the right-hand side.

The other, very ingenious way of circumventing the quadratic nature of the equation $p^2 - m^2 = 0$ is to try to factorize the left-hand side. Obviously for any scalar quantity a one can factorize $a^2 - m^2$ as $(a - m)(a + m)$, but p is not a scalar, but rather a 4-vector, and what the equation really means is $p^\mu p_\mu - m^2 = 0$. However, suppose one multiplies p_μ by a set of four matrices γ^μ to form $\gamma{\cdot}p$ and then requires that

$$p^2 - m^2 = (\gamma \cdot p - m)(\gamma \cdot p + m) \tag{11.27}$$

i.e.

$$\begin{aligned} p^2 - m^2 &= \gamma_\mu \gamma_\nu p^\mu p^\nu - m^2 \\ &= \tfrac{1}{2}(\gamma_\mu \gamma_\nu + \gamma_\nu \gamma_\mu) p^\mu p^\nu - m^2 \end{aligned}$$

using the symmetry of $p^\mu p^\nu$. So the factorization of equation (11.27) can be achieved provided the γ matrices satisfy the anticommutation relations

$$\{\gamma_\mu, \gamma_\nu\} = 2\eta_{\mu\nu} \tag{11.28}$$

defining what is known as a *Clifford algebra*.

It is indeed possible to find γ_μ which satisfy this condition. It turns out that they must be at least 4×4 matrices. A possible set are

$$\gamma_0 = \begin{pmatrix} \mathbb{1} & 0 \\ 0 & -\mathbb{1} \end{pmatrix} \qquad \gamma_i = \begin{pmatrix} 0 & \sigma_i \\ -\sigma_i & 0 \end{pmatrix} \tag{11.29}$$

(see problem 10.8)which necessarily act on a 4-component wavefunction ψ_α. Given the factorization (11.27) one can use only one of the factors to give a linear wave equation. With the choice of γ matrices adopted in equation (11.29) the appropriate factor is the first one, $\gamma \cdot p - m$. According to the correspondence principle $p_\mu \to \mathrm{i}\partial_\mu$ we then obtain the free *Dirac equation*

$$(\mathrm{i}\gamma \cdot \partial - m)\psi = 0 \tag{11.30}$$

This can be cast in Hamiltonian form by multiplying through by $\beta := \gamma_0$ to obtain

$$\begin{aligned} \mathrm{i}\frac{\partial \psi}{\partial t} &= (-\mathrm{i}\gamma_0 \boldsymbol{\gamma} \cdot \boldsymbol{\nabla} + m)\psi \\ &= (-\mathrm{i}\boldsymbol{\alpha} \cdot \boldsymbol{\nabla} + m)\psi \end{aligned}$$

where $\boldsymbol{\alpha} = \beta\boldsymbol{\gamma}$. Thus we can identify the Hamiltonian operator as $\hat{H} = \boldsymbol{\alpha} \cdot \hat{\boldsymbol{p}} + m$.

As far as group theory is concerned, the wavefunction ψ carries the representation $(\frac{1}{2}, 0) + (0, \frac{1}{2})$ of the homogeneous Lorentz group, whereby the Lorentz generators $L_{\mu\nu}$ are represented by $\frac{1}{2}\sigma_{\mu\nu}$, with $\sigma_{\mu\nu} = \frac{1}{2}\mathrm{i}[\gamma_\mu, \gamma_\nu]$, as the dedicated reader will have shown in problem10.6. In the non-relativistic limit ψ effectively reduces to a 2-component object, an ordinary SU(2) spinor, if one adopts the Pauli set of γ matrices given in equation (11.29). For suppose we have a momentum eigenstate $\psi_\alpha(x) = u_\alpha(p)\mathrm{e}^{-\mathrm{i}p \cdot x}$. Separating the 4-component object $u(p)$

into the two upper and the two lower components, which we call $u_L(p)$ and $u_S(p)$ respectively, for reasons that will soon become apparent, the free Dirac equation $(\gamma \cdot p - m)u = 0$ becomes

$$\begin{pmatrix} E - m & -\boldsymbol{\sigma}\cdot\boldsymbol{p} \\ \boldsymbol{\sigma}\cdot\boldsymbol{p} & -E - m \end{pmatrix} \begin{pmatrix} u_L \\ u_S \end{pmatrix} = 0$$

i.e.

$$(E - m)u_L = \boldsymbol{\sigma}\cdot\boldsymbol{p}\; u_S$$

and (11.31)

$$(E + m)u_S = \boldsymbol{\sigma}\cdot\boldsymbol{p}\; u_L$$

From the second of these equations, the ratio of the magnitude of the lower components to those of the upper components is of order $|\boldsymbol{p}|/(E + m)$, which is small in the non-relativistic limit. Hence the nomenclature L and S for 'large' and 'small' respectively. In the rest-frame the lower components vanish and u becomes just a 2-component spinor $\varphi := u_L$, an eigenstate of the projection operator $\frac{1}{2}(1 + \gamma_0)$, for which the rotation generators are just $\frac{1}{2}\sigma_i$.

Again, however, the Dirac trick has not really resolved the problem of negative energies. In the above we have taken E to be the positive square root $(\boldsymbol{p}^2 + m^2)^{1/2}$, but if one took $E = -(\boldsymbol{p}^2 + m^2)^{1/2}$ the role of upper and lower components would be reversed. Although the two transform separately under rotations, they are connected by boosts, for which the generators K are

$$\begin{pmatrix} 0 & \frac{1}{2}\mathrm{i}\boldsymbol{\sigma} \\ \frac{1}{2}\mathrm{i}\boldsymbol{\sigma} & 0 \end{pmatrix}$$

The fact that electrons are fermions satisfying Fermi statistics allowed Dirac to make a further ingenious proposal in order to exorcize the negative energy states, namely that the physical vacuum consisted of an infinite 'sea' of occupied negative energy states with $E < -m$. Then a positive energy particle could not make a transition to an occupied negative energy state because of the exclusion principle! However, given enough energy a negative energy state could jump to a positive energy state with $E > m$, leaving behind a 'hole', which would behave like a positive energy state with opposite charge to the electron. This process, the creation of electron–positron pairs, does indeed occur, but cannot be adequately described in the context of a single-particle wave equation. That is, the problem can again only be properly resolved by second quantization, whereby ψ becomes an operator capable of annihilating electrons or creating positrons. However, the single-particle equation does have a limited domain of applicability, particularly in atomic physics. There the typical kinetic energies are very much less than $2mc^2 \approx 1$ MeV, the threshold for pair creation.

Let us take this opportunity to review these and other wave equations in terms of the representations of the Lorentz group, classified in the previous chapter as (j_1, j_2). The Klein–Gordon equation, equation (11.26), is the equation usually adopted for a spinless particle, represented by the scalar field $\varphi(x)$, while the Dirac equation corresponds to the 4-component reducible representation $(0, \frac{1}{2}) \oplus (\frac{1}{2}, 0)$. That reducibility is not apparent in the Pauli representation of the γ matrices, but becomes so in the Weyl representation of problem 10.6. Indeed, the two representations can be isolated by application of the projection operators $\frac{1}{2}(1 \pm i\gamma_5)$, where $\gamma_5 = \gamma_0\gamma_1\gamma_2\gamma_3$, satisfying $\gamma_5^2 = -1$ and $\{\gamma_5, \gamma_\mu\} = 0$. They are linked together in the massive Dirac equation, but the field operator of a fermion of zero mass can be taken as an eigenstate of γ_5 consistent with the massless equation $\gamma\cdot\partial\psi = 0$. Such 2-component representations are relevant for the electron neutrino and its antiparticle, if indeed they are truly massless.

A massive spin-1 particle can be represented by a vector field operator $\varphi_\mu(x)$ transforming according to the representation $(\frac{1}{2}, \frac{1}{2})$. That means that it has 4 degrees of freedom, which does not agree with the 3 degrees of freedom we expect for spin 1. It must therefore be constrained by an auxiliary condition, $\partial^\mu\varphi_\mu = 0$, in addition to the Klein–Gordon equation. For a momentum eigenfunction of the form $\varphi_\mu(x) = \xi_\mu(p)e^{-ip\cdot x}$ this means that $p\cdot\xi = 0$, which reduces in the rest-frame to $\xi_0 = 0$.

The peculiar nature of the gauge field A_μ is connected with the very different nature of the representation theory of the Poincaré group for massless, as opposed to massive, particles. As we have seen in the previous chapter, the relevant Lorentz generators are then J_3, $L_1 := J_1 + K_2$ and $L_2 := J_2 - K_1$, rather than J_3, J_1 and J_2. Moreover, physical states with canonical momentum $\omega_\mu = \omega(1;\, 0,\, 0,\, 1)$ must be annihilated by the operators L_1 and L_2. Correspondingly the *polarization vector* $\xi_\mu(\omega)$ should satisfy $(L_1)_\mu{}^\nu\xi_\nu = (L_2)_\mu{}^\nu\xi_\nu = 0$. Now in the $(\frac{1}{2}, \frac{1}{2})$ representation $(J_1)_{\mu\nu} = i(\eta_{\mu2}\eta_{\nu3} - \eta_{\mu3}\eta_{\nu2})$ and $(K_2)_{\mu\nu} = i(\eta_{\mu0}\eta_{\nu2} - \eta_{\mu2}\eta_{\nu0})$, so that

$$\begin{aligned}(L_1)_\mu{}^\nu\xi_\nu &= i(-\eta_{\mu3}\xi_2 + \eta_{\mu0}\xi_2)\\ &= i\xi_2\omega_\mu/\omega \end{aligned} \tag{11.32}$$

where we have used $\xi_3 = \xi_0$, from $\omega\cdot\xi = 0$. Similarly $(L_2)_\mu{}^\nu\xi_\nu = -i\xi_1\omega_\mu/\omega$. In a general frame the corresponding expressions are proportional to p_μ, instead of being zero, as the group theory requires. This is another aspect of the fact that A_μ does not really transform as a 4-vector, but may have an accompanying gauge transformation. The situation can be saved if in any physical result $\xi_\mu(p)$ is contracted with an object $M_\mu(p)$ which itself satisfies $p \cdot M = 0$. This is in fact the case, and is a consequence of gauge invariance and the conservation of the electromagnetic current j_μ.

11.5 Quantum Fields and Their Interactions

As mentioned in the previous section, the problems associated with the negative energy solutions of the Klein–Gordon and Dirac equations can be overcome within the framework of second quantization. This takes account of the fact that relativity inevitably implies the possibility of the creation and annihilation of particles, so that one needs to go beyond a single-particle approach. Accordingly, $\varphi(x)$ and $\psi(x)$ are taken as *operators* which can annihilate particles or create their antiparticles.

It would take us too far afield to develop the whole framework of quantum field theory. We are primarily interested in the group theoretical aspects, and in particular in gauge groups, but in order to appreciate the structure of the field theories which arise by the imposition of a local group invariance we will need to be familiar with the Lagrangian formulation of relativistic field theory, which is explained in Appendix F, and to recognize the interactions implied by a particular Lagrangian.

As far as the first item is concerned, we learn from Appendix F that for a general field $\varphi_\alpha(x)$ the Lagrangian L is written as the spatial integral of a Lagrange density $\mathscr{L}$, which is a function of φ_α and its space–time derivatives $\partial_\mu \varphi$:

$$L = \int \mathrm{d}^3 x \mathscr{L}(\varphi_\alpha(x), \partial_\mu \varphi_\alpha(x)) \tag{11.33}$$

The Euler–Lagrange equations of motion are then

$$\partial_\mu\left(\frac{\partial \mathscr{L}}{\partial(\partial_\mu \varphi_\alpha)}\right) = \frac{\partial \mathscr{L}}{\partial \varphi_\alpha} \tag{11.34}$$

We now discuss in turn the form of the appropriate Lagrangians for the scalar, Dirac and Maxwell fields, both for the free fields and with the incorporation of interactions.

Scalar Field φ

A neutral spinless particle is described by a Hermitian scalar field φ. The canonical Lagrange density for the free field is

$$\mathcal{L}_0 = \tfrac{1}{2}(\partial_\mu\varphi)^2 - \tfrac{1}{2}m^2\varphi^2 \tag{11.35}$$

which, as is readily verified, leads to the free Klein–Gordon equation. Actually, any multiple of (11.35) would give the same equation of motion: the normalization is fixed by going over to the Hamiltonian formalism and expressing H in terms of the Fourier components of φ:

$$\varphi = \int (\mathrm{d}\boldsymbol{p})(a(\boldsymbol{p})\mathrm{e}^{-\mathrm{i}p\cdot x} + a^\dagger(\boldsymbol{p})\mathrm{e}^{\mathrm{i}p\cdot x}) \tag{11.36}$$

where $(\mathrm{d}\boldsymbol{p})$ is the invariant integration measure of equation (D1). Up to an infinite (!) constant one then obtains

$$H = \int (\mathrm{d}\boldsymbol{p})E(\boldsymbol{p})a^\dagger(\boldsymbol{p})a(\boldsymbol{p}) \tag{11.37}$$

The $a^\dagger(\boldsymbol{p})$ and $a(\boldsymbol{p})$ have commutation relations just like the raising and lowering operators of the simple harmonic oscillator, and in quantum field theory are interpreted as particle creation and annihilation operators respectively. Thus H can be recognized as the integral over all momenta of the energy times the number operator appropriate to that momentum. The infinite constant we have discarded is the integral of the ground state energy of all the oscillators.

If we wish to include self-interactions of the φ field, a possible term to add would be

$$\mathcal{L}_\mathrm{I} = -\lambda\varphi^4 \tag{11.38}$$

Applying the Euler–Lagrange equation to $\mathcal{L} = \mathcal{L}_0 + \mathcal{L}_\mathrm{I}$ then gives the equation

$$(\partial^2 + m^2)\varphi = -4\lambda\varphi^3 \tag{11.39}$$

This interacting theory is normally treated by perturbation theory in the coupling constant λ. In the covariant perturbation theory developed by Feynman, scattering amplitudes are calculated by means of momentum space Feynman diagrams, whose ingredients are 4-point vertices, arising from $\mathcal{L}_\mathrm{I}$, connected together by propagators, which are essentially the inverse of the operator appearing on the left-hand side of equation (11.39), which in momentum space becomes $1/(p^2 - m^2)$. Figure 11.1 shows one of the lowest-order Feynman diagrams for the elastic scattering of two particles.

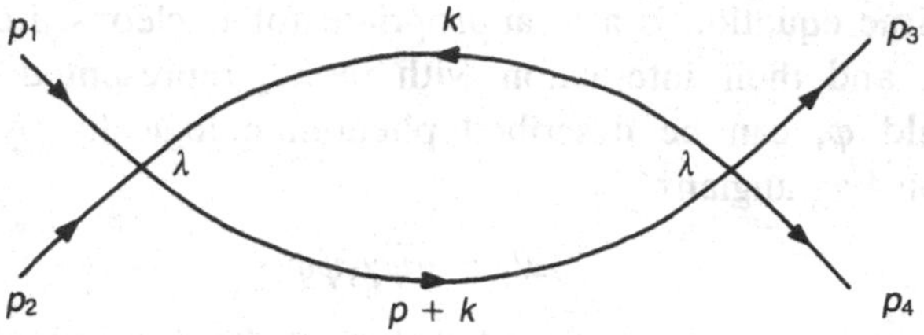

Figure 11.1 Lowest-order Feynman diagram for the scattering of two scalar particles with $\lambda\varphi^4$ interaction.

Dirac Field

The standard Lagrange density which gives rise to the free Dirac equation for the quantum field $\psi_\alpha(x)$ is

$$\mathcal{L}_0 = \overline{\psi}(\mathrm{i}\gamma\cdot\partial - m)\psi \tag{11.40}$$

where $\overline{\psi} := \psi^\dagger\gamma_0$. The Dirac equation $(\mathrm{i}\gamma\cdot\partial - m)\psi = 0$ is indeed just the Euler–Lagrange equation for $\overline{\psi}$, where ψ and $\overline{\psi}$ are regarded as independent. Again, however, the normalization of (11.40) is determined by the resulting Hamiltonian expressed as an integral over the Fourier components of ψ. The momentum space decomposition of the field ψ_α now involves electron annihilation operators $b(\boldsymbol{p}, \lambda)$ and positron creation operators $d^\dagger(\boldsymbol{p}, \lambda)$:

$$\psi_\alpha(x) = \int (\mathrm{d}\boldsymbol{p}) \sum_\lambda (u^{(\lambda)}(\boldsymbol{p})b(\boldsymbol{p}, \lambda)\mathrm{e}^{-\mathrm{i}p\cdot x} + v^{(\lambda)}(\boldsymbol{p})d^\dagger(\boldsymbol{p}, \lambda)\mathrm{e}^{\mathrm{i}p\cdot x}) \tag{11.41}$$

Here $u(\boldsymbol{p})$ is a plane wave solution of the momentum space Dirac equation $(\gamma\cdot p - m)u(\boldsymbol{p}) = 0$. Referring to problem 10.7, it can be expressed in terms of a 2-component rest-frame Pauli spinor φ as $u(\boldsymbol{p}) = (\gamma\cdot p + m)\varphi$. The superscript λ is inherited from φ and refers to the helicity. The other spinor $v^{(\lambda)}(\boldsymbol{p})$ is a solution of the negative energy equation $(\gamma\cdot p + m)v^{(\lambda)}(\boldsymbol{p}) = 0$ and can be taken proportional to $\gamma_5 u^{(-\lambda)}(\boldsymbol{p})$. After a certain amount of manipulation one obtains

$$H = \int (\mathrm{d}\boldsymbol{p})E(\boldsymbol{p}) \sum_\lambda (b^\dagger(\boldsymbol{p}, \lambda)b(\boldsymbol{p}, \lambda) + d^\dagger(\boldsymbol{p}, \lambda)d(\boldsymbol{p}, \lambda)) \tag{11.42}$$

again up to an infinite constant, *provided* that the b's and d's are taken to satisfy *anticommutation* relations, corresponding to the fact that electrons satisfy Fermi–Dirac rather than Bose–Einstein statistics. Then H can be recognized as the integral over all momenta of the energy times the sum of the number operators for electrons and positrons.

The Dirac equation is also appropriate for nucleons, i.e. protons and neutrons, and their interaction with pions, represented by a pseudo-scalar field φ, can be described phenomenologically by the Yukawa interaction Lagrangian†

$$\mathcal{L}_{\mathrm{I}} = g\bar{\psi}\gamma_5\psi\varphi \tag{11.43}$$

Here the γ_5 is necessary in order that $\mathcal{L}_{\mathrm{I}}$ be a true scalar under the parity transformation (see problem 11.6). The Feynman rules for the complete Lagrangian

$$\mathcal{L} = \tfrac{1}{2}[(\partial\varphi)^2 - \mu^2\varphi^2] + \bar{\psi}(\mathrm{i}\gamma\cdot\partial - m)\psi + g\bar{\psi}\gamma_5\psi\varphi \tag{11.44}$$

involve the pion propagator $1/(p^2 - \mu^2)$, the nucleon propagator $1/(\gamma\cdot p - m)$, which really means $(\gamma\cdot p + m)/(p^2 - m^2)$, and the 3-point vertex proportional to g. Figure 11.2 shows the lowest-order diagram for neutron–proton scattering. However, calculations with the Lagrangian of (11.44) face the fundamental problem that for this strong interaction the dimensionless coupling constant $g^2/4\pi$ is not small, so that perturbation theory is not a useful method of calculation.

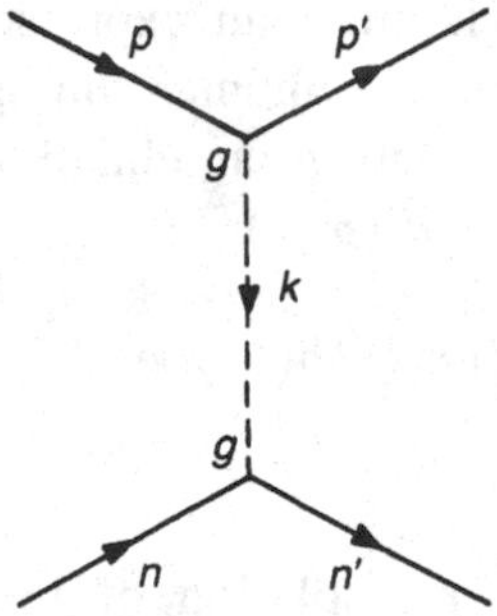

Figure 11.2 Lowest-order diagram for neutron–proton scattering via the exchange of a virtual pion.

Maxwell Field A_μ

As we have already seen at the end of §11.4, the use of the gauge field A_μ to describe the photon leads to special problems which have to be treated with some care. Related problems occur again in setting up the Lagrange formalism for photons.

One's first guess at a Lagrangian would be the gauge-invariant object

$$\mathcal{L}_0 = -\tfrac{1}{4}F_{\mu\nu}F^{\mu\nu} \tag{11.45}$$

†At this point we give up the unequal struggle to distinguish between the Lagrangian (L) and the Lagrange density ($\mathcal{L}$).

which leads to the Euler–Lagrange equation

$$\partial^2 A_\nu - \partial_\nu(\partial \cdot A) = 0 \tag{11.46}$$

Classically one could impose the Lorentz condition $\partial \cdot A = 0$, and then (11.46) would be the desired free equation of motion for A_μ. However, it turns out that it is not consistent in the second-quantized theory to impose $\partial \cdot A = 0$ as an operator condition. In fact, in order to quantize the theory consistently one is forced to break the gauge invariance by adding to $\mathscr{L}_0$ a so-called gauge-fixing term of the form

$$\mathscr{L}_{\text{gf}} = (\partial \cdot A)^2/2\xi \tag{11.47}$$

although this still leaves a residual gauge invariance, of the form $A_\mu \to A_\mu - \partial_\mu \Omega$, where $\partial^2 \Omega = 0$. The choice of the parameter ξ is arbitrary and can be shown not to affect the calculation of any physical quantity. A convenient choice is $\xi = 1$, which leads to the Fermi Lagrange density

$$\mathscr{L}_{\text{F}} = -\tfrac{1}{2}(\partial_\mu A_\nu)(\partial^\mu A^\nu) \tag{11.48}$$

up to a total derivative. The Euler–Lagrange equation derived from (11.48) is just $\partial^2 A_\mu = 0$, as required, and the photon propagator $D_{\mu\nu}(k)$ is $-\eta_{\mu\nu}/k^2$.

This does not quite exhaust the special measures we have to take in order to quantize the photon field successfully. Going over to the Hamiltonian formalism and expressing A_μ in terms of creation and annihilation operators as

$$A_\mu(x) = \int (\mathrm{d}\boldsymbol{k}) \sum_l \xi_\mu^{(l)}(\boldsymbol{k})(a(\boldsymbol{k}, l)\mathrm{e}^{-ik\cdot x} + a^\dagger(\boldsymbol{k}, l)\mathrm{e}^{ik\cdot x}) \tag{11.49}$$

we obtain

$$H = -\int (\mathrm{d}\boldsymbol{k}) \sum_l \eta_{ll} a^\dagger(\boldsymbol{k}, l) a(\boldsymbol{k}, l) \tag{11.50}$$

up to the usual infinite constant. In (11.49) the $\xi_\mu^{(l)}$ $(l = 0 \ldots 3)$ are a set of real polarization vectors normalized to $\xi^{(l)} \cdot \xi^{(l')} = \eta_{ll'}$. $\xi^{(1)}$ and $\xi^{(2)}$ are taken as transverse, $k \cdot \xi^{(1)} = k \cdot \xi^{(2)} = 0$, and correspond to the physical degrees of linear polarization, while the unphysical timelike and longitudinal polarization vectors may be taken as $\xi^{(0)} = (1; \mathbf{0})$ and $\xi^{(3)} = (0; \hat{\boldsymbol{k}})$. In order for H to make sense, the contributions over the unphysical degrees of freedom must cancel. This can be ensured by imposing the subsidiary condition

$$(a(\boldsymbol{k}, 0) - a(\boldsymbol{k}, 3))|\text{phys}\rangle = 0 \tag{11.51}$$

on all physical states. The latter is equivalent to the condition $(\partial \cdot A^{(+)})|\text{phys}\rangle = 0$, where $A^{(+)}$ denotes the positive frequency part of A. This is the nearest we can get to imposing $\partial \cdot A = 0$.

The interactions of the photon field with charged matter can be derived by the application of the gauge principle, as described in the following section.

11.6 Gauge Field Theories

We are finally in a position to describe the application of the gauge principle to relativistic field theories. The case of quantum electrodynamics, involving a local U(1) invariance, is of long standing, but the application to non-Abelian groups such as SU(2) or SU(3) is more recent, and results in theories with some remarkable properties.

(i) U(1)—Quantum Electrodynamics

Quantum electrodynamics (QED) is the theory which describes the relativistic interactions of photons and electrons. In contrast to the discussions of earlier sections of this chapter, both the electromagnetic and Dirac fields are now quantized, enabling the theory to encompass processes involving the creation and annihilation of photons and electrons.

With an eye to the non-Abelian case it will prove very useful to construct the Lagrangian for QED by applying the gauge principle to the free Dirac Lagrangian, although this is not the way that it was originally derived. The Lagrange density of equation (11.40),

$$\mathcal{L}_0(\psi, \bar{\psi}) = \bar{\psi}(\mathrm{i}\gamma \cdot \partial - m)\psi \tag{11.40}$$

is invariant under the global U(1) transformation $\psi \to \psi' = \mathrm{e}^{\mathrm{i}e\Lambda}\psi$. However, if we try to make this a local transformation, with Λ a function of x, the derivative acting on Λ produces an extra term $-e\bar{\psi}\gamma \cdot (\partial\Lambda)\psi$. As will now be familiar, the way to remedy this is to introduce an additional field A_μ, with the transformation $A_\mu \to A'_\mu = A_\mu + \partial_\mu \Lambda$, and change the ordinary derivative in (11.40) to the covariant derivative $D_\mu := \partial_\mu - \mathrm{i}eA_\mu$, to produce

$$\mathcal{L}' = \bar{\psi}(\mathrm{i}\gamma \cdot D - m)\psi \tag{11.40'}$$

invariant under the combined transformations

$$\begin{aligned} \psi(x) &\to e^{ie\Lambda(x)}\psi(x) \\ A_\mu(x) &\to A_\mu(x) + \partial_\mu\Lambda(x) \end{aligned} \tag{11.52}$$

as you should check.

Equation (11.40′) is no longer a free Lagrangian, but contains the interaction term $\mathscr{L}_\mathrm{I} = e\bar{\psi}\gamma\cdot A\psi$, which in Feynman diagram language gives rise to the photon–electron vertex. In some contexts, when one is dealing with a macroscopic external electromagnetic field, as in the earlier sections of this chapter, that is as far as one need go. However, if we are to treat the electromagnetic field as a dynamical quantum field we need to complete the Lagrangian by adding the appropriate free Lagrangian for the field A_μ we have introduced. The requirement of gauge invariance leads us to

$$\mathscr{L}_0(A) = -\tfrac{1}{4}F_{\mu\nu}F^{\mu\nu} \tag{11.53}$$

with the normalization fixed by the considerations of(11.5). From those considerations we also know that an additional gauge-fixing term $\mathscr{L}_\mathrm{gf} = -(\partial\cdot A)^2/2\xi$ has to be added in order to quantize the theory successfully. Altogether, then, we arrive at the full Lagrangian of QED:

$$\mathscr{L}_\mathrm{QED} = \bar{\psi}(\mathrm{i}\gamma\cdot D - m)\psi - \tfrac{1}{4}F_{\mu\nu}F^{\mu\nu} - (\partial\cdot A)^2/2\xi \tag{11.54}$$

by insisting that the original global U(1) invariance of the free Dirac Lagrangian should become a local one. It is this procedure which is known as 'gauging' a symmetry.

(ii) SU(2)

As the first example of a non-Abelian gauge symmetry let us consider the case of SU(2). That group is relevant for a Dirac Lagrangian of the form

$$\mathscr{L}_0 = \bar{\psi}(\mathrm{i}\gamma\cdot\partial - M)\psi \tag{11.55}$$

where ψ is a doublet of Dirac fields, $\psi = (\psi_1, \psi_2)$, and $M = m\delta_{ab}$ is a diagonal matrix in the (1–2) space. To be more explicit, (11.55) reads

$$\mathscr{L}_0 = \sum_a \bar{\psi}_a(\mathrm{i}\gamma\cdot\partial - m)\psi_a$$

which is invariant under the transformation $\psi_a \to U_a^b\psi_b$, where U is a constant SU(2) matrix. To the extent that their masses are equal, such a Lagrangian is applicable to the u and d quarks discussed in Chapter 8.

The symmetry in that case would be the SU(2) of isospin.

If we wish to promote the global symmetry to a local one by letting U become a function of x, we will in the first instance produce extra terms, which will have to be cancelled by the introduction of gauge fields. In analogy with the Abelian case, it is convenient to write U in exponential form:

$$U = \mathrm{e}^{\mathrm{i}g\boldsymbol{T}\cdot\boldsymbol{\Lambda}} \tag{11.56}$$

where the $\boldsymbol{T}$ are the three Hermitian generators, which in the fundamental representation are essentially the Pauli matrices: $\boldsymbol{T} = \frac{1}{2}\boldsymbol{\sigma}$. When we now make the substitution $\psi \to U\psi$, the derivative acting on U will produce the additional term $-g\bar{\psi}U^{-1}\gamma^\mu(\partial_\mu\boldsymbol{\Lambda})\cdot\boldsymbol{T}\psi$. In order to cancel this term, which depends on the three independent components of $\boldsymbol{\Lambda}$, we need to introduce *three* additional gauge fields, $\boldsymbol{W}_\mu$ say, one for each generator. So, in analogy with the electromagnetic case we introduce the covariant derivative

$$D_\mu := \partial_\mu - \mathrm{i}g\boldsymbol{T}\cdot\boldsymbol{W}_\mu \tag{11.57}$$

and go over to the new Lagrangian

$$\mathcal{L}' = \bar{\psi}(\mathrm{i}\gamma\cdot D - M)\psi \tag{11.58}$$

We now need to establish the transformation property of the $\boldsymbol{W}_\mu$ in order that (11.58) should be invariant when $\psi \to U(x)\psi$. What is required is that D_μ should transform without any additional terms:

$$U^{-1}D'_\mu U = D_\mu \tag{11.59}$$

Let us first confine ourselves to an infinitesimal transformation, so that U can be expanded as $1 + \mathrm{i}g\boldsymbol{T}\cdot\boldsymbol{\Lambda}$. Then (11.59) reads

$$(1 - \mathrm{i}g\boldsymbol{T}\cdot\boldsymbol{\Lambda})(\partial_\mu - \mathrm{i}g\boldsymbol{T}\cdot\boldsymbol{W}'_\mu)(1 + \mathrm{i}g\boldsymbol{T}\cdot\boldsymbol{\Lambda}) = \partial_\mu - \mathrm{i}g\boldsymbol{T}\cdot\boldsymbol{W}_\mu$$

giving

$$\mathrm{i}g\boldsymbol{T}\cdot\partial_\mu\boldsymbol{\Lambda} - \mathrm{i}g\boldsymbol{T}\cdot\boldsymbol{W}'_\mu - g^2[\boldsymbol{T}\cdot\boldsymbol{\Lambda}, \boldsymbol{T}\cdot\boldsymbol{W}'_\mu] = \mathrm{i}g\boldsymbol{T}\cdot\boldsymbol{W}_\mu$$

Working to first order in $\boldsymbol{\Lambda}$, we can replace $\boldsymbol{W}'_\mu$ by $\boldsymbol{W}_\mu$ in the commutator term on the LHS. In that term we use the commutation relations of the T_i, namely $[T_i, T_j] = \mathrm{i}\varepsilon_{ijk}T_k$, to write $[\boldsymbol{T}\cdot\boldsymbol{\Lambda}, \boldsymbol{T}\cdot\boldsymbol{W}_\mu] = \mathrm{i}\varepsilon_{ijk}\Lambda_i W_{j\mu} T_k = \mathrm{i}\boldsymbol{T}\cdot(\boldsymbol{\Lambda}\times\boldsymbol{W}_\mu)$. Then a factor of $\boldsymbol{T}\cdot$ can be extracted from the entire equation to give

$$\boldsymbol{W}'_\mu = \boldsymbol{W}_\mu + \partial_\mu\boldsymbol{\Lambda} - g\boldsymbol{\Lambda}\times\boldsymbol{W}_\mu \tag{11.60}$$

Actually this equation should not surprise us. The term $\partial_\mu\boldsymbol{\Lambda}$ is a gauge piece, closely analogous to the corresponding term in QED. However, the last term is just the normal transformation of an isovector, which would occur even if $\boldsymbol{\Lambda}$ were constant. The finite version of (11.60) can be cast in the very elegant form

$$W'_\mu = \frac{\mathrm{i}}{g} U D_\mu U^{-1} \tag{11.61}$$

where W_μ denotes the 2×2 matrix $\boldsymbol{T}\cdot\boldsymbol{W}_\mu$.

Having introduced the gauge fields $\boldsymbol{W}_\mu$ in an interaction term, it remains to construct their free Lagrangian, which must be invariant under (11.60) or (11.61). The appropriate generalization of $F_{\mu\nu}$ turns out to be

$$\boldsymbol{F}_{\mu\nu} = \partial_\mu \boldsymbol{W}_\nu - \partial_\nu \boldsymbol{W}_\mu + g\boldsymbol{W}_\mu \times \boldsymbol{W}_\nu \tag{11.62}$$

whose infinitesimal change under $\boldsymbol{W}_\mu \to \boldsymbol{W}'_\mu$ is

$$\begin{aligned}\delta\boldsymbol{F}_{\mu\nu} &= \partial_\mu(\partial_\nu\boldsymbol{\Lambda} - g\boldsymbol{\Lambda} \times \boldsymbol{W}_\nu) + g(\partial_\mu\boldsymbol{\Lambda} - g\boldsymbol{\Lambda} \times \boldsymbol{W}_\mu) \times \boldsymbol{W}_\nu \\ &\quad - \partial_\nu(\partial_\mu\boldsymbol{\Lambda} - g\boldsymbol{\Lambda} \times \boldsymbol{W}_\mu) + g\boldsymbol{W}_\mu \times (\partial_\nu\boldsymbol{\Lambda} - g\boldsymbol{\Lambda} \times \boldsymbol{W}_\nu) \\ &= -\, g\boldsymbol{\Lambda} \times (\partial_\mu \boldsymbol{W}_\nu - \partial_\nu \boldsymbol{W}_\mu + g\boldsymbol{W}_\mu \times \boldsymbol{W}_\nu)\end{aligned}$$

after various cancellations and use of the Jacobi identity

$$\boldsymbol{A} \times (\boldsymbol{B} \times \boldsymbol{C}) = (\boldsymbol{A} \times \boldsymbol{B}) \times \boldsymbol{C} + \boldsymbol{B} \times (\boldsymbol{A} \times \boldsymbol{C}) \tag{11.63}$$

with $\boldsymbol{A} = \boldsymbol{\Lambda}$, $\boldsymbol{B} = \boldsymbol{W}_\mu$ and $\boldsymbol{C} = \boldsymbol{W}_\nu$. In other words

$$\delta\boldsymbol{F}_{\mu\nu} = -g\boldsymbol{\Lambda} \times \boldsymbol{F}_{\mu\nu} \tag{11.64}$$

which is the transformation of a true isovector, with no gauge terms. A more elegant way of obtaining (11.62) is to write $F_{\mu\nu} = \boldsymbol{T}\cdot\boldsymbol{F}_{\mu\nu}$ and note that

$$F_{\mu\nu} = \frac{\mathrm{i}}{g}[D_\mu, D_\nu] \tag{11.65}$$

which transforms as $UF_{\mu\nu}U^{-1}$ by virtue of (11.59).

In any case the point is that $\boldsymbol{F}_{\mu\nu}\cdot\boldsymbol{F}^{\mu\nu}$ is a scalar, and the required Lagrangian for the $\boldsymbol{W}_\mu$ is

$$\mathscr{L}(\boldsymbol{W}_\mu) = -\tfrac{1}{4}\boldsymbol{F}_{\mu\nu}\cdot\boldsymbol{F}^{\mu\nu} \tag{11.66}$$

to which a gauge-fixing term of the form $\mathscr{L}_{\mathrm{gf}} = -(\partial\cdot W)^2/2\xi$ must also be added. Actually there are yet more terms which have to be added in order to ensure correct quantization, but these are technicalities which need not concern us here. The crucial feature of (11.66) as compared with (11.53) is that it is not a free Lagrangian, but contains self-interactions, precisely because of the additional term in $\boldsymbol{F}_{\mu\nu}$. This can be understood from the fact that the $\boldsymbol{W}_\mu$ couple to the SU(2) current, and they themselves carry non-trivial SU(2) quantum numbers. The vertices implied by (11.66) consist of a 3-point vertex proportional to g and a 4-point vertex proportional to g^2, as illustrated in figure 11.3.

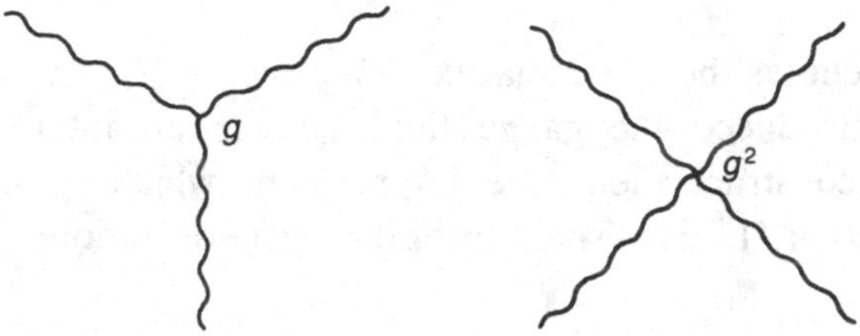

Figure 11.3 Self-interactions of gauge mesons.

A field theory of this sort does in fact occur in the so-called Weinberg–Salam model of weak and electromagnetic interactions. There are, however, several important ramifications of that theory. In the first place the gauge group is not simply SU(2), but SU(2) × U(1), so that there are four gauge mesons, one for each generator. Secondly, in a pure gauge theory, of the kind discussed above, the gauge particles are massless, just like the photon, because gauge invariance does not allow a mass term of the form $\frac{1}{2}\mu^2 \boldsymbol{W}_\mu \cdot \boldsymbol{W}^\mu$ to be included in the Lagrangian. The Weinberg–Salam theory gets around this difficulty by the phenomenon of spontaneous symmetry breaking, which may occur when the theory is coupled to scalar fields in a special way. The net result is that three of the gauge bosons acquire mass, of about 80 GeV/c^2, and have subsequently been identified experimentally as the W^+, W^- and Z^0 bosons. There is a remaining unbroken U(1) symmetry mediated by the still massless photon.

(iii) SU(3)—QCD

It is a relatively simple matter to generalize the case of SU(2) to a general SU(n) group. The special feature of SU(2) is that it has three generators, which naturally lends itself to a vector notation, and the ε_{ijk} occurring in their commutation relations $[T_i, T_j] = \mathrm{i}\varepsilon_{ijk} T_k$ produces the cross-products of (11.60) and subsequent equations. For a general SU(n) group we have $n^2 - 1$ generators T_a, with commutation relations

$$[T_a, T_b] = \mathrm{i}f^c_{ab} T_c \tag{11.67}$$

and the corresponding equations will contain the structure constants f^c_{ab} explicitly.

As in the case of SU(2), we expect a gauge field W^a_μ for each generator T_a, and the covariant derivative analogous to (11.57) is

$$D_\mu = \partial_\mu - igT_aW^a_\mu \tag{11.68}$$

Here, as in equation (11.67), there is an implicit summation over the repeated index. For some purposes it is useful to denote the $n \times n$ matrix $T_aW^a_\mu$ occurring in equation (11.68) by W_μ. To obtain the required transformation property of the W^a_μ, we follow the same arguments that led up to equation (11.60), to obtain

$$W'^a_\mu = W^a_\mu + \partial_\mu\Lambda^a - gf^a_{bc}\Lambda^bW^c_\mu \tag{11.69}$$

for the change in W^a_μ induced by the infinitesimal transformation $U = \exp(\mathrm{i}T_a\Lambda^a)$. For a finite transformation equation (11.61) stands.

The covariant curl $F^a_{\mu\nu}$, transforming according to the adjoint representation, is most easily derived from equation (11.65), which gives

$$F^a_{\mu\nu} = \partial_\mu W^a_\nu - \partial_\nu W^a_\mu + gf^a_{bc}W^b_\mu W_{\nu c} \tag{11.70}$$

It is from $F^a_{\mu\nu}$ that the gauge Lagrangian $\mathcal{L}(W^a_\mu)$ is constructed:

$$\mathcal{L}(W^a_\mu) = -\tfrac{1}{4}F^a_{\mu\nu}F^{\mu\nu a} \tag{11.71}$$

together with various gauge-fixing terms. Again this has the crucial feature that it contains self-interactions of the gauge fields.

In the particular case $n = 3$, such a theory turns out to be of very great relevance for the strong interactions of elementary particles. Apart from their 'flavour' quantum numbers, such as charge, strangeness, etc., quarks are now believed to have another, 3-fold, quantum number, rather frivolously termed 'colour'. Rather than labelling as 'red', 'green' and 'yellow', let us more prosaically give them an extra index a, running from 1 to 3. Thus, for example, there will be three copies of the up quark, $u_a = (u_1, u_2, u_3)$, which are identical as far as the weak and electromagnetic interactions are concerned, and in particular have the same mass. Thus the free quark Lagrangian has a global SU(3) symmetry, which we will call $\mathrm{SU}(3)_C$ to distinguish it from the 'flavour' SU(3) of Chapter 8, which is to do with transformations between the u, d and s quarks.

In the spirit of the present chapter it is natural to try to make this global symmetry a local one, i.e. to gauge the SU(3) of colour. This implies the existence of massless gauge bosons W^a_μ, known as gluons, which couple to the colour quantum number, and have self-interactions among themselves in the usual way. The resulting theory has been termed 'quantum chromodynamics' (QCD) and has been found to give a very good account of the strong interactions of quarks, and hence ultimately of elementary particles such as pions, nucleons, etc. Colour is hidden, in the sense that all such ordinary particles are singlets under the colour group. As we know from Chapter 8, the products $3 \otimes \overline{3}$ and $3 \otimes 3 \otimes 3$ both contain the trivial representation.

QCD has three very remarkable properties, all deriving from the self-interactions inherent in (11.71). In the first place it is 'renormalizable', which means that the divergences which arise in perturbation theory when loop diagrams are evaluated can be absorbed by redefinitions of the coupling constant and rescaling of the field. In general this is not true of field theories including spin-1 particles, but gauge theories are the exception. (This, incidentally, was the rationale for the construction of the Weinberg–Salam theory.) Secondly, non-Abelian gauge theories, and only such theories, have the property of 'asymptotic freedom'. After renormalization the effective coupling constant is dependent on the momentum scale at which one is working. In asymptotically free theories the effective coupling constant becomes *smaller* at higher momenta, or equivalently at shorter distances, which means that perturbation theory again becomes possible. Finally, at large distances the effective coupling becomes so large as to produce 'confinement', which means that quarks can only exist as individual entities in small, confined volumes. If one attempts to separate them, the vacuum produces quark–antiquark pairs in such a way as to neutralize the colour charge, and all one sees at large distance are the singlet bound states such as the pion or nucleon.

These, then, are some extremely important physical aspects of the application of the gauge principle to non-Abelian symmetries. The mathematical side of gauge theories is equally rich. To understand them fully one needs to go beyond groups to consider objects called 'fibre bundles'. In this framework the gauge fields have a very natural geometrical interpretation in terms of 'connections', which effect parallel transport from one fibre to another. This would be the topic of another book, but serves as a further illustration of the fascinating interplay between mathematical constructs and the physical world.

Problems for Chapter 11

11.1 In electrostatics the potential φ can be written as the line integral

$$\varphi(\boldsymbol{x}) = -\int_{\boldsymbol{x}_0}^{\boldsymbol{x}} \boldsymbol{E}(\boldsymbol{x}')\cdot \mathrm{d}\boldsymbol{x}'$$

along any path from $\boldsymbol{x}_0$ to $\boldsymbol{x}$. Verify that in the special gauge $\boldsymbol{x}\cdot\boldsymbol{A} = 0$ the vector potential $\boldsymbol{A}$ can be written as the integral along the straight line from the origin to $\boldsymbol{x}$:

$$\boldsymbol{A}(\boldsymbol{x}) = -\int_0^1 \boldsymbol{x}' \times \boldsymbol{B}(\boldsymbol{x}')\,\mathrm{d}t$$

where $\boldsymbol{x}' = \boldsymbol{x}t$. (Hint: use equation (6.29) and note that $\partial B_k(\boldsymbol{x}')/\partial x_i = t\partial B_k/\partial x'_i$, while $\mathrm{d}B_k(\boldsymbol{x}')/\mathrm{d}t = x_i \partial B_k/\partial x'_i$.)

11.2 If in equation (11.18) we choose

$$\alpha(\boldsymbol{x}) = (e/\hbar)\int_{\boldsymbol{x}_0}^{\boldsymbol{x}} \boldsymbol{A}(\boldsymbol{x}')\cdot \mathrm{d}\boldsymbol{x}'$$

we can transform $\boldsymbol{A}$ to zero and so reduce equation (11.19) to the free particle equation, at the cost of giving ψ a phase $\alpha(\boldsymbol{x})$. One can carry out an interference experiment with beams of slow neutrons which follow two different paths from $\boldsymbol{x}_0$ to $\boldsymbol{x}$, as indicated below.

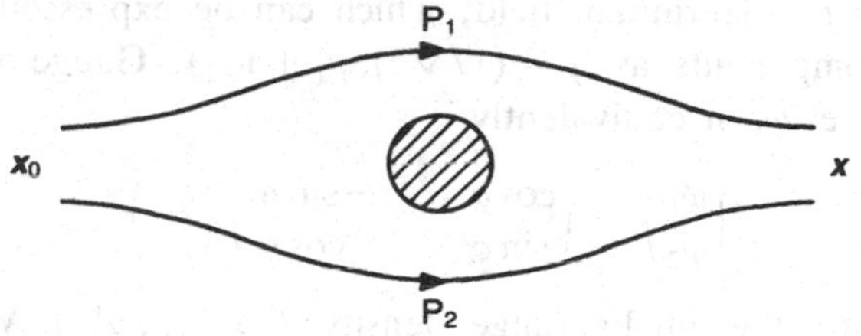

Show that a phase difference is induced, given by $(e/\hbar)\Phi$, where Φ is the magnetic flux enclosed between the two paths. (This phenomenon is known as the Aharonov–Bohm effect.)

11.3 Verify that the four Maxwell equations can be written in covariant form as

$$\partial^{\mu} F_{\mu\nu} = \mu j_{\nu}$$

$$\varepsilon_{\mu\nu\rho\sigma}\partial^{\nu} F^{\rho\sigma} = 0$$

11.4 Show that in the Weyl representation $(\frac{1}{2}, 0)$ for massless spin-$\frac{1}{2}$ particles, in which $\boldsymbol{J} = \frac{1}{2}\boldsymbol{\sigma}$, $\boldsymbol{K} = -\frac{1}{2}\mathrm{i}\boldsymbol{\sigma}$ acting on 2-component spinors φ, the condition

$$L_1\varphi = L_2\varphi = 0$$

is satisfied if

$$\varphi \propto \begin{pmatrix} 0 \\ 1 \end{pmatrix}$$

with $\lambda = -\frac{1}{2}$.

11.5 Show that the commutation relations of γ_μ, $\Sigma_{\rho\sigma} := \frac{1}{2}\sigma_{\rho\sigma}$ exactly parallel those of P_μ, $L_{\rho\sigma}$ (equation (10.60)), and hence that the Dirac bilinear $\bar{\psi}(x)\gamma_\mu\psi(x)$ transforms like a 4-vector density under $\psi(x) \to \exp(\frac{1}{2}\mathrm{i}a_{\rho\sigma}\Sigma^{\rho\sigma})\psi(\Lambda x)$.

11.6 Under parity the components of a general 4-vector V_μ become $(\Pi V)_\mu = (V_0;\ -\boldsymbol{V})$. Noting that $\gamma_0\gamma_\mu\gamma_0 = (\gamma_0;\ -\boldsymbol{\gamma})$, show that $\bar{\psi}\gamma_5\psi$ transforms like a pseudoscalar density under $\psi(x) \to \gamma_0\psi(\Pi x)$.

11.7 A charged scalar field has the Lagrange density

$$\mathcal{L}(\varphi, \varphi^\dagger) = (\partial_\mu\varphi)(\partial^\mu\varphi^\dagger) - m^2\varphi\varphi^\dagger$$

where φ is a non-Hermitian field, which can be expressed in terms of Hermitian components as $\varphi = (1/\sqrt{2})(\varphi_1 + \mathrm{i}\varphi_2)$. Gauge the local invariance $\varphi \to \mathrm{e}^{\mathrm{i}\alpha}\varphi$, or equivalently

$$\begin{pmatrix} \varphi_1 \\ \varphi_2 \end{pmatrix} \to \begin{pmatrix} \cos\alpha & -\sin\alpha \\ \sin\alpha & \cos\alpha \end{pmatrix} \begin{pmatrix} \varphi_1 \\ \varphi_2 \end{pmatrix}$$

and write down the full Lagrange density $\mathcal{L}(\varphi, \varphi^\dagger, A_\mu)$. What interactions with photons does this imply?

11.8 Check that equation (11.59) is indeed satisfied by the finite transformation of W_μ given in equation (11.61).

A

Dirac Notation in Quantum Mechanics

A1 State Vectors

In wave mechanics we typically work with the wavefunction $\psi_\alpha(\boldsymbol{x})$ of a system in a state specified by the labelling α. For example, in the case of a free particle α could be the momentum $\boldsymbol{p}$, and $\psi_\alpha(\boldsymbol{x})$ would then be the de Broglie wavefunction, $\psi_\alpha(\boldsymbol{x}) = u_{\boldsymbol{p}}(\boldsymbol{x}) = e^{i\boldsymbol{p}\cdot\boldsymbol{x}/\hbar}$. The electronic state of a hydrogen atom is specified by the quantum numbers n, l and m, so in this case $\psi_\alpha(\boldsymbol{x}) = u_{nlm}(\boldsymbol{x}) = R_{nl}(r)Y_{lm}(\theta, \varphi)$. If we also want to specify the spin state of the electron we have to multiply this latter wavefunction by a column vector χ_{m_s}, where $\chi_{1/2} = \binom{1}{0}$ and $\chi_{-1/2} = \binom{0}{1}$ represent 'spin-up' and 'spin-down' respectively, and now α comprises the four numbers $(nlmm_s)$.

We should notice that the wavefunction is a mathematical construct, with interpretational rules for obtaining the answers to physical questions, but the *physical specification* of the state is given by the labels α, which are usually the *eigenvalues of commuting observables*, as in all the above examples. The last example serves to show that in order to incorporate a specification of the electron spin we are forced into an uneasy mixture of wavefunctions and matrices. (Note that the strict analogue of $\psi_\alpha(\boldsymbol{x})$ for spin would be the *components* $(\chi_{m_s})_i$ of the column vectors.) In the Dirac formulation of quantum mechanics, wavefunctions and spin vectors are put on a common footing. The distinguishing feature of this formulation is that it concentrates on the essentials, namely the state specification α, and does not initially tie one down to any particular calculational scheme. For that reason it begins at a somewhat abstract level, but once the formalism has been developed one can then introduce the physical interpretation and make contact with the two apparently different ways of doing quantum mechanics, namely wave mechanics and matrix mechanics.

State Vectors (kets)

With the physical state specified by the labelling α is associated a vector $|\alpha\rangle$ in an abstract vector space (see §3.4). Thus states can be added and multiplied by (complex) scalars, and these operations satisfy certain rules. Wavefunctions provide a specific example of such a vector space under ordinary addition and multiplication.

Actually the correspondence between physical state and state vector $|\alpha\rangle$ is not 1:1. In wave mechanics we are familiar with the fact that the wavefunction is undefined up to an overall multiplicative constant. Even a normalized wavefunction is still undetermined up to a phase $e^{i\varphi}$. Thus a physical state actually corresponds to a *ray*, consisting of all possible non-zero multiples $c|\alpha\rangle$ of a given ket $|\alpha\rangle$ (which in fact form an equivalence class—see §2.1).

The space is provided with an inner product (§3.5), which in Dirac notation is written as $\langle\alpha|\beta\rangle$. It is the moduli or real parts of such complex numbers which will eventually provide the answers to physical questions. It is useful, but not essential, to give a meaning to the symbol $\langle\alpha|$ standing on its own. It is taken to be a vector in a conjugate vector space of 'bra' vectors. The scalar product $\langle\alpha|\beta\rangle$ is then regarded as a product defined between the two different spaces—a bra(c)ket! In order that the axioms (S) hold, the mapping between $|\alpha\rangle$ and $\langle\alpha|$ must be such that linear combinations of ket vectors correspond to the *complex conjugate* linear combination of bra vectors: $a|\alpha\rangle + b|\beta\rangle \leftrightarrow a^*\langle\alpha| + b^*\langle\beta|$.

A2 Orthonormal Basis

We recall (§3.4) that a basis is a set of vectors, which in the present case we can write as $\{|e_i\rangle\}$, which span the space, so that a general ket $|\alpha\rangle$ can be expanded as a linear combination of those vectors:

$$|\alpha\rangle = \sum_i a_i|e_i\rangle \tag{A1}$$

In quantum mechanics the number of independent basis vectors $|e_i\rangle$ is usually infinite, so that the vector space is infinite-dimensional. In that case we have to worry about the convergence of such expansions. In order for us to be able to carry over properties and theorems from finite-dimensional vector spaces, the space of state vectors must be what is technically known as a *separable Hilbert space*. A Hilbert space is a complete normed vector space, 'complete' in the sense that it contains the limit point of any Cauchy sequence $\{|n\rangle\}$, with $\|(|n\rangle - |m\rangle)\| \to 0$ as $n, m \to \infty$, and it is 'separable' if the basis is countably infinite, i.e. if it can be put into 1:1 correspondence with the integers. In many physical problems, however, we can restrict our attention to a finite-dimensional subspace, for example the states of a given energy or angular momentum, in which case the theorems of finite-dimensional vector spaces apply directly.

It is usually very convenient to choose an *orthonormal* basis, satisfying

$$\langle e_i|e_j\rangle = \delta_{ij} \tag{A2}$$

If the original basis is not orthonormal, one can always be constructed by the Gram–Schmidt procedure explained in §3.5. Then the scalar product $\langle\alpha|\beta\rangle$ can be expressed in terms of the respective components a_i, b_i as

$$\langle\alpha|\beta\rangle = \sum_i a_i^* b_i \tag{A3}$$

With such a basis these components are just the scalar products

$$a_i = \langle e_i|\alpha\rangle \tag{A4}$$

Substituting back into (A1) gives

$$|\alpha\rangle = \sum_i \langle e_i|\alpha\rangle|e_i\rangle \tag{A5}$$

If we expand the notation so that multiplication by scalars $a|\alpha\rangle$ can equivalently be written as $|\alpha\rangle a$, equation (A5) leads to an identity which is perhaps the most useful identity in all of quantum mechanics. We first rewrite it as

$$|\alpha\rangle = \sum_i |e_i\rangle\langle e_i|\alpha\rangle \tag{A5$'$}$$

and then insert brackets thus:

$$|\alpha\rangle = \sum_i (|e_i\rangle\langle e_i|)|\alpha\rangle \tag{A5$''$}$$

The quantity $|A\rangle\langle B|$ is a linear operator. Acting on an arbitrary vector $|\alpha\rangle$ it gives the vector $|A\rangle\langle B|\alpha\rangle$, and on the linear combination $a|\alpha\rangle + b|\beta\rangle$ the vector $|A\rangle\langle B|(a|\alpha\rangle + b|\alpha\rangle) = a|A\rangle\langle B|\alpha\rangle +$

$b|A\rangle\langle B|\alpha\rangle$, in accordance with equation (3.42). The linear operator in (A5″) is a very special linear operator in that it leaves $|\alpha\rangle$ unchanged. Since this is true for an *arbitrary* $|\alpha\rangle$, it must in fact be the identity operator. Thus

$$\boxed{\sum_i |e_i\rangle\langle e_i| = \mathbb{1}} \tag{A6}$$

expresses in operator form the completeness of the orthonormal basis $\{|e_i\rangle\}$.

A3 Linear Operators

In general a linear operator A is a mapping of the vector space into itself, associating with any vector $|\alpha\rangle$ another $A|\alpha\rangle$ in such a way that $A(a|\alpha\rangle + b|\beta\rangle) = a(A|\alpha\rangle) + b(B|\beta\rangle)$. The matrix elements of A with respect to an orthonormal basis are

$$A_{ij} = \langle e_i|(A|e_j\rangle) \tag{A7}$$

In Dirac notation the brackets are usually omitted, so that A_{ij} is written as

$$A_{ij} = \langle e_i|A|e_j\rangle \tag{A7'}$$

but it should always be remembered that A acts to the right on the ket $|e_j\rangle$. By double use of equation (A6) a more general overlap $\langle\alpha|A|\beta\rangle$ can be expressed in terms of coordinates as

$$\langle\alpha|A|\beta\rangle = \sum_{i,j} a_i^* A_{ij} b_j \tag{A8}$$

In wave mechanics one is not used to specifying an operator in this way. It is rather the approach used in matrix mechanics. However, even the familiar operators $\hat{x}$ and $\hat{p} = -i\hbar\partial/\partial x$ can be defined by an infinite set of matrix elements $(x)_{ij}$ and $(p)_{ij}$ with respect to some orthonormal basis of wavefunctions $\{u_i(x)\}$.

For example, we could use as an orthonormal basis of wavefunctions the energy eigenfunctions $u_n(x) = A_n H_n \exp(-\frac{1}{2}y^2)$ of the simple harmonic oscillator. Here $y = x/a$, where a is the classical amplitude $\sqrt{(\hbar/m\omega)}$, $H_n(y)$ is the Hermite polynomial and $A_n = (2^n n! a\sqrt{\pi})^{-1/2}$ is the appropriate normalization constant. The H_n satisfy the recursion relation

$$2yH_n = H_{n+1} + 2nH_{n-1}$$

so that

$$yu_n = (1/\sqrt{2})[\sqrt{(n+1)}u_{n+1} + \sqrt{n}u_{n-1}]$$

Hence

$$\begin{aligned}(x)_{mn} = a \int u_m^* y u_n \, \mathrm{d}x &= (a/\sqrt{2}) \int u_m^* [\sqrt{(n+1)}u_{n+1} + \sqrt{n}u_{n-1}] \, \mathrm{d}x \\ &= (a/\sqrt{2})[\sqrt{(n+1)}\delta_{m,n+1} + \sqrt{n}\delta_{m,n-1}] \quad \text{(A9)}\end{aligned}$$

which can be thought of as an infinite matrix with non-zero elements immediately above and below the main diagonal. A similar representation can be found for $\hat{p}$ using the recursion relation $H'_n = 2nH_{n-1}$.

A4 Hermitian Operators

Hermitian operators are very important in quantum mechanics because of the interpretative formalism whereby the eigenvalues of operators represent the results of physical measurements. These clearly have to be real, which is assured if the operators are Hermitian.

The *adjoint* $A^\dagger$ of a linear operator A is defined as follows. Consider the scalar product $\langle\beta|\{A|\alpha\rangle\}$. If its complex conjugate can be written as the scalar product of $\langle\alpha|$ with a ket $|\beta'\rangle$,

$$(\langle\beta|\{A|\alpha\rangle\})^* = \langle\alpha|\beta'\rangle \quad \text{(A10)}$$

we write

$$|\beta'\rangle = A^\dagger|\beta\rangle$$

i.e.

$$(\langle\beta|\{A|\alpha\rangle\})^* = \langle\alpha|\{A^\dagger|\beta\rangle\} \quad \text{(A10')}$$

Then a *Hermitian* operator H is one for which $H^\dagger = H$, so that

$$(\langle\beta|\{H|\alpha\rangle\})^* = \langle\alpha|\{H|\beta\rangle\} \quad \text{(A11)}$$

In what follows we shall assume that equation (A11) holds for all $|\beta\rangle$ in the space, which strictly speaking is the definition of a *self-adjoint* operator.

Reality of Eigenvalues

Let $|\lambda_i\rangle$ represent the eigenvectors of H, with eigenvalues λ_i:

$$H|\lambda_i\rangle = \lambda_i|\lambda_i\rangle \qquad \text{(no summation)} \tag{A12}$$

Taking the scalar product of this equation with $\langle\lambda_i|$ gives

$$\langle\lambda_i|H|\lambda_i\rangle = \lambda_i\langle\lambda_i|\lambda_i\rangle \tag{A13}$$

But

$$\langle\lambda_i|H|\lambda_i\rangle = \langle\lambda_i|\{H|\lambda_i\rangle\} = (\langle\lambda_i|\{H|\lambda_i\rangle\})^*$$

using hermiticity, equation (A11). Thus in equation (A13) $\langle\lambda_i|H|\lambda_i\rangle$ is real, as is $\langle\lambda_i|\lambda_i\rangle$. Hence the eigenvalue λ_i must also be real.

Orthogonality of Eigenvectors

Now consider two eigenvectors $|\lambda_i\rangle$, $|\lambda_j\rangle$ corresponding to *distinct* eigenvalues:

$$\begin{aligned} H|\lambda_i\rangle &= \lambda_i|\lambda_i\rangle \\ H|\lambda_j\rangle &= \lambda_j|\lambda_j\rangle \end{aligned} \tag{A14}$$

Multiplying the first of these equations by $\langle\lambda_j|$ and the second by $\langle\lambda_i|$ we get

$$\begin{aligned} \langle\lambda_j|H|\lambda_i\rangle &= \lambda_i\langle\lambda_j|\lambda_i\rangle \\ \langle\lambda_i|H|\lambda_j\rangle &= \lambda_j\langle\lambda_i|\lambda_j\rangle \end{aligned} \tag{A15}$$

Taking the complex conjugate of the second equation, using hermiticity and the already established reality of λ_j gives

$$\langle\lambda_j|H|\lambda_i\rangle = \lambda_j\langle\lambda_j|\lambda_i\rangle \tag{A16}$$

Subtracting this from equation (A15) gives $0 = (\lambda_i - \lambda_j)\langle\lambda_i|\lambda_j\rangle$. Hence, since we have assumed $\lambda_i \neq \lambda_j$, $\langle\lambda_i|\lambda_j\rangle = 0$.

In the case of degeneracy, where there is more than one eigenvector of H corresponding to a given eigenvalue, they can be *chosen* to be orthogonal by the Gram–Schmidt procedure. In practice one usually specifies the states further by making them eigenstates of additional Hermitian operators which commute with the first, as for example in the specification $|nlm\rangle$, which refers to the eigenvalues of the Hamiltonian H, the square of the orbital angular momentum operator $\boldsymbol{L}^2$, and its z component L_z.

Completeness of Eigenvectors

In a finite-dimensional space it can be proved that the eigenvectors of a Hermitian operator span the space; that is, that the space $\mathscr{H}$ spanned by the eigenvectors is the whole space. The proof proceeds by showing that $\mathscr{H}_\perp$, the orthogonal complement of $\mathscr{H}$, is in fact the null space. We do not give it here, since in quantum mechanics we are concerned with an infinite-dimensional space, where the proof is much more difficult, and the completeness of the eigenvectors is usually assumed without proof.

Operationally this completeness is most usefully expressed in the form of equation (A6):

$$\sum_i |\lambda_i\rangle\langle\lambda_i| = \mathbb{1} \tag{A17}$$

choosing the eigenvectors to be orthogonal.

Representation of H in Terms of Eigenvectors

If we know the eigenvalues λ_i and (orthonormal) eigenvectors $|\lambda_i\rangle$ of a Hermitian operator H we can write it in the form

$$H = \sum_i \lambda_i|\lambda_i\rangle\langle\lambda_i| \tag{A18}$$

This representation, also known as the spectral decomposition of H, is easily proved by showing that it gives the correct result when acting on any of the $|\lambda_j\rangle$ which span the space. (Thus $(\Sigma_i \lambda_i|\lambda_i\rangle\langle\lambda_i|)|\lambda_j\rangle = \Sigma_i\lambda_i|\lambda_i\rangle\delta_{ij} = \lambda_j|\lambda_j\rangle = H|\lambda_j\rangle$, as required.) The way to understand equation (A18) is that each individual operator $|\lambda_i\rangle\langle\lambda_i|$ in the sum is a projection operator onto the subspace spanned by $|\lambda_i\rangle$, and in this subspace H is effectively equal to $\lambda_i\mathbb{1}$.

Given the spectral decomposition (A18), we can define a function $f(H)$ by

$$f(H) := \sum_i f(\lambda_i)|\lambda_i\rangle\langle\lambda_i| \tag{A19}$$

Of particular interest is the exponential function. Indeed, when H is in fact the Hamiltonian, the time development of the system is given by the operator

$$\mathrm{e}^{-\mathrm{i}Ht/\hbar} = \sum_i \mathrm{e}^{-\mathrm{i}E_it/\hbar}|E_i\rangle\langle E_i| \tag{A20}$$

Example

Consider the matrix operator

$$\sigma_2 = \begin{pmatrix} 0 & -i \\ i & 0 \end{pmatrix}$$

Its eigenvalues are ± 1, with corresponding eigenvectors,

$$\chi_+ = \frac{1}{\sqrt{2}}\begin{pmatrix} 1 \\ i \end{pmatrix} \text{ and } \chi_- = \frac{1}{\sqrt{2}}\begin{pmatrix} 1 \\ -i \end{pmatrix}$$

In matrix form the completeness relation, equation (A17), reads

$$\chi_+(\chi_+)^\dagger + (\chi_-)^\dagger\chi_- = \frac{1}{2}\begin{pmatrix} 1 \\ i \end{pmatrix}(1 \quad -i) + \frac{1}{2}\begin{pmatrix} 1 \\ -i \end{pmatrix}(1 \quad i)$$

$$= \frac{1}{2}\left[\begin{pmatrix} 1 & -i \\ i & 1 \end{pmatrix} + \begin{pmatrix} 1 & i \\ -i & 1 \end{pmatrix}\right] = \begin{pmatrix} 1 & 0 \\ 0 & 1 \end{pmatrix}$$

Similarly equation (A18) reads

$$\chi_+(\chi_+)^\dagger - (\chi_-)^\dagger\chi_- = \frac{1}{2}\left[\begin{pmatrix} 1 & -i \\ i & 1 \end{pmatrix} - \begin{pmatrix} 1 & i \\ -i & 1 \end{pmatrix}\right] = \begin{pmatrix} 0 & -i \\ i & 0 \end{pmatrix} = \sigma_2$$

Finally

$$e^{-i\varphi\sigma_2} = \frac{1}{2}\left[e^{-i\varphi}\begin{pmatrix} 1 & -i \\ i & 1 \end{pmatrix} + e^{+i\varphi}\begin{pmatrix} 1 & i \\ -i & 1 \end{pmatrix}\right] = \begin{pmatrix} \cos\varphi & -\sin\varphi \\ \sin\varphi & \cos\varphi \end{pmatrix}$$

$$= \mathbb{1}\cos\varphi - i\sigma_2\sin\varphi$$

which is the value obtained in §6.1 by summation of the exponential series.

A5 Unitary Operators

Another class of linear operators which feature prominently in quantum mechanics consists of *unitary* operators. These can be defined as operators U whose adjoint is their inverse (again assuming that the adjoint is defined for all vectors in the space), i.e.

$$U^\dagger = U^{-1} \tag{A21}$$

Their importance lies in the fact that overlaps of states and matrix elements of operators remain invariant under the simultaneous transformations

$$\begin{aligned} |\psi\rangle \to |\psi'\rangle &= U|\psi\rangle \\ A &\to UAU^{-1} \end{aligned} \tag{A22}$$

for all states $|\psi\rangle$ and all linear operators A.

As a preliminary it is useful to remark in general that the bra vector corresponding to the ket $B|\varphi\rangle$ is $\langle\varphi|B^{\dagger}$. Thus, writing $|\zeta\rangle = B|\varphi\rangle$ we have

$$\langle\zeta|\psi\rangle = \langle\psi|\zeta\rangle^* = \langle\psi|B|\varphi\rangle^* = \langle\varphi|B^{\dagger}|\psi\rangle$$

for an arbitrary ket $|\psi\rangle$. Thus

$$\begin{aligned} \langle\varphi'|\psi'\rangle &= \langle\varphi|U^{\dagger}U|\psi\rangle = \langle\varphi|\psi\rangle \\ \langle\varphi'|A'|\psi'\rangle &= \langle\varphi|U^{\dagger}(UAU^{-1})U|\psi\rangle = \langle\varphi|A|\psi\rangle \end{aligned} \tag{A23}$$

as claimed. The first of these two equations corresponds to the definition of unitary transformations given in Chapter 3 (equation (3.55)).

The complex exponential of the Hamiltonian, as in equation (A20), is just such a unitary operator. Similarly the exponentials of the momentum and angular momentum operators are unitary operators generating respectively spatial translations and rotations.

Another important example of a unitary operator occurs when one changes from one orthonormal basis to another. Suppose that $\{|\alpha_i\rangle\}$ and $\{|\beta_i\rangle\}$ are two orthonormal bases. Then the linear operator

$$U := \sum_i |\beta_i\rangle\langle\alpha_i| \tag{A24}$$

when acting on an arbitrary ket $|\varphi\rangle$ has the effect

$$U|\varphi\rangle = \sum_i |\beta_i\rangle\langle\alpha_i|\varphi\rangle$$

That is, the transformed vector has the components of $|\varphi\rangle$ in the decomposition $|\varphi\rangle = \Sigma_i|\alpha_i\rangle\langle\alpha_i|\varphi\rangle$, but now referred to the new basis $\{|\beta_i\rangle\}$. The adjoint of U is easily seen to be

$$U^{\dagger} = \sum_i |\alpha_i\rangle\langle\beta_i| \tag{A25}$$

which satisfies $UU^{\dagger} = U^{\dagger}U = \mathbb{1}$ by virtue of the orthonormality and completeness of both bases.

Conversely, a given unitary operator U, such as $\exp(-\mathrm{i}Ht/\hbar)$, transforms an orthonormal basis $\{|\alpha_i\rangle\}$ into another one, say $\{|\beta_i\rangle\}$, by $|\beta_i\rangle = U|\alpha_i\rangle$.

One frequently refers to 'diagonalizing' an operator. In the case of a Hermitian or unitary operator A, this simply means choosing as orthonormal basis the eigenvectors $|\lambda_i\rangle$ of A. Referred to an arbitrary basis $\{|\beta_i\rangle\}$ the matrix elements of A are $A_{ij} = \langle\beta_i|A|\beta_j\rangle$. But with respect to the basis $\{|\alpha_i\rangle\} = \{|\lambda_i\rangle\}$ we have $A'_{ij} = \langle\lambda_i|A|\lambda_j\rangle = \lambda_i\delta_{ij}$ (no summation). The transformation to the basis of eigenvectors is effected by the unitary operator $U = \Sigma_i|\lambda_i\rangle\langle\beta_i|$.

For example, the matrix σ_2 above actually consists of the matrix elements $X_{ij} = \langle\beta_i|X|\beta_j\rangle$ of an operator X with respect to two states $|\beta_1\rangle$, $|\beta_2\rangle$ (which are in fact eigenstates of another operator X_3), on which X has the effect

$$X|\beta_1\rangle = \mathrm{i}|\beta_2\rangle$$
$$X|\beta_2\rangle = -\mathrm{i}|\beta_1\rangle$$

so that $X_{11} = X_{22} = 0$ and $X_{21} = -X_{12} = \mathrm{i}$. The eigenstates of X are the linear combinations

$$|\lambda_\pm\rangle = (1/\sqrt{2})(|\beta_1\rangle \pm \mathrm{i}|\beta_2\rangle)$$

and referred to these vectors as basis the matrix elements of X are just $\langle\lambda_\pm|X|\lambda_\pm\rangle = \pm 1$, $\langle\lambda_\pm|X|\lambda_\mp\rangle = 0$. With respect to the original basis the matrix representing the operator U is

$$U_{ij} = \frac{1}{\sqrt{2}}\begin{pmatrix}1\\0\end{pmatrix}(1 \quad -\mathrm{i}) + \frac{1}{\sqrt{2}}\begin{pmatrix}0\\1\end{pmatrix}(1 \quad \mathrm{i}) = \frac{1}{\sqrt{2}}\begin{pmatrix}1 & -\mathrm{i}\\1 & \mathrm{i}\end{pmatrix}$$

Similarly

$$(U^{-1})_{ij} = \frac{1}{\sqrt{2}}\begin{pmatrix}1\\\mathrm{i}\end{pmatrix}(1 \quad 0) + \frac{1}{\sqrt{2}}\begin{pmatrix}1\\-\mathrm{i}\end{pmatrix}(0 \quad 1) = \frac{1}{\sqrt{2}}\begin{pmatrix}1 & 1\\\mathrm{i} & -\mathrm{i}\end{pmatrix} = (U^\dagger)_{ij}$$

and

$$\begin{aligned}(UXU^{-1})_{ij} &= \frac{1}{2}\begin{pmatrix}1 & -\mathrm{i}\\1 & \mathrm{i}\end{pmatrix}\begin{pmatrix}0 & -\mathrm{i}\\\mathrm{i} & 0\end{pmatrix}\begin{pmatrix}1 & 1\\\mathrm{i} & -\mathrm{i}\end{pmatrix}\\ &= \frac{1}{2}\begin{pmatrix}1 & -\mathrm{i}\\1 & \mathrm{i}\end{pmatrix}\begin{pmatrix}1 & -1\\\mathrm{i} & \mathrm{i}\end{pmatrix} = \begin{pmatrix}1 & 0\\0 & -1\end{pmatrix}\end{aligned}$$

the diagonal matrix of the eigenvalues of X.

A6 Continuous Eigenvalues

So far we have been dealing with situations where a Hermitian operator has a discrete spectrum, as is the case in a finite-dimensional space or for bound states in quantum mechanics. However, to be useful, the theory needs to be extended to the case when an operator can have a continuous range of eigenvalues, as for the scattering states of the hydrogen atom, or for the momentum or energy of free particles. The theory can be extended in a mathematically rigorous way (see e.g. Jordan *Linear Operators for Quantum Mechanics* Wiley), but in the context of this appendix we shall be content to follow the more heuristic approach of Dirac.

The signs of difficulty are already there in elementary wave mechanics, where the 'eigenfunction' of the momentum operator $\hat{p} = -i\hbar\partial/\partial x$ is $u_p(x) = \exp(-ipx/\hbar)$, which is not a properly normalizable wavefunction since $|u_p(x)| = 1$ for all x. So we will somehow have to accommodate states with infinite norm.

In going over from discrete to continuous normalization one can really start almost anywhere in the circle, but perhaps the most convenient place is the completeness relation, which for discrete eigenvalues takes the form of equation (A17). For continuous eigenvalues λ, we will want to replace the summation by an integration, so that

$$\mathbb{1} = \int d\lambda\, |\lambda\rangle\langle\lambda| \tag{A26}$$

which is equivalent to the statement that $|\varphi\rangle = \int d\lambda\, |\lambda\rangle\langle\lambda|\varphi\rangle$ for any state $|\varphi\rangle$. Then necessarily

$$\langle\lambda'|\varphi\rangle = \int d\lambda \langle\lambda'|\lambda\rangle\langle\lambda|\varphi\rangle \tag{A27}$$

The only way that this relation can be true for any $|\varphi\rangle$ is if $\langle\lambda'|\lambda\rangle$ is the Dirac delta function $\delta(\lambda' - \lambda)$. It is not really a function at all, but rather a *distribution*, which is defined by its property under an integral sign:

$$f(\lambda') = \int d\lambda\ \delta(\lambda' - \lambda) f(\lambda) \tag{A28}$$

for all reasonably behaved functions $f(\lambda)$.

One can think of many different representations of $\delta(x)$ as limits of ordinary functions, for example $\lim_{g\to\infty}[(\sin gx)/\pi x]$, but all that is relevant is equation (A28), from which one can deduce a number of useful identities, such as

(i) $\delta(ax) = (1/|a|)\delta(x)$ for a real,
(ii) $f(x)\delta(x) = f(0)\delta(x)$,
(iii) $\delta(x^2 - a^2) = (1/2|a|)[\delta(x - a) + \delta(x + a)]$ for a real,

or more generally

(iv) $\delta(f(x)) = \Sigma_i \delta(x - x_i)/|f'(x_i)|$, where the x_i are the real roots of $f(x)$.

We will also encounter the *derivative* of the delta function. This again is defined by its effect under an integral sign. There the idea is to integrate by parts to convert the integral to the form of equation (A28). In this way one can prove the identity

(v) $x\delta'(x) = -\delta(x)$,

which also follows by naïve differentiation of $x\delta(x) = 0$ (cf. (ii)).

Replacing the previous formula (A18) we have

$$H = \int \mathrm{d}\lambda\, \lambda\, |\lambda\rangle\langle\lambda| \tag{A29}$$

and the statement of orthonormality is now

$$\langle\lambda'|\lambda\rangle = \delta(\lambda' - \lambda) \tag{A30}$$

In general an operator will have both a discrete and a continuous part of its spectrum, in which case equations (A28) and (A29) will contain a mixture of integrations and summations, and (A30) a mixture of delta functions and Kronecker deltas.

A7 Interpretative Formalism; Wavefunctions

Remarkably this abstract mathematical construction of vectors and linear operators seems to provide the precise framework we need to describe the quantum world. The interpretative formalism which we have already adumbrated several times, and has been found by experience to be the one that 'works', is as follows:

(i) With every physical observable $\mathcal{A}$ is associated a Hermitian operator A.
(ii) With every physical state of the system is associated a ket vector $|\varphi\rangle$.
(iii) The possible results a_i of a measurement of $\mathcal{A}$ are the eigenvalues

of the corresponding operator A. Immediately after the measurement the system is in the state represented by the ket $|a_i\rangle$.

(iv) If the system is in the state described by the ket $|\varphi\rangle$, the probability that a measurement of $\mathcal{A}$ will give the value of a_i is $|\langle a_i|\varphi\rangle|^2$. (If the eigenvalue a is part of a continuum rather than discrete then the probability of a value in the interval $(a, a + \delta a)$ is $|\langle a|\varphi\rangle|^2\delta a$.)

(v) Therefore the expectation value of $\mathcal{A}$ (the average value of measurements on an ensemble of identical systems) in the normalized state specified by $|\varphi\rangle$ is

$$\langle a\rangle_\varphi = \begin{cases} \sum_i a_i|\langle a_i|\varphi\rangle|^2 \\ \int \mathrm{d}a\,|\langle a|\varphi\rangle|^2 \end{cases} \tag{A31}$$

By virtue of the spectral decomposition, equation (A18) or (A29), this can be written as

$$\langle a\rangle_\varphi = \left\{ \begin{array}{l} \sum_i \langle\varphi|a_i\rangle a_i\langle a_i|\varphi\rangle \\ \int \mathrm{d}a\,\langle\varphi|a\rangle a\langle a|\varphi\rangle \end{array} \right\} = \langle\varphi|A|\varphi\rangle \tag{A32}$$

This is the form in which the expectation value is often first given. Note, however, that equation (A32) is not an independent assumption, but follows logically from (iv).

Wavefunctions

Finally we come to the connection of the Dirac formalism with wave mechanics. In that version of quantum mechanics the wavefunction $\psi(x)$ of a particle has the interpretation that $|\psi(x)|^2\delta x$ is the probability of finding the particle in the range $(x, x + \delta x)$. According to (iv) above this would be $|\langle x|\psi\rangle|^2\delta x$. This does not establish a unique correspondence, but by choice of phase of the states $|x\rangle$ we can take

$$\boxed{\psi(x) = \langle x|\psi\rangle} \tag{A33}$$

Wave mechanics is carried out in the framework where the operator $\hat{x}$ is diagonal, which means that

$$\langle x'|\hat{x}|x\rangle = x\delta(x' - x) \tag{A34}$$

In the evaluation of overlaps and matrix elements we always insert a complete set of x states, using the completeness relation $\mathbb{1} = \int dx\, |x\rangle\langle x|$. Thus the overlap $\langle a_i|\varphi\rangle$ needed in (iv) above is evaluated as

$$\begin{aligned}\langle a_i|\varphi\rangle &= \int dx\, \langle a_i|x\rangle\langle x|\varphi\rangle \\ &= \int dx\, u_i^*(x)\varphi(x)\end{aligned} \tag{A35}$$

where we have identified $\langle x|a_i\rangle = u_i(x)$, the wavefunction of the eigenstate $|a_i\rangle$.

The matrix elements of any operator function of $\hat{x}$ will again be diagonal:

$$\langle x|A(\hat{x})|x'\rangle = A(x)\delta(x' - x) \tag{A36}$$

by the continuum version of equation (A19). Then the x representative of the state $A(\hat{x})|\varphi\rangle$ is

$$\begin{aligned}\langle x|A(\hat{x})|\varphi\rangle &= \int dx' \langle x|A(\hat{x})|x'\rangle\langle x'|\varphi\rangle \\ &= A(x)\varphi(x)\end{aligned} \tag{A37}$$

As far as the momentum operator $\hat{p}$ is concerned, its matrix elements must be taken to be the *derivative* of a delta function:

$$\langle x|\hat{p}|x'\rangle = i\hbar(d/dx')\delta(x' - x) \tag{A38}$$

To check that this corresponds with the familiar form of the operator in wave mechanics we must evaluate the x representation of $\hat{p}|\varphi\rangle$, which is

$$\begin{aligned}\langle x|\hat{p}|\varphi\rangle &= \int dx' [i\hbar(d/dx')\delta(x' - x)]\langle x'|\varphi\rangle \\ &= -i\hbar \int dx' \delta(x' - x)(d/dx')\langle x'|\varphi\rangle\end{aligned}$$

upon integration by parts. Thus

$$\langle x|\hat{p}|\varphi\rangle = -i\hbar \frac{d\varphi(x)}{dx} \tag{A39}$$

The time-independent Schrödinger equation is just the x coordinate representation of the energy eigenvalue equation

$$H|E\rangle = E|E\rangle \tag{A40}$$

with

$$H = \hat{p}^2/2m + V(\hat{x})$$

Thus on taking the overlap with $\langle x|$, equation (A40) becomes the familiar

$$\left(\frac{-\hbar^2}{2m}\frac{\mathrm{d}^2}{\mathrm{d}x^2} + V(x)\right)\psi(x) = E\psi(x) \tag{A40$'$}$$

or in 3 dimensions

$$\left(\frac{-\hbar^2}{2m}\nabla^2 + V(\boldsymbol{x})\right)\psi(\boldsymbol{x}) = E\psi(\boldsymbol{x}) \tag{A40$''$}$$

B

Eigenstates of Angular Momentum in Quantum Mechanics

In books on quantum mechanics the commutation relations (6.28) are usually derived by generalizing those of the orbital angular momentum operators $\hat{\boldsymbol{L}} := \hat{\boldsymbol{x}} \times \hat{\boldsymbol{p}}$. From the fundamental commutation relations $[\hat{x}_i, \hat{p}_j] = \delta_{ij}$ it is straightforward to deduce that

$$[\hat{L}_i, \hat{L}_j] = \mathrm{i}\hbar\varepsilon_{ijk}\hat{L}_k \tag{B1}$$

which are the same as (6.28) up to a factor $\hbar$.

In the context of wave mechanics the connection is via equation (3.10). For an infinitesimal rotation the change in the argument of the wavefunction from $\boldsymbol{x}$ to $R^{-1}\boldsymbol{x} = \boldsymbol{x} - \varphi(\boldsymbol{n} \times \boldsymbol{x})$ (cf. equation (6.21)) can be effected by the application of the operator $\boldsymbol{n}\cdot\hat{\boldsymbol{L}}$, which in wave mechanics is realized as $-\mathrm{i}\hbar\boldsymbol{n}\cdot\boldsymbol{x} \times \boldsymbol{\nabla}$. Thus, by use of Maclaurin's theorem,

$$\begin{aligned}\psi'(\boldsymbol{x}) &= \psi(\boldsymbol{x} - \varphi\boldsymbol{n} \times \boldsymbol{x}) \\ &= (1 - \varphi\boldsymbol{n} \times \boldsymbol{x}\cdot\boldsymbol{\nabla})\psi(\boldsymbol{x}) \\ &= [1 - \mathrm{i}\varphi\boldsymbol{n}\cdot(\hat{\boldsymbol{L}}/\hbar)]\psi(\boldsymbol{x})\end{aligned}$$

The problem in quantum mechanics is to find the possible eigenvalues and eigenvectors of the angular momentum operators $\boldsymbol{J}$ satisfying the commutation relations (B1) in the form $[J_i, J_j] = \mathrm{i}\hbar\varepsilon_{ijk}J_k$. We do not insist that the operators be realizable like the $\hat{L}_i$ as differential operators acting on wavefunctions, but rather take the commutation relations as fundamental. The latter admit to half-integral eigenvalues which have no classical analogue, but which are certainly needed to describe the physical world, in particular the intrinsic angular momentum, or spin, of particles such as the proton or the electron.

Since the J's do not commute with each other, it is not possible simultaneously to assign definite values. Conventionally J_z is singled out, with eigenstates $|m\rangle$ such that $J_z|m\rangle = m\hbar|m\rangle$. The non-commutativity of J_x and J_y with J_z is an expression of the fact that $\boldsymbol{J}$ transforms as a vector. By the same token the square of the total angular momentum operator $\boldsymbol{J}^2 = J_x^2 + J_y^2 + J_z^2$ is a rotational scalar, which commutes with any component of $\boldsymbol{J}$. States $|\beta, m\rangle$ can therefore be defined which are simultaneous eigenstates of $\boldsymbol{J}^2$ and J_z:

$$\begin{aligned} \boldsymbol{J}^2|\beta, m\rangle &= \beta^2\hbar^2|\beta, m\rangle \\ J_z|\beta, m\rangle &= m\hbar|\beta, m\rangle \end{aligned} \tag{B2}$$

In order to find the possible values of β, m it is useful to work with the combinations $J_\pm := J_x \pm \mathrm{i}J_y$. These also commute with $\boldsymbol{J}^2$, and their commutation relations with J_z are given by

$$[J_z, J_\pm] = \pm\hbar J_\pm \tag{B3}$$

$J_\pm$ turn out to be respectively *raising* and *lowering* operators for J_z. That is, $J_\pm|\beta, m\rangle$ is an eigenstate of J_z with eigenvalue $(m \pm 1)\hbar$, while that of $\boldsymbol{J}^2$ is unchanged. For consider the effect of J_z on $J_\pm|\beta, m\rangle$:

$$\begin{aligned} J_z(J_\pm|\beta, m\rangle) &= (J_\pm J_z + [J_z, J_\pm])|\beta, m\rangle \\ &= (J_\pm m\hbar \pm \hbar J_\pm)|\beta, m\rangle \\ &= (m \pm 1)\hbar(J_\pm|\beta, m\rangle) \end{aligned} \tag{B4}$$

Thus the eigenvalues of J_z are spaced apart by units of $\hbar$, forming a sort of ladder. The raising and lowering operators J_+, J_- take the system up and down one rung respectively, as illustrated in figure B1.

Figure B1 Raising and lowering operators for angular momentum.

For a given value of β this raising or lowering process cannot go on indefinitely, for the value of m is limited. Thus

$$\beta^2\hbar^2 = \langle\beta, m|\boldsymbol{J}^2|\beta, m\rangle$$

$$= \langle \beta, m|(J_x^2 + J_y^2 + m^2\hbar^2)|\beta, m\rangle$$
$$\geqslant m^2\hbar^2 \qquad \text{(B5)}$$

since J_x and J_y are both Hermitian operators, so that $\langle J_x^2 + J_y^2\rangle \geqslant 0$. Thus at some stage the raising and lowering processes must come to an end. Concentrating first on the raising operator, the only way for this to happen is if, instead of giving a proper state, $J_+|\beta, m\rangle = 0$ for some value of m, say j. This value is closely related to β. Going back to equation (B5) let us re-express $\boldsymbol{J}^2$ in terms of J_z, $J_\pm$:

$$\boldsymbol{J}^2 = (J_x - iJ_y)(J_x + iJ_y) - i[J_x, J_y] + J_z^2$$
$$= J_-J_+ + J_z(J_z + \hbar) \qquad \text{(B6)}$$

Thus acting on the state $|\beta, j\rangle$ we have

$$\boldsymbol{J}^2|\beta, j\rangle = j\hbar(j\hbar + \hbar)|\beta, j\rangle$$

i.e.

$$\beta^2 = j(j + 1) \qquad \text{(B7)}$$

Alternatively we can rewrite (B6) as

$$\boldsymbol{J}^2 = J_+J_- + J_z(J_z - \hbar) \qquad \text{(B6')}$$

so that the value of m for which $J_-|\beta, m\rangle = 0$ is given by $\beta^2 = m(m - 1)$. Thus $j(j + 1) = m(m - 1)$, with solutions $m = -j$ or $m = j + 1$. The latter is inadmissible since j was already the maximum possible value of m.

As illustrated in figure B2, we have thus shown that for a given value of j, related to β^2 by $\beta^2 = j(j + 1)$, the spectrum of m ranges from $-j$ to $+j$ in integer steps. The total number of eigenvalues thus obtained, the multiplicity of the representation, is $2j + 1$. This must be an integer, so that from the present point of view j can be either integral or half-integral.

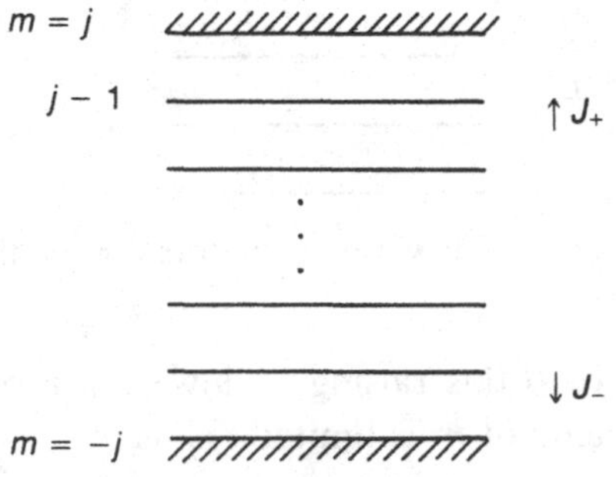

Figure B2 The range of J_3 is bounded above by j and below by $-j$.

The action of the operators J^2, J_z on the states $|\beta, m\rangle$, now usually rewritten as $|j, m\rangle$, is specified by equations (B2). To complete the specification we need to be more precise about the action of the raising and lowering operators. That is, $J_\pm|j, m\rangle \propto |j, m \pm 1\rangle$, but what is the constant of proportionality? To fix this we need to consider the norm of the state $J_\pm|j, m\rangle$, since $|j, m \pm 1\rangle$ and $|j, m\rangle$ are implicitly assumed to be normalized. Now since J_x and J_y are Hermitian, the Hermitian conjugate of J_+ is J_- and vice versa. Hence

$$\begin{aligned}||J_+|j, m\rangle||^2 &= \langle j, m|J_-J_+|j, m\rangle \\ &= \hbar^2[j(j+1) - m(m+1)]\end{aligned}$$

by virtue of (B6). Thus $J_+|j, m\rangle = N|j, m+1\rangle$, where $|N|^2 = \hbar^2[j(j+1) - m(m+1)] = \hbar^2(j-m)(j+m+1)$. There is an arbitrary choice of phase involved in fixing N itself: it is always chosen to be real, and the standard (Condon–Shortley) phase convention takes it to be *positive*. Within this convention, then,

$$J_+|j, m\rangle = \hbar[(j-m)(j+m+1)]^{1/2}|j, m+1\rangle \tag{B8}$$

Similarly

$$J_-|j, m\rangle = \hbar[(j+m)(j-m+1)]^{1/2}|j, m-1\rangle$$

Hence the matrix elements of J_z, $J_\pm$ are as given in equations (6.33), taking account of the factor of $\hbar$ between $\boldsymbol{J}$ and $\boldsymbol{X}$. These specify an irreducible representation of the SU(2) algebra by $(2j+1) \times (2j+1)$ matrices.

Addition of Angular Momenta

In quantum mechanics the problem of finding the decomposition of the direct product of representations is usually cast in the language of addition of angular momenta, the angular momentum operators being the generators which give finite rotations upon exponentiation. The Clebsch–Gordan coefficients of equation (6.44), and indeed the Clebsch–Gordan series itself, can be constructed by a systematic application of raising and lowering operators.

In general when two systems originally specified by the eigenvalues of their individual angular momentum operators $\boldsymbol{J}_1$, $\boldsymbol{J}_2$ are combined, the 'magnetic' quantum numbers m_1 and m_2 are no longer separately conserved. For example, $\boldsymbol{J}_1$ could refer to the orbital angular momentum $\boldsymbol{L}$, and $\boldsymbol{J}_2$ to the spin $\boldsymbol{S}$ of the electron in a hydrogen atom. At the first level of approximation the Hamiltonian is spin-independent and invariant under rotations of the spatial coordinates, so that $[H, J_1] = [H, J_2] = 0$. However, when the spin–orbit interaction, proportional to $\boldsymbol{L}\cdot\boldsymbol{S}$, is taken into account, H is only invariant under simultaneous rotations of the spatial and spin states, so that $[H, \boldsymbol{J}] = 0$, where $\boldsymbol{J}$ is the total angular momentum operator $\boldsymbol{J} = \boldsymbol{J}_1 + \boldsymbol{J}_2$. Thus it is appropriate to reassemble the product state vectors $|j_1m_1; j_2m_2\rangle$ into linear combinations which, as eigenstates of $\boldsymbol{J}^2$, $\boldsymbol{J}_z$, reflect this reduced invariance. Since $\boldsymbol{L}\cdot\boldsymbol{S}$ commutes with both $\boldsymbol{L}^2$ and $\boldsymbol{S}^2$, the quantum numbers j_1, j_2 remain good quantum numbers and may be retained. The appropriate states are therefore labelled as $|(j_1j_2)jm\rangle$.

First a few words about notation. When we write $\boldsymbol{J} = \boldsymbol{J}_1 + \boldsymbol{J}_2$ we are referring to an operator which acts on the combined state $|j_1m_1; j_2m_2\rangle$, often written as $|j_1m_1\rangle|j_2m_2\rangle$. However, $\boldsymbol{J}_1$ only acts on the first factor, and $\boldsymbol{J}_2$ on the second. It is useful to make this explicit by writing instead $\boldsymbol{J} = \boldsymbol{J}_1\mathbb{1}_2 + \mathbb{1}_1\boldsymbol{J}_2$, where $\mathbb{1}$ is the identity operator. The raising and lowering operators constructed from $\boldsymbol{J}_1$ will be written as $J_{1\pm}$, and similarly for $\boldsymbol{J}_2$.

The procedure is constructive and begins with the state $|j_1j_1; j_2j_2\rangle$, that is, the 'top' state of both systems, with eigenvalue $m = j_1 + j_2$ for the z coordinate of the total angular momentum:

$$J_z|j_1j_1; j_2j_2\rangle = (j_1 + j_2)\hbar|j_1j_1; j_2j_2\rangle \tag{B9}$$

As the top state it also satisfies $J_{1+}|j_1j_1\rangle = 0$ and $J_{2+}|j_2j_2\rangle = 0$, and therefore

$$J_+|j_1j_1; j_2j_2\rangle = 0 \tag{B10}$$

where $J_+ := J_{1+}\mathbb{1}_2 + \mathbb{1}_1J_{2+}$ is the raising operator for the total angular momentum. Treating $\boldsymbol{J}$ as a single entity satisfying the commutation relations of angular momentum, we can deduce from the arguments of the first part of the appendix that this state is an eigenstate of $\boldsymbol{J}^2$ with $j = j_1 + j_2$. Thus we can identify

$$|(j_1j_2)j_1 + j_2, j_1 + j_2\rangle = |j_1j_1; j_2j_2\rangle \tag{B11}$$

From this state all the other states with $j = j_1 + j_2$ and with m lying between $j_1 + j_2 - 1$ and $-(j_1 + j_2)$ can be constructed by repeated application of the lowering operator J_-, as shown in figure B3. In applying J_- to the left-hand side we treat it as a single entity, but in acting on the states on the right-hand side we recognize its decomposition into the individual operators $J_{1-}\mathbb{1}_2 + \mathbb{1}_1J_{2-}$.

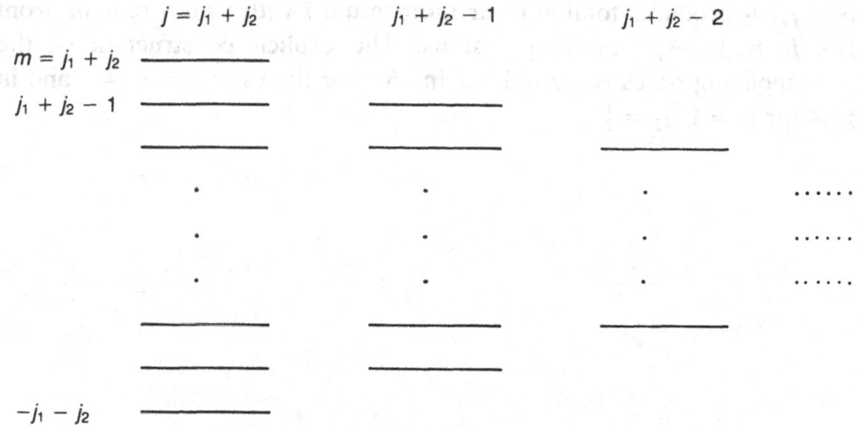

Figure B3 Addition of angular momenta j_1 and j_2.

There are two states with $m = j_1 + j_2 - 1$, namely $|j_1\, j_1 - 1;\, j_2\, j_2\rangle$ and $|j_1\, j_1;\, j_2\, j_2 - 1\rangle$. One linear combination of these must be the state $|(j_1\, j_2)\, j_1 + j_2,\ j_1 + j_2 - 1\rangle$ obtained by lowering (B11). However, another combination $\alpha|j_1\, j_1 - 1;\, j_2\, j_2\rangle + \beta|j_1\, j_1;\, j_2\, j_2 - 1\rangle$ can be constructed which is annihilated by the raising operator J_+. Thus

$$J_+(\alpha|j_1\, j_1 - 1;\, j_2\, j_2\rangle + \beta|j_1\, j_1;\, j_2\, j_2 - 1\rangle)$$
$$= [\alpha(2j_1)^{1/2} + \beta(2j_2)^{1/2}]|j_1\, j_1;\, j_2\, j_2\rangle$$

which is zero if $\alpha = -\beta(j_1/j_2)^{1/2}$. The state thus obtained, being annihilated by J_+ at an m value of $j_1 + j_2 - 1$, can be identified with $|(j_1\, j_2)\, j_1 + j_2 - 1,\ j_1 + j_2 - 1\rangle$, the top state with $j = j_1 + j_2 - 1$. All the states with lower values of m can again be generated by J_-, as shown in figure B3.

We then progress to the level $m = j_1 + j_2 - 2$, construct the state which is annihilated by J_+ and generate all the other states with $j = j_1 + j_2 - 2$ by the action of J_-. The process continues until we run out of states, which occurs at the level $j = |j_1 - j_2|$. For each j value the multiplicity of states is $2j + 1$. Thus the total multiplicity is

$$\sum_{|j_1-j_2|}^{j_1+j_2} (2j + 1) = \sum_{|j_1-j_2|}^{j_1+j_2} [(j + 1)^2 - j^2]$$
$$= (j_1 + j_2 + 1)^2 - (j_1 - j_2)^2$$
$$= (2j_1 + 1)(2j_2 + 1)$$

which was the original multiplicity of the states $|j_1 m_1;\, j_2 m_2\rangle$.

Thus the combination of two angular momenta with quantum numbers j_1, j_2 produces total angular momentum j with values ranging from $j_1 + j_2$ to $|j_1 - j_2|$ in integer steps. The explicit construction of the corresponding states is carried out in §6.3 for the case $j_1 = j_2 = \frac{1}{2}$ and in §7.4 for $j_1 = 1$, $j_2 = \frac{1}{2}$.

C

Group-invariant Measure for SO(3)

In the case of finite groups we made frequent use of the sum Σ_g over all members of the group. This has the almost trivial property that $\Sigma_{g'} = \Sigma_g$, where $g' = hg$, h being any fixed element of G. For compact Lie groups, whose continuous parameters range over a finite interval, it is not so trivial to find the integral $\int \mathrm{d}\mu(g)$ which replaces Σ_g and has the analogous property that $\int \mathrm{d}\mu(g') = \int \mathrm{d}\mu(g)$. In this appendix we find the required measure $\mathrm{d}\mu(g)$ for the rotation group SO(3).

The first question to be addressed is how best to parametrize the elements of the group. One possibility is to use the Euler angles α, β, γ, whereby any general rotation is expressed as $R(\alpha, \beta, \gamma) = R_3(\alpha)R_2(\beta)R_3(\gamma)$. However, for our present purposes it turns out to be better to use the specification $R_{\boldsymbol{n}}(\psi)$, identifying the rotation by the unit vector $\boldsymbol{n}$ along its axis and by ψ, the angle of rotation. We have changed the notation from that of Chapter 6 in order to leave θ and φ to denote the polar and azimuthal angles respectively of the vector $\boldsymbol{n}$ (i.e. $\boldsymbol{n} = (\sin\theta\cos\varphi, \sin\theta\sin\varphi, \cos\theta)$). This vector can be represented by a point N, say, on the surface of a unit sphere.

Now suppose we have two rotations R_{A} and R_{B} through angles ψ_{a}, ψ_{b}, with their respective axes represented by points A and B on the unit sphere, as in figure C1. There is a very elegant geometrical construction whereby we can find the point C and the angle ψ_{c} which specify the product rotation $R_{\mathrm{B}}R_{\mathrm{A}}$.

If we apply this product rotation to the point A, the first rotation R_{A} leaves it invariant and the second rotation R_{B} moves it to the point A′ shown in figure C1. The lines BA and BA′ are arcs of great circles, of equal length, and the angle between them is ψ_{b}. A moment's thought should convince you that the end of the axis of any rotation taking A into A′ must lie on the dotted line, which is the great circle bisecting

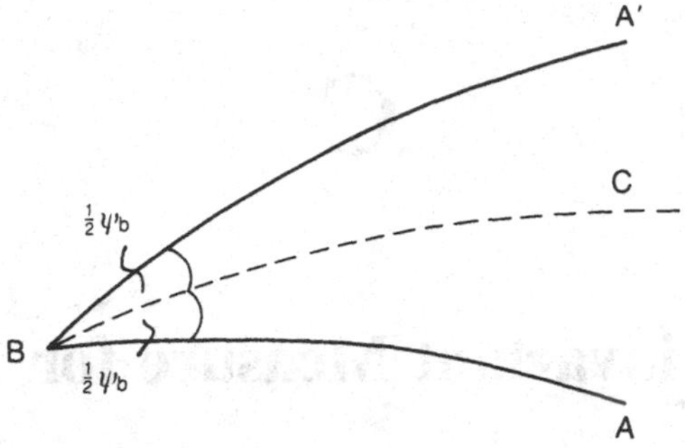

Figure C1 Geometrical construction for the product rotation $R_B R_A$. C must lie on the great circle bisecting AB and A′B.

BA and BA′. We can completely fix the position of C by one further condition, namely that it should remain invariant under the combined rotation $R_B R_A$. It takes perhaps more than a moment's thought to spot the solution, which is that C is the third vertex of the spherical triangle with angles $\frac{1}{2}\psi_a$ and $\frac{1}{2}\psi_b$ at A and B respectively, as shown in figure C2. The point is that the first rotation R_A takes C into the mirror image point C′, and the second rotation R_B takes it back again.

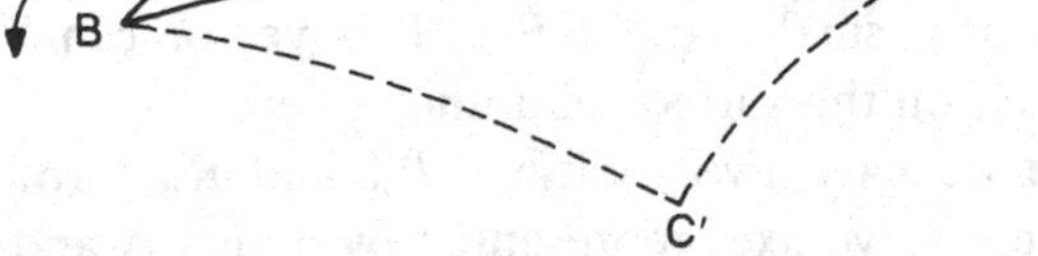

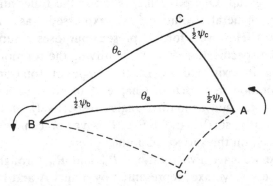

Figure C2 C is the third vertex of the spherical triangle shown.

Not only does this construction determine the position of the axis, it also gives the angle ψ_c of the product rotation! In fact $\frac{1}{2}\psi_c$ is just the external angle at C of the spherical triangle ABC, as is clear by considering again the transformation of the point A.

This geometric construction now enables us to relate the angular parameters θ_c, φ_c and ψ_c to the corresponding parameters of the individual rotations. We are considering R_{B} to be the fixed rotation, in analogy to h in the case of a finite group, and it is convenient to choose the z axis along $\boldsymbol{n}_{\mathrm{b}}$. The relation between the azimuthal angles is then just

$$\varphi_{\mathrm{c}} = \varphi_{\mathrm{a}} + \tfrac{1}{2}\psi_{\mathrm{b}} \tag{C1}$$

The sine rule for the spherical triangle ABC gives

$$\sin\theta_{\mathrm{c}} \sin\tfrac{1}{2}\psi_{\mathrm{c}} = \sin\theta_{\mathrm{a}} \sin\tfrac{1}{2}\psi_{\mathrm{a}} \tag{C2}$$

while the cosine rule gives

$$\cos\tfrac{1}{2}\psi_{\mathrm{c}} = \cos\tfrac{1}{2}\psi_{\mathrm{a}} \cos\tfrac{1}{2}\psi_{\mathrm{b}} - \sin\tfrac{1}{2}\psi_{\mathrm{a}} \sin\tfrac{1}{2}\psi_{\mathrm{b}} \cos\theta_{\mathrm{a}} \tag{C3}$$

Another trigonometric relation can be derived from equations (C2) and (C3), namely

$$\cos\theta_{\mathrm{c}} \sin\tfrac{1}{2}\psi_{\mathrm{c}} = \cos\tfrac{1}{2}\psi_{\mathrm{a}} \sin\tfrac{1}{2}\psi_{\mathrm{b}} + \sin\tfrac{1}{2}\psi_{\mathrm{a}} \cos\tfrac{1}{2}\psi_{\mathrm{b}} \cos\theta_{\mathrm{a}} \tag{C4}$$

From these equations we wish to find the Jacobian $\partial(\psi_{\mathrm{c}}, \cos\theta_{\mathrm{c}})/\partial(\psi_{\mathrm{a}}, \cos\theta_{\mathrm{a}})$. This can be done in two stages using the identity

$$\frac{\partial(\psi_{\mathrm{c}}, \cos\theta_{\mathrm{c}})}{\partial(\psi_{\mathrm{a}}, \cos\theta_{\mathrm{a}})} = \frac{\partial(\psi_{\mathrm{c}}, \cos\theta_{\mathrm{c}})}{\partial(\cos\frac{1}{2}\psi_{\mathrm{c}}, \sin\theta_{\mathrm{c}} \sin\frac{1}{2}\psi_{\mathrm{c}})} \; \frac{\partial(\cos\frac{1}{2}\psi_{\mathrm{c}}, \sin\theta_{\mathrm{c}} \sin\frac{1}{2}\psi_{\mathrm{c}})}{\partial(\psi_{\mathrm{a}}, \cos\theta_{\mathrm{a}})} \tag{C5}$$

The first factor is just the inverse of

$$\frac{\partial(\cos\frac{1}{2}\psi_{\mathrm{c}}, \sin\theta_{\mathrm{c}} \sin\frac{1}{2}\psi_{\mathrm{c}})}{\partial(\psi_{\mathrm{c}}, \cos\theta_{\mathrm{c}})} = \begin{vmatrix} -\frac{1}{2}\sin\frac{1}{2}\psi_{\mathrm{c}} & \frac{1}{2}\sin\theta_{\mathrm{c}}\cos\frac{1}{2}\psi_{\mathrm{c}} \\ 0 & -\cot\theta_{\mathrm{c}}\sin\frac{1}{2}\psi_{\mathrm{c}} \end{vmatrix}$$

$$= \tfrac{1}{2}\sin^2\tfrac{1}{2}\psi_{\mathrm{c}} \cot\theta_{\mathrm{c}}$$

while the second is

$$\frac{\partial(\cos\frac{1}{2}\psi_{\mathrm{c}}, \sin\theta_{\mathrm{c}})}{\partial(\psi_{\mathrm{a}}, \cos\theta_{\mathrm{a}})}$$

$$= \begin{vmatrix} -\frac{1}{2}\sin\frac{1}{2}\psi_{\mathrm{a}}\cos\frac{1}{2}\psi_{\mathrm{b}} - \frac{1}{2}\cos\frac{1}{2}\psi_{\mathrm{a}}\sin\frac{1}{2}\psi_{\mathrm{b}}\cos\theta_{\mathrm{a}} & -\sin\frac{1}{2}\psi_{\mathrm{a}}\sin\frac{1}{2}\psi_{\mathrm{b}} \\ \frac{1}{2}\sin\theta_{\mathrm{a}}\cos\frac{1}{2}\psi_{\mathrm{a}} & -\cot\theta_{\mathrm{a}}\sin\frac{1}{2}\psi_{\mathrm{a}} \end{vmatrix}$$

$$= (\sin\tfrac{1}{2}\psi_{\mathrm{a}}/2\sin\theta_{\mathrm{a}})(\cos\tfrac{1}{2}\psi_{\mathrm{b}}\sin\tfrac{1}{2}\psi_{\mathrm{a}}\cos\theta_{\mathrm{a}} + \sin\tfrac{1}{2}\psi_{\mathrm{b}}\cos\tfrac{1}{2}\psi_{\mathrm{a}})$$

$$= \tfrac{1}{2}\sin\tfrac{1}{2}\psi_{\mathrm{a}}\sin\tfrac{1}{2}\psi_{\mathrm{c}}\cos\theta_{\mathrm{c}}/\sin\theta_{\mathrm{a}}$$

by virtue of (C4). So altogether

$$\frac{\partial(\psi_{\mathrm{c}}, \cos\theta_{\mathrm{c}})}{\partial(\psi_{\mathrm{a}}, \cos\theta_{\mathrm{a}})} = \frac{\sin\frac{1}{2}\psi_{\mathrm{a}}\sin\theta_{\mathrm{c}}}{\sin\frac{1}{2}\psi_{\mathrm{c}}\sin\theta_{\mathrm{a}}} = \frac{\sin^2\frac{1}{2}\psi_{\mathrm{a}}}{\sin^2\frac{1}{2}\psi_{\mathrm{c}}} \tag{C6}$$

Thus we have shown that the measure $\sin^2 \frac{1}{2}\psi \, d\psi \, d(\cos\theta) \, d\varphi$ is group-invariant. Note that the θ, φ part must be the differential solid angle $d\Omega = d(\cos\theta)\, d\varphi$ to ensure invariance under similarity transformations such as the one we effectively used to align $\boldsymbol{n}_b$ with the z axis. Normalizing it to 1, the measure is

$$d\mu(g) = \frac{1}{4\pi^2} \sin^2 \tfrac{1}{2}\psi \, d\psi \, d(\cos\theta) \, d\varphi \qquad \text{(C7)}$$

When it comes to integrating over characters, which depend only on the angle of rotation and not on the orientation of the axis, the integration over θ and φ just gives the total solid angle 4π, and we are left with the measure $d\mu(\psi) = (1/\pi)\sin^2 \frac{1}{2}\psi \, d\psi$, as in equation (6.40).

D

Calculation of roots for SO(n) and Sp($2r$)

SO($2r$)

The generators of SO(n) are antisymmetric Hermitian matrices, generalizations of σ_2, like equation (6.15):

$$(L_{ab})_{kl} = -\mathrm{i}(\delta_{ak}\delta_{bl} - \delta_{al}\delta_{bk}) \tag{D1}$$

whose commutation relations can be evaluated, after some algebra, as

$$[L_{ab}, L_{cd}] = \mathrm{i}(\delta_{ac}L_{bd} - \delta_{ad}L_{bc} + \delta_{bd}L_{ac} - \delta_{bc}L_{ad}). \tag{D2}$$

This is zero if the two L's have no indices in common. So if n is even, with $n = 2r$, we can take as the r elements of the Cartan subalgebra the generators $L_{12}, L_{34}, \ldots, L_{n-1,n}$, i.e.

$$H_i = L_{2i-1,2i} \qquad i = 1 \ldots r. \tag{D3}$$

If the indices are restricted to the range $1 \ldots 3$, the generators L_{12}, L_{13} and L_{23} are just the generators L_3, $-L_2$ and L_1 of SO(3), with commutation relations

$$\begin{aligned}[L_{12}, L_{13}] &= \mathrm{i}L_{23}\\ [L_{12}, L_{23}] &= -\mathrm{i}L_{13}.\end{aligned} \tag{D4}$$

Thus we know that $L_{13} + \mathrm{i}L_{23}$, which is analogous to $\mathrm{i}(L_1 + \mathrm{i}L_2)$, is a raising operator for L_{12}. Unfortunately it also has an effect on L_{34} which is not that of a raising or lowering operator, namely $[L_{34}, L_{13} + \mathrm{i}L_{23}] = \mathrm{i}(L_{14} + \mathrm{i}L_{24})$. However, by taking appropriate combinations of $L_{13} \pm \mathrm{i}L_{23}$ and $L_{14} \pm \mathrm{i}L_{24}$, namely $L(\varepsilon, \eta) = (L_{13} + \mathrm{i}\varepsilon L_{23}) + \mathrm{i}\eta(L_{14} + \mathrm{i}\varepsilon L_{24})$, where ε and η are independently ± 1, we can indeed construct four step operators, with commutation relations:

$$\begin{aligned}[L_{12}, L(\varepsilon, \eta)] &= \varepsilon L(\varepsilon, \eta)\\ [L_{12}, L(\varepsilon, \eta)] &= \eta L(\varepsilon, \eta)\end{aligned} \tag{D5}$$

while they commute with the other members of the Cartan subalgebra H_i for $i > 2$. Thus we have four roots $(\pm 1, \pm 1, 0, \ldots 0)$, or $\pm e_1 \pm e_2$ in the notation used previously for SU(n). Clearly this process can be extended, to produce the set of roots $\pm e_i \pm e_j$ for $i < j$, each in the range $1 \ldots r$. All the roots are accounted for in this way, since they number $4 \times \frac{1}{2} r(r-1) = 2r(r-1)$. Together with the r generators of the Cartan subalgebra, this gives $r(2r-1) = \frac{1}{2} n(n-1)$, the number of independent real antisymmetric $n \times n$ matrices.

In this case the simple roots can be taken as

$$\begin{cases} \alpha_i = e_i - e_{i+1} & i = 1 \ldots r-1 \\ \alpha_r = e_{r-1} + e_r . & \end{cases} \tag{D6}$$

To verify that this is correct we need to show that all the roots $\pm e_i \pm e_j$ can be expressed as integral linear combinations of the α_k with all positive or all negative coefficients. First consider $e_i - e_j$ for $i < j$. As in the case of SU(n), this can be expressed in the form (9.64). On the other hand, the sum $e_i + e_j$ can be written as

$$\begin{aligned} e_i + e_j &= (e_i - e_j) + 2(e_j - e_{r-1}) + (e_{r-1} - e_r) + (e_{r-1} + e_r) \\ &= \sum_{k=i}^{j-1} \alpha_k + 2 \sum_{j}^{r-1} \alpha_k + \alpha_{r-1} + \alpha_r . \end{aligned} \tag{D7}$$

These are the positive roots. The negative roots are $e_j - e_i$ for $i < j$ and $-(e_i + e_j)$.

SO($2r + 1$)

In the case of SO($2r + 1$) we have the additional index $q \equiv 2r + 1$ and an additional r generators, L_{aq}, together with their opposite numbers, L_{qa}. These do not commute with the elements of the Cartan subalgebra, so the rank is still r, but they make up the total number of generators to $r(2r-1) + 2r = r(2r+1) = \frac{1}{2} n(n-1)$, as required. Considering first L_{1q} and L_{2q}, it is clear that the combinations $L_{1q} \pm \mathrm{i} L_{2q}$ act as raising and lowering operators for L_{12}, but commute with the other H's. Similar combinations can be formed from L_{3q} and L_{4q}, etc. The corresponding roots are $\pm e_i$, $i = 1, \ldots r$. The simple roots α_i

must be such that the positive roots $e_i - e_j$ for $i < j$, $e_i + e_j$ and e_i can all be expressed as positive integer combinations of them. In this case it turns out that we can choose

$$\begin{cases} \alpha_i = e_i - e_{i+1} & i = 1, \ldots r-1 \\ \alpha_r = e_r \end{cases} \tag{D8}$$

rather than $e_{r-1} + e_r$, as was the case in equation (9.66). The required combination for $e_i - e_j$ is the same as before (equation (9.64)). We now express $e_i + e_j$ in the form

$$\begin{aligned} e_i + e_j &= (e_i - e_j) + 2(e_j - e_r) + 2e_r \\ &= \sum_{k=i}^{j-1} \alpha_k + 2\sum_{j}^{r-1} \alpha_k + 2\alpha_r. \end{aligned} \tag{D9}$$

Finally,

$$e_i = (e_i - e_r) + \alpha_r = \sum_{k=i}^{r} \alpha_k. \tag{D10}$$

Sp(2*r*)

In equation (9.69) we have shown the general form of a generator G. From this we have to identify the elements of the Cartan subalgebra and the step operators. The calculation simplifies if G is written in direct product notation, namely $G = \mathbb{1} \times A + \sigma_3 \times S_3 + \sigma_+ \times S_+ + \sigma_- \times S_-$.

The elements of the Cartan subalgebra can be taken as $\sigma_3 \times \mathbb{1}_i$, where $(\mathbb{1}_i)_{kl} = \delta_{ik}\delta_{il}$. Unlike the case of SU($n$), these $r \times r$ matrices do not need to be traceless: the overall tracelessness of the complete $2r \times 2r$ matrix is ensured by the factor of σ_3. Then, one set of step operators are the E_{ab} of equation (9.60). As for SU(n), these give $r(r-1)$ roots, $e_a - e_b$. An additional set of step operators are, in direct product form, $\sigma_+ \times (\sigma_1)_{ab}$, where $(\sigma_1)_{ab}$ is a σ_1 matrix in the $a - b$ subspace, with matrix elements $((\sigma_1)_{ab})_{kl} = \delta_{ak}\delta_{bl} + \delta_{al}\delta_{bk}$. Thus, for a general element $X = x \cdot H$ of the Cartan subalgebra

$$
\begin{aligned}
[\sigma_3 \times X, \sigma_+ \times (\sigma_1)_{ab}] &= \sigma_+ \times \left[\begin{pmatrix} x_a & 0 \\ 0 & x_b \end{pmatrix} \begin{pmatrix} 0 & 1 \\ 1 & 0 \end{pmatrix} \right. \\
&\quad \left. + \begin{pmatrix} 0 & 1 \\ 1 & 0 \end{pmatrix} \begin{pmatrix} x_a & 0 \\ 0 & x_b \end{pmatrix} \right] \\
&= \sigma_+ \times \left[\begin{pmatrix} 0 & x_a \\ x_b & 0 \end{pmatrix} + \begin{pmatrix} 0 & x_b \\ x_a & 0 \end{pmatrix} \right] \\
&= (x_a + x_b)\ (\sigma_+ \times (\sigma_1)_{ab}) \\
&= x \cdot (e_a + e_b)\ (\sigma_+ \times (\sigma_1)_{ab})
\end{aligned} \tag{D11}
$$

Thus we have another $r(r-1)$ roots $\pm(e_a + e_b)$ for $b \neq a$, together with $2r$ roots $\pm 2e_a$. Altogether we have $2r^2$ step operators, which together with the r diagonal elements make up $r(2r+1)$, the order of the group. For Sp($2r$) the simple roots may be taken as

$$
\left\{ \begin{aligned} \alpha_i &= e_i - e_{i+1} \qquad i = 1, \ldots r-1 \\ \alpha_r &= 2e_r . \end{aligned} \right. \tag{D12}
$$

The proof is on the lines of previous ones, and is left to the reader. The reason that the last root cannot be taken as e_r is that then we would have the root $2e_r = 2\alpha_r$, which is not allowed.

E

Covariant Normalization and Relativistic Scattering

When we come to consider the scattering of relativistic particles it is appropriate to use a formalism in which Lorentz invariance is made manifest as far as possible. In this spirit it is desirable to replace the non-relativistic normalization of plane wave states, equation (10.82), by a covariant normalization, which takes the same form in all frames of reference. The 3-dimensional delta function does not have this property since it is defined by

$$\int d^3\boldsymbol{p}\, \delta^3(\boldsymbol{p}-\boldsymbol{p}')g(\boldsymbol{p}) = g(\boldsymbol{p}')$$

for a suitable test function $g(\boldsymbol{p})$, and the volume element $d^3\boldsymbol{p}$ is not a Lorentz invariant. Consider for definiteness a Lorentz transformation along the z axis whereby $p_1' = p_1$, $p_2' = p_2$ and $p_3' = p_3 \cosh\zeta + E\sinh\zeta$. The Jacobian is just

$$\partial p_3'/\partial p_3 = \cosh\zeta + (p_3/E)\sinh\zeta = E'/E$$

Thus the invariant volume element in p space is $d^3\boldsymbol{p}/E$. A more elegant way of deriving this results from considering the manifestly invariant volume element $d^4p\, \delta^{(+)}(p^2 - m^2)$, where $\delta^{(+)}(p^2 - m^2) := \theta(p_0)\delta(p^2 - m^2)$ is a delta function which imposes $p_0 = E(\boldsymbol{p}) = +\surd(\boldsymbol{p}^2 + m^2)$. Referring to Appendix A we see that $\delta(p^2 - m^2) = [\delta(p_0 - E(\boldsymbol{p})) + \delta(p_0 + E(\boldsymbol{p}))]/2E(\boldsymbol{p})$, from which the step function $\theta(p_0)$ picks out the positive energy part ($\theta(x) = 1$ for $x > 0$, zero otherwise). Performing the p_0 integration we can therefore write $d^4p\, \delta^{(+)}(p^2 - m^2) = d^3\boldsymbol{p}/2E$.

In performing the Fourier transform from momentum space to configuration space one inevitably encounters factors of $(2\pi)^{1/2}$. A very elegant way of keeping track of these factors is always to associate a momentum integration with a factor of $1/2\pi$ for each component of momentum involved. This can be indicated in shorthand notation by replacing the differential d symbol by đ, in analogy with the relation between h and $\hbar$. We will therefore take the invariant volume element as

$$(\mathrm{d}\boldsymbol{p}) := \frac{\mathrm{d}^3\boldsymbol{p}}{2E(2\pi)^3} = \frac{đ^3\boldsymbol{p}}{2E} \tag{E1}$$

Correspondingly, every momentum delta function is naturally associated with a factor of 2π for each component involved, and it is useful to indicate this by a 'crossed' delta $\eth$. Thus $\eth^3(\boldsymbol{p}' - \boldsymbol{p}) = (2\pi)^3\delta^3(\boldsymbol{p}' - \boldsymbol{p})$, for example. If, indeed, we wish the completeness relation for momentum eigenstates to be

$$\sum_\lambda \int (\mathrm{d}\boldsymbol{p})|\boldsymbol{p}, \lambda\rangle\langle\boldsymbol{p}, \lambda| = \mathbb{1} \tag{E2}$$

the corresponding normalization is

$$\langle\boldsymbol{p}', \lambda'|\boldsymbol{p}, \lambda\rangle = 2E\eth^3(\boldsymbol{p}' - \boldsymbol{p})\delta_{\lambda'\lambda} \tag{E3}$$

rather than equation (10.82).

Scattering from an initial state $|i\rangle$ to a final state $|f\rangle$ is governed by the S operator, defined in analogy with equation (7.44), so that

$$^{\text{out}}\langle f|i\rangle^{\text{in}} = {}^{\text{out}}\langle f|S|i\rangle^{\text{out}} \tag{E4}$$

The initial state will always be a two-particle state $|\boldsymbol{p}_1\lambda_1; \boldsymbol{p}_2\lambda_2\rangle$, but the essence of relativity is that energy can be exchanged for mass, so the final state may contain more than two particles, as, for example, in $\pi\mathrm{N} \rightarrow \pi\pi\mathrm{N}$.

Translational invariance of the interaction implies momentum conservation, and the S matrix elements will always contain a delta function $\eth^4(P_f - P_i)$ conserving the overall 4-momentum. Exhibiting this factor explicitly we write

$$^{\text{out}}\langle f|S|i\rangle^{\text{out}} = \eth^4(P_f - P_i)[\mathbb{1} + \mathrm{i}T]_{fi} \tag{E5}$$

which is the relativistic analogue of equation (7.43). To calculate the transition probability one must take the squared modulus of this expression, which formally leads to a factor of $\eth^4(0)$. A more careful analysis using wave packets rather than plane wave states shows that this factor is to be interpreted as VT, the space–time volume. It is natural that the transition probability should be proportional to the time, and we can eliminate the factor of T by considering instead the transition *rate*. As for the factor of V, we must recognize that the spatial density corresponding to equation (E3) is $2E$ per unit volume. The *cross-section* is defined as the transition rate divided by the incident flux, which itself is proportional to the volume. Thus the cross-section $d\sigma$ is given by

$$d\sigma = \frac{1}{4F}(\Sigma\bar{\Sigma}|T_{fi}|^2)\,d\rho \tag{E6}$$

where $4F$ is the initial flux per unit volume, namely $(2E_1)(2E_2)v_{12}$, where v_{12} is the relative speed (in a frame where $\boldsymbol{p}_1$ and $\boldsymbol{p}_2$ are parallel). An explicitly Lorentz-invariant form of F is

$$F = ((p_1.p_2)^2 - m_1^2 m_2^2)^{1/2} \tag{E7}$$

In the rest-frame of particle 1, where $p_1 = (m_1, \mathbf{0})$, this reduces to $m_1 p_2$, which is precisely what we expect, namely $F = \frac{1}{4}(2m_1)(2E_2)(p_2/E_2)$. It is left as an exercise to the reader to show that in the centre-of-mass frame F becomes qW, where $W := E_1 + E_2$ is the total energy and q the magnitude of the momentum of either particle.

The second factor in equation (E6) is just the modulus squared of T_{fi}, summed and averaged over final and initial quantum numbers respectively. The final factor is the density-of-states factor

$$d\rho = \left(\prod_r (d\boldsymbol{p}_r)\right) \delta^4(P_f - P_i) \tag{E8}$$

where the product extends over the momenta $\boldsymbol{p}_r$ of the final state particles. This again is explicitly Lorentz-invariant and can be evaluated in any convenient frame. For a two-particle final state $|\boldsymbol{p}_3\lambda_3; \boldsymbol{p}_4\lambda_4\rangle$ in the centre-of-mass frame, where $P_i = (W; \mathbf{0})$, we have

$$d\rho = \frac{1}{4E_3E_4}\frac{1}{(2\pi)^2}(q')^2\,dq'\,d\Omega\,\delta(E_3 + E_4 - W)$$

The 3-dimensional delta function has been used to eliminate $\boldsymbol{p}_4$, setting $\boldsymbol{p}_4 = -\boldsymbol{p}_3 = -\boldsymbol{q}'$, say, so that now $E_4 = \sqrt{[(q')^2 + m_4]}$. The remaining delta function may be used to fix q' at the cost of a factor of $[d(E_3 + E_4)/dq']^{-1} = E_3E_4/Wq'$. Thus

$$d\rho_{CM} = \frac{q'}{16\pi^2 W}\,d\Omega \tag{E9}$$

Putting this all together, the differential cross-section in the centre-of-mass frame for the process $p_1 + p_2 \rightarrow p_3 + p_4$ is

$$\frac{d\sigma}{d\Omega} = \frac{1}{(8\pi W)^2}\frac{q'}{q}\Sigma\bar{\Sigma}|T_{fi}|^2 \tag{E10}$$

F

Lagrangian Mechanics

Lagrange's Equations

Suppose we have an assembly of N particles, with individual masses $m^{(r)}$ and positions $\boldsymbol{x}^{(r)}$, interacting via a potential V. The total kinetic energy is

$$T = \tfrac{1}{2} \sum_{r=1}^{N} m^{(r)} (\dot{\boldsymbol{x}}^{(r)})^2 \tag{F1}$$

and the potential V is usually a function of the positions, $V = V(\{\boldsymbol{x}^{(r)}\})$. Then the ith component of the force acting on particle r is $F_i^{(r)} = -\partial V / \partial x_i^{(r)}$. By taking the derivative of T with respect to $\dot{x}_i^{(r)}$, we obtain $m^{(r)} \dot{x}_i^{(r)}$. Hence Newton's equations, force = mass × acceleration, can be written in the form

$$\frac{\mathrm{d}}{\mathrm{d}t} \left(\frac{\partial T}{\partial \dot{x}_i^{(r)}} \right) = - \frac{\partial V}{\partial x_i^{(r)}} \tag{F2}$$

The Lagrangian L is defined as the *difference* of the total kinetic and potential energies:

$$L = T - V \tag{F3}$$

If V is indeed a function of position only, equation (F2) can be rewritten in terms of L as

$$\boxed{\frac{\mathrm{d}}{\mathrm{d}t} \left(\frac{\partial L}{\partial \dot{x}_i^{(r)}} \right) = \frac{\partial L}{\partial x_i^{(r)}}} \tag{F4}$$

which constitute Lagrange's equations.

Hamilton's Principle

The importance of Lagrange's equations is that they can be derived from a variational principle; that is, they emerge as the condition that a certain quantity, S, be stationary with respect to possible variations in the motions of the particles. S is defined as the time integral of L:

$$S(t_1, t_2) = \int_{t_1}^{t_2} L(\{\boldsymbol{x}^{(r)}\}, \{\dot{\boldsymbol{x}}^{(r)}\})\,\mathrm{d}t \tag{F5}$$

and *Hamilton's principle* states that the actual motion is a stationary point of S with respect to small variations $\delta\boldsymbol{x}^{(r)}(t)$ of the particle positions which vanish at the end points: $\delta\boldsymbol{x}^{(r)}(t_1) = \delta\boldsymbol{x}^{(r)}(t_2) = 0$. Thus

$$\delta S = \int_{t_1}^{t_2} \sum_r \left(\frac{\partial L}{\partial x_i^{(r)}}\,\delta x_i^{(r)} + \frac{\partial L}{\partial \dot{x}_i^{(r)}}\,\delta \dot{x}_i^{(r)} \right) \mathrm{d}t$$

with a summation over the repeated Cartesian index i understood. It is important to note that the variations of position and velocity are not independent, but are related by $\delta\dot{x}_i = \mathrm{d}(\delta x_i)/\mathrm{d}t$. Then we can integrate the second term by parts, to obtain

$$\delta S = \int_{t_1}^{t_2} \sum_r \left[\frac{\partial L}{\partial x_i^{(r)}} - \frac{\mathrm{d}}{\mathrm{d}t}\left(\frac{\partial L}{\partial \dot{x}_i^{(r)}} \right) \right] \delta x_i^{(r)}\,\mathrm{d}t$$

with no boundary terms because of the condition $\delta\boldsymbol{x}^{(r)}(t_1) = \delta\boldsymbol{x}^{(r)}(t_2) = 0$.

Because all the $x_i^{(r)}$ are allowed to vary independently, as arbitrary functions of time subject only to the above condition at the end points, δS is stationary only if each term in square brackets vanishes. But these are precisely the Lagrange equations for the $3N$ coordinates.

The great advantage of a variational principle like Hamilton's principle is that it is coordinate-independent, i.e. independent of the variables we choose as arguments of L. In the above, we took L as a function of the Cartesian position components and their time derivatives, but we could equally well have used various linear combinations, for example centre-of-mass and relative coordinates. There is no need to restrict ourselves to Cartesian coordinates: we could use spherical polar coordinates, cylindrical coordinates, indeed any set of $3N$ independent quantities $q_\alpha(t)$. By exactly similar steps, Hamilton's principle then leads to the *Euler–Lagrange* equations

$$\boxed{\frac{\mathrm{d}}{\mathrm{d}t}\left(\frac{\partial L}{\partial \dot{q}_\alpha} \right) = \frac{\partial L}{\partial q_\alpha}} \tag{F6}$$

for the generalized coordinates q_α.

Hamilton's Equations

When the kinetic energy is a quadratic function of the $\dot{q}_\alpha$, the Euler–Lagrange equations are a set of $3N$ second-order differential equations. For some purposes it is useful to recast these as a set of $6N$ first-order equations. To this end we define the *canonical momenta* p_α as

$$p_\alpha := \frac{\partial L}{\partial \dot{q}_\alpha} \tag{F7}$$

The Euler–Lagrange equations then become

$$\dot{p}_\alpha = \frac{\partial L}{\partial q_\alpha} \tag{F8}$$

We then imagine that the $3N$ equations of (F7) are inverted, to express the $\{\dot{q}_\alpha\}$ in terms of the p's and q's, and define the Hamiltonian $H(\{p\}, \{q\})$ as

$$H := \sum_\alpha p_\alpha \dot{q}_\alpha - L \tag{F9}$$

Then

$$\begin{aligned} \mathrm{d}H &= \sum_\alpha \left[(p_\alpha \, \mathrm{d}\dot{q}_\alpha + \dot{q}_\alpha \, \mathrm{d}p_\alpha) - \left(\frac{\partial L}{\partial q_\alpha} \, \mathrm{d}q_\alpha + \frac{\partial L}{\partial \dot{q}_\alpha} \, \mathrm{d}\dot{q}_\alpha \right) \right] \\ &= \sum_\alpha (\dot{q}_\alpha \, \mathrm{d}p_\alpha - \dot{p}_\alpha \, \mathrm{d}q_\alpha) \end{aligned}$$

by virtue of equations (F7) and (F8). Hence

$$\boxed{\frac{\partial H}{\partial p_\alpha} = \dot{q}_\alpha \qquad \frac{\partial H}{\partial q_\alpha} = -\dot{p}_\alpha} \tag{F10}$$

These are Hamilton's equations, which, as advertised above, give a pair of first-order equations for each index α.

If there is no explicit time dependence in T or V, the Hamiltonian is a conserved quantity, for then

$$\frac{\mathrm{d}H}{\mathrm{d}t} = \sum \left(\frac{\partial H}{\partial p_\alpha} \dot{p}_\alpha + \frac{\partial H}{\partial q_\alpha} \dot{q}_\alpha \right) = 0 \tag{F11}$$

In the normal situation when the kinetic energy is a homogeneous quadratic function of the $\{\dot{q}_\alpha\}$ and V is independent of them, T can be written as

$$T = \tfrac{1}{2} \sum_{\alpha} \frac{\partial T}{\partial \dot{q}_{\alpha}} \dot{q}_{\alpha} = \tfrac{1}{2} \sum_{\alpha} p_{\alpha} \dot{q}_{\alpha} \tag{F12}$$

Thus

$$\begin{aligned} H &= 2T - (T - V) \\ &= T + V \end{aligned} \tag{F13}$$

the total energy. The Hamiltonian formulation of classical mechanics is very important in setting up the quantum mechanical equations via the correspondence principle.

Lagrangian for Lorentz Force

A non-relativistic particle of charge e in a static electromagnetic field is subject to the Lorentz force, and consequently has the equation of motion

$$m\dot{\boldsymbol{v}} = e(\boldsymbol{E} + \boldsymbol{v} \times \boldsymbol{B}) \tag{F14}$$

In the absence of a magnetic field everything is simple: $L = T - U$, where the potential function U is just $U = e\varphi$. We can argue how this should generalize to $\boldsymbol{B} \neq 0$ by consideration of Lorentz invariance. Under a Lorentz transformation a static potential φ gives rise to a vector potential $\boldsymbol{A}$. Indeed we know that $(\varphi/c; \boldsymbol{A})$ is a 4-vector, as also is $(c; \boldsymbol{v}) \times \mathrm{d}t/\mathrm{d}\tau$ (equation(10.22)). In order for $S(-\infty, \infty)$ to be an invariant, the correct generalization of U, transforming like $\mathrm{d}\tau/\mathrm{d}t$, must therefore be $U = e(\varphi - \boldsymbol{A}\cdot\boldsymbol{v})$, a velocity-dependent potential. Thus the Lagrangian is

$$L = \tfrac{1}{2}m\boldsymbol{v}^2 - e\varphi + e\boldsymbol{A}\cdot\boldsymbol{v} \tag{F15}$$

Let us verify that the Euler–Lagrange equations

$$\frac{\mathrm{d}}{\mathrm{d}t}\left(\frac{\partial L}{\partial v_i}\right) = \frac{\partial L}{\partial x_i}$$

indeed reproduce the Lorentz force equation (F14). The left-hand side is just

$$\frac{\mathrm{d}}{\mathrm{d}t}\left(\frac{\partial L}{\partial v_i}\right) = m\dot{v}_i + e\frac{\mathrm{d}A_i}{\mathrm{d}t}$$

The right-hand side is ($\partial_i \equiv \partial/\partial x_i$)

$$\begin{aligned}\partial_i L &= e(-\partial_i\varphi + v_l\partial_i A_l)\\ &= e[(E_i + \partial A_i/\partial t)\\ &\qquad + v_l\partial_l A_i + v_l(\partial_i A_l - \partial_l A_i)]\end{aligned}$$

Here we have added and subtracted a term $v_l\partial_l A_i$. The reason for so doing is that $\partial A_i/\mathrm{d}t + (\boldsymbol{v}\cdot\boldsymbol{\nabla})A_i$ is just the total time derivative $\mathrm{d}A_i/\mathrm{d}t$, which will cancel with the identical term arising on the left-hand side. The combination $v_l(\partial_i A_l - \partial_l A_i)$ which remains is the ith component of $\boldsymbol{v} \times (\boldsymbol{\nabla} \times \boldsymbol{A})$, i.e. of $\boldsymbol{v} \times \boldsymbol{B}$. So altogether we have

$$m\dot{v}_i = e[E_i + (\boldsymbol{v} \times \boldsymbol{B})_i]$$

that is, equation (F14) in component form.

We are now in a position to justify the principle of minimal substitution used in §11.2. Going over to the Hamiltonian formulation, the canonical momentum $\boldsymbol{p}$ is

$$\boldsymbol{p} = \frac{\partial L}{\partial \boldsymbol{v}} = m\boldsymbol{v} + e\boldsymbol{A} \tag{F16}$$

rather than the usual $\boldsymbol{p} = m\boldsymbol{v}$. Equivalently

$$\boldsymbol{v} = \frac{1}{m}(\boldsymbol{p} - e\boldsymbol{A}) \tag{F17}$$

The Hamiltonian H is

$$\begin{aligned}H &= \boldsymbol{p}\cdot\boldsymbol{v} - L\\ &= \frac{(\boldsymbol{p} - e\boldsymbol{A})^2}{2m} + e\varphi\end{aligned} \tag{F18}$$

Thus the magnetic field is taken into account by the minimal substitution $\boldsymbol{p} \rightarrow \boldsymbol{p} - e\boldsymbol{A}$.

Lagrangian Field Theory

The dynamics of a field, such as the electromagnetic fields $\boldsymbol{E}(\boldsymbol{x})$ and $\boldsymbol{B}(\boldsymbol{x})$, can also be formulated in terms of Lagrangians. To take a scalar field $\varphi(\boldsymbol{x})$ for simplicity, it is a system with an infinite number of degrees of freedom, the value of the field at each spatial position $\boldsymbol{x}$. The Lagrangian formalism carries over to this situation if we replace the discrete labelling α by the continuous index $\boldsymbol{x}$, sums by integrals, and ordinary derivatives by functional derivatives, of which a word of explanation is in order.

In the most usual situation, where there is no preferred origin in space, the Lagrangian will obtain contributions of the same form from each point $\boldsymbol{x}$, and will in fact be expressible as the spatial integral of a Lagrange density $\mathcal{L}$:

$$L = \int \mathrm{d}^3\boldsymbol{x}\, \mathcal{L}(\varphi(\boldsymbol{x}), \dot{\varphi}(\boldsymbol{x})) \tag{F19}$$

L is a *functional* of φ, meaning that it depends on the entire form of φ rather than its value at any particular value of $\boldsymbol{x}$. When φ is varied, to become $\varphi + \delta\varphi$, with $\delta\varphi(\boldsymbol{x}) = \eta(\boldsymbol{x})$, say, a completely independent function, the change δL in L correspondingly depends on the entire form of η. Indeed an infinitesimal change can be written as an integral over η with some weight function g, say:

$$\delta L = \int \mathrm{d}^3\boldsymbol{x}\, g(\boldsymbol{x})\, \eta(\boldsymbol{x}) \tag{F20}$$

That function $g(\boldsymbol{x})$ is called the functional derivative of L with respect to φ and written $g(\boldsymbol{x}) = \delta L/\delta\varphi(\boldsymbol{x})$. Then (F20) can be rewritten in the suggestive form

$$\delta L = \int \mathrm{d}^3\boldsymbol{x}\, \frac{\delta L}{\delta\varphi(\boldsymbol{x})}\, \delta\varphi(\boldsymbol{x}) \tag{F20$'$}$$

The Euler–Lagrange equations arising from (F19) are the obvious generalization of equation (F6), namely

$$\frac{\mathrm{d}}{\mathrm{d}t}\left(\frac{\delta L}{\delta\dot{\varphi}(\boldsymbol{x})}\right) = \frac{\delta L}{\delta\varphi(\boldsymbol{x})} \tag{F21}$$

It is more convenient, particularly in relativistic field theories, to recast this last equation in terms of ordinary partial derivatives of the Lagrange density $\mathcal{L}$. In a relativistic field theory, $\dot{\varphi}$ will never occur on its own, but always in the combination $\partial_\mu\varphi = (\dot{\varphi}; -\boldsymbol{\nabla}\varphi)$, in natural units. So then $L = \int \mathrm{d}^3\boldsymbol{x}\, \mathcal{L}(\varphi, \partial_\mu\varphi)$ and

$$\delta L = \int \mathrm{d}^3\boldsymbol{x} \left(\frac{\partial\mathcal{L}}{\partial\varphi}\,\delta\varphi + \frac{\partial\mathcal{L}}{\partial(\partial_i\varphi)}\,\delta(\partial_i\varphi) + \frac{\partial\mathcal{L}}{\partial(\partial_0\varphi)}\,\delta(\partial_0\varphi)\right) \tag{F22}$$

Here $\delta(\partial_i\varphi) = \partial_i(\delta\varphi)$, which can be integrated by parts to give

$$\delta L = \int \mathrm{d}^3\boldsymbol{x} \left[\left(\frac{\partial\mathcal{L}}{\partial\varphi} - \partial_i\,\frac{\partial\mathcal{L}}{\partial(\partial_i\varphi)}\right)\delta\varphi + \frac{\partial\mathcal{L}}{\partial(\partial_0\varphi)}\,\delta(\partial_0\varphi)\right] \tag{F23}$$

We can now identify

$$\frac{\delta L}{\delta \varphi} = \frac{\partial \mathcal{L}}{\partial \varphi} - \partial_i \frac{\partial \mathcal{L}}{\partial(\partial_i \varphi)}$$
$$\frac{\delta L}{\delta \dot{\varphi}} = \frac{\partial \mathcal{L}}{\partial \dot{\varphi}} \tag{F24}$$

Hence equation (F21) becomes

$$\partial_0 \frac{\partial \mathcal{L}}{\partial(\partial_0 \varphi)} = \frac{\partial \mathcal{L}}{\partial \varphi} - \partial_i \frac{\partial \mathcal{L}}{\partial(\partial_i \varphi)}$$

which can be written in the manifestly covariant form

$$\boxed{\partial_\mu \frac{\partial \mathcal{L}}{\partial(\partial_\mu \varphi)} = \frac{\partial \mathcal{L}}{\partial \varphi}} \tag{F25}$$

a form which can alternatively be obtained by considering the variation of $S(-\infty, \infty) = \int d^4x\, \mathcal{L}(\varphi, \partial_\mu \varphi)$ directly.

Glossary of Mathematical Symbols

Sets and Groups

$\mathbb{Z}$	integers
$\mathbb{R}$	real numbers
$\mathbb{C}$	complex numbers
$\{\}$	set of elements
$\exists$	there exist(s)
$\vert$	such that
$\in$	member of
$\forall$	for all
$\cup$	union
$\subset$	subset
$<$	proper subgroup
$\sim$	equivalent (conjugate)
$(\ )$	equivalence (conjugacy) class
$[\ ]$	order of finite group
$\oplus$	direct sum
$\otimes$	direct product

Equality

$\equiv$	identical to
$:=$	definition
$\cong$	isomorphic
$\simeq$	approximately equal
$\approx$	of the order of
$\sim$	asymptotic behaviour

Mappings

$f: A \to B$	map from A to B
$a \mapsto b$	mapping of individual elements
E	identity map

Vector Space

$[n]$	n-dimensional
$\lvert\ \rvert$	norm of ordinary 3-vector
$\hat{\boldsymbol{v}}$	unit vector
$\lVert\ \rVert$	generalized norm

Operators and Matrices

$\hat{A}$	operator
$\mathbb{1}$	identity operator or matrix
$A^{\dagger}$	adjoint, Hermitian conjugate
A^{T}	transpose of matrix
$[\ ,\ ]$	commutator
$\{\ ,\ \}$	anticommutator

Miscellaneous

iff	if and only if
$\otimes$	contradiction
δ_{ij}	Kronecker delta
ε_{ab}	antisymmetric 2-index symbol
ε_{ijk}	antisymmetric 3-index symbol
$\varepsilon_{\mu\nu\rho\sigma}$	antisymmetric 4-index symbol
$\delta(x)$	Dirac delta function
$\bar{\delta}(k)$	$2\pi\delta(k)$
$\bar{d}k$	$\mathrm{d}k/2\pi$
$\theta(x)$	theta function
$\partial/\partial x$	partial derivative
$\delta/\delta\varphi(x)$	functional derivative

Bibliography

This is a very selective list of books which may be useful for background reading or further study.

Primarily Mathematical

Finite Groups

W Ledermann	*Introduction to Group Theory*	Longman
W Ledermann	*Introduction to Group Characters*	Cambridge University Press

SO(3)

D M Brink G R Satchler	*Angular Momentum*	Oxford University Press

SU(N)

M Hamermesh	*Group Theory and its Application to Physical Problems*	Addison-Wesley (out of print)

Physics and Mathematics

General

J P Elliott P G Dawber	*Symmetry in Physics* Vols 1 and 2	Macmillan

A W Joshi	*Elements of Group Theory for Physicists*	Wiley Eastern
V Heine	*Group Theory in Quantum Mechanics*	Pergamon

Solid State Physics

M Lax	*Symmetry Principles in Solid State and Molecular Physics*	Wiley

Atomic Physics

M Weissbluth	*Atoms and Molecules*	Academic Press
E P Wigner	*Group Theory and its Application to Atomic Spectra*	Academic Press

Elementary Particles

D B Lichtenberg	*Unitary Symmetry and Elementary Particles*	Academic Press
S Gasiorowicz	*Elementary Particle Physics*	Wiley
I J R Aitchison	*An Informal Introduction to Gauge Field Theories*	Cambridge University Press
H Georgi	*Lie Algebras in Particle Physics*	Benjamin/Cummings

Quantum Mechanics

P T Matthews	*Introduction to Quantum Mechanics*	McGraw-Hill
L I Schiff	*Quantum Mechanics* (3rd Edition)	McGraw-Hill
E Merzbacher	*Quantum Mechanics*	Wiley Toppan

Problem Solutions

Chapter 1

1.1 The products a_1a_2 are given by the table

$a_1\backslash a_2$	e	c	c^2	b	bc	bc^2
e	e	c	c^2	b	bc	bc^2
c	c	c^2	e	bc^2	b	bc
c^2	c^2	e	c	bc	bc^2	b
b	b	bc	bc^2	e	c	c^2
bc	bc	bc^2	b	c^2	e	c
bc^2	bc^2	b	bc	c	c^2	e

1.2

$$\begin{pmatrix} 1 & 2 & 3 & 4 & 5 & 6 & 7 & 8 \\ 6 & 1 & 4 & 8 & 5 & 7 & 2 & 3 \end{pmatrix} = (1\ 6\ 7\ 2)(3\ 4\ 8)(5)$$

$$\begin{pmatrix} 1 & 2 & 3 & 4 & 5 & 6 & 7 & 8 & 9 \\ 3 & 5 & 4 & 1 & 8 & 9 & 6 & 7 & 2 \end{pmatrix} = (1\ 3\ 4)(2\ 5\ 8\ 7\ 6\ 9)$$

1.3 (1 6 7 2)(3 4 8)(5) has order 12 (the LCM of 4 and 3).
(1 3 4)(2 5 8 7 6 9) has order 6 (the LCM of 3 and 6).
In general $(a_1a_2 \ldots a_{2r})(b_1b_2 \ldots b_r)$ has order $2r$.

1.4 As permutations of the vertices $A_1 \equiv A$, $A_2 \equiv B$, $A_3 \equiv C$, $A_4 = D$, the generators are $b = (1\ 3)(2\ 4)$, with $b^2 = e$, and $c = (2\ 3\ 4)$, with $c^3 = e$. Then $bc = (1\ 2\ 3)$, satisfying $(bc)^3 = e$. The 12 elements of the group are $\{e, c, c^2; b, bc, bc^2; cb, c^2b, cbc, c^2bc^2; cbc^2, c^2bc\}$.

As elements of S_4, b and c are both even permutations. Hence they generate only even permutations, and since there are 12 of these, $T \cong A_4$.

1.5

(i) is $C_4 = gp\{c\}$, with $c = i$.
(ii) is also C_4, with $c = 2$.
(iii) is $D_2 = gp\{a, b\}$, with $a = (1\ 2)$, $b = (3\ 4)$.
(iv) is again C_4, with $c = (1\ 2\ 3\ 4)$.
(v) is D_2, with $a = \begin{pmatrix} 1 & 0 \\ 0 & -1 \end{pmatrix}$, $b = \begin{pmatrix} -1 & 0 \\ 0 & 1 \end{pmatrix}$.

1.6 $z_1, z_2 \in Z \Rightarrow z_1 z_2 g = z_1 g z_2 = g z_1 z_2$. So Z is closed under multiplication and is therefore a subgroup. Moreover it is Abelian, since we can take $g = z_2$ in the defining relation $z_1 g = g z_1$.

1.7 Labelling the elements as $\{g_1, g_2, g_3, g_4, g_5, g_6\} = \{e, c, c^2, b, bc, bc^2\}$ in that order, the isomorphism for the generators is $\Pi(c) = (1\ 2\ 3)(4\ 5\ 6)$, $\Pi(b) = (1\ 4)(2\ 5)(3\ 6)$ from which all the others follow. For example, $\Pi(bc) = \Pi(b)\Pi(c) = (1\ 5)(2\ 6)(3\ 4)$, as can be checked directly.

Chapter 2

2.1 The double cosets for D_4 are $AeA = A = (e, b)$, $AcA = (c, bc, c^3, bc^3)$ and $Ac^2A = (c^2, bc^2)$.

2.2

(i) As permutations, $c = (1\ 2\ 3\ 4)$ and $b = (1\ 2)(3\ 4)$. Thus $c^2 = (1\ 3)(2\ 4)$, $c^3 = (1\ 4\ 3\ 2)$, $bc = (2\ 4)$, $bc^2 = (1\ 4)(2\ 3)$ and $bc^3 = (1\ 3)$. According to the cycle structure the conjugacy classes would be (e), (c, c^3), (c^2, b, bc^2), (bc, bc^3). However, *within the subgroup* D_4, there is no element which effects the conjugation of c^2 into b or bc^2. Hence the conjugacy classes are actually (e), (c, c^3), c^2, (b, bc^2) and (bc, bc^3).
(ii) The full consequences of the defining relations are $bc = c^3 b$, $bc^2 = c^2 b$ and $bc^3 = cb$. Hence c^2 is self-conjugate and c is

conjugate to c^3. By conjugating with all the elements of the group, the conjugacy class of b is found to be (b, bc^2), and similarly that of bc to be (bc, bc^3).

(iii) Geometrically (c, c^3) form a class since they are rotations through $\pi/2$ about axes which can be rotated into one another by b, for example. Similarly (c^2) forms a class by itself. (b, bc^2) are π rotations about the x and y axes respectively, which can be rotated into one another by c. (bc, bc^3) are π rotations about the diagonals. They are not conjugate to b, bc^2, since to rotate a diagonal into an axis would require a $\pi/4$ rotation $\notin D_4$.

2.3 The conjugacy classes are (e), (b, cbc^2, c^2bc), (c, bc, cb, c^2bc^2) and (c^2, c^2b, bc^2, cbc).

(i) In S_4 the permutations of the last two classes are all 3-cycles and would combine into a single class. However, the required conjugating elements are not elements of T.

(ii) Basically we use the defining relation to reduce any product to a standard form containing only one factor of b. Thus $(bc)^3 = e \Rightarrow bcbc = c^2b$. Hence $bcb = c^2bc^2$ and $bc^2b = cbc$. Then it is a question of laboriously performing all the conjugations.

(iii) Geometrically the elements $\{b, cbc^2, c^2bc\}$, of order 2, represent π rotations about the three coordinate axes. These axes can be turned into one another by a $2\pi/3$ rotation about a cube diagonal. The other eight rotations are $2\pi/3$ and $4\pi/3$ rotations about the four cube diagonals. The directed diagonals $(1, 1, 1)$, $(1, -1, -1)$, $(-1, 1, -1)$, and $(-1, -1, 1)$, but not their negatives, can be rotated into one another by the π rotations above.

2.4 Evenness and oddness are multiplicative. Thus A_n is closed under the action of the group, and if $P \in A_n$, any conjugate permutation $Q^{-1}PQ$ also $\in A_n$. One coset is $eA_n = A_n$ itself. Then pick any element Q not in A_n, which of necessity will be odd, and form the coset QA_n. This exhausts the group. For consider any other odd permutation Q'. Then $Q^{-1}Q'$ is even, so that $Q^{-1}Q' = P$ and $Q' = QP$. So S_n is made up of just two cosets, of equal order, and hence the order of A_n is $\frac{1}{2}n!$. The group S_n/A_n is isomorphic to C_2.

2.5 If the conjugating element g is in H then $gHg^{-1} = H$ because H is a subgroup. Otherwise, take $g = g_1h_1$, say. Then $ghg^{-1} = g_1h_2g_1^{-1}$, which is supposed to be in H. If not, it must be in g_1H, with

$g_1 h_2 g_1^{-1} = g_1 h_3$, say. But this implies $g_1 = h_3^{-1} h_2 \in H$, which is a contradiction.

2.6 The different conjugacy classes of S_4 are determined by the cycle structure, or the different partitions $l_1\, l_2 \ldots$ of 4. These are:

1^4	corresponding to the single element	(e);
$1^2.2$	corresponding to the six 2-cycles	(ab);
1.3	corresponding to the eight 3-cycles	(abc);
2^2	corresponding to the three double 2-cycles	$(ab)(cd)$;
4	corresponding to the six 4-cycles	$(abcd)$.

A normal subgroup must be made up of entire classes, including (e), and its order must be a divisor of 4!, namely 12, 8, 6, 4, 3 or 2. Apart from the normal subgroup A_4, of order 12, made up of the even permutations 1^4, 1.3, 2^2, the only other possibility is a subgroup of order 4 made up of the classes 1^4, 2^2. This is indeed closed, and is isomorphic to D_2, or $C_2 \times C_2$.

2.7 Since $bc^2 = c^2 b$, the centre is $Z = (e,\ c^2)$, made up of the complete conjugacy classes (e), (c^2). The cosets are $eZ = (e,\ c^2)$, $cZ = (c,\ c^3)$, $bZ = (b,\ bc^2)$ and $bcZ = (bc,\ bc^3)$, with the structure of $C_2 \times C_2$, i.e. $D_4/C_2 \cong C_2 \times C_2$. However, D_4 is not isomorphic to $C_2 \times (C_2 \times C_2)$, since the latter is Abelian, which D_4 is not.

2.8 Let $b = c^2$, generating the group $C_3 = \{e,\ b,\ b^2\} = B$, say, and $a = c^3$, generating the group $C_2 = \{e,\ a\} = A$, say. b and a commute, and every element of C_6 can be written uniquely as $a^r b^s$. Thus $c = ab^2$, $c^2 = b$, $c^3 = a$, $c^4 = b^2$ and $c^5 = ab$. Hence $C_6 = A \times B = C_3 \times C_2$.

2.9 The mapping is clearly (1:1). The only question is whether it preserves multiplication. But $f(g_1 g_2) = (g_1 g_2)^{-1} = g_2^{-1} g_1^{-1} = g_1^{-1} g_2^{-1}$, since G is Abelian. Thus $f(g_1 g_2) = f(g_1) f(g_2)$, as required.

2.10 The possible homomorphic images are restricted to the quotient groups D_4/N, where N is a normal subgroup. There are four proper invariant subgroups of D_4; namely:

$N = C_2 = \mathrm{gp}\{c^2\}$, with $D_4/N \cong C_2 \times C_2$ (see problem 2.7);
$N = D_2 = \mathrm{gp}\{b,\ c^2\}$, with $D_4/N \cong C_2$ (cosets eN, cN);
$N = C_4 = \mathrm{gp}\{c\}$, with $D_4/N \cong C_2$ (cosets eN, bN).

A non-trivial mapping of D_4 into D_3 is possible in the last two cases.

2.11 $D_n = \mathrm{gp}\{c, b\}$, with $c^n = b^2 = (bc)^2 = e$. Clearly $C_n = \mathrm{gp}\{c\}$ is a subgroup of D_n. It is invariant (element by element) under conjugation by any power of c, while under conjugation by b, $bcb = c^{n-1}$. Hence C_n is a normal subgroup. The cosets are just eC_n, bC_n, with the structure of $\{e, b^2\}$, i.e. C_2.

2.12 We have to check the group axioms for $I(G) = \{f_g\}$.

G0 $f_a f_b$ is the map $g \mapsto a(bgb^{-1})a^{-1} = (ab)g(ab)^{-1}$, which is again a conjugation.
G2 The associative law $f_a(f_b f_c) = (f_a f_b)f_c$ follows from that of G.
G2 f_e is the unit conjugation.
G3 $(f_a)^{-1} = f_{a^{-1}}$.

Consider the homomorphism Θ: $G \to I(G)$, whereby $g \mapsto f_g$. The kernel K is the centre of the group, for then f_z is the map $g \mapsto zgz^{-1} = g$, the identity map. Hence, by the isomorphism theorem, $I(G) \cong G/Z$.

Chapter 3

3.1 Under the rotation (3.4), $x' \pm iy' = e^{\pm i\theta}(x \pm iy)$. So with respect to the spherical basis the matrix representative becomes $\mathrm{diag}(e^{i\theta}, 1, e^{-i\theta})$, which is completely reducible. The character is $\chi(\theta) = e^{i\theta} + 1 + e^{-i\theta} = 1 + 2\cos\theta$, which is the same as that of $R(\theta)$ (equation (3.5)).

3.2 $D(c)$ and $D(c^2)$ are already given in equation (3.6). If b is a π rotation about the x axis, it is easy to see that $D(b) = \mathrm{diag}(1, -1, -1)$. By multiplication,

$$D(bc) = \begin{pmatrix} -1/2 & -\sqrt{3}/2 & 0 \\ -\sqrt{3}/2 & 1/2 & 0 \\ 0 & 0 & -1 \end{pmatrix}$$

$$D(bc^2) = \begin{pmatrix} -1/2 & \sqrt{3}/2 & 0 \\ \sqrt{3}/2 & 1/2 & 0 \\ 0 & 0 & -1 \end{pmatrix}$$

The traces are $\chi(e) = 3$, $\chi(c) = \chi(c^2) = 0$, $\chi(b) = \chi(bc) = \chi(bc^2) = -1$. If two elements g, g' are in the same conjugacy class they are related by $g' = aga^{-1}$. Then $\chi(g') \equiv \mathrm{Tr}(D(g')) =$

$\mathrm{Tr}(D(a)D(g)D^{-1}(a)) = \chi(g)$, by the cyclic property of the trace.

3.3 All that is necessary is to verify that $(D(c))^3 = 1$. Suppose it is possible to bring $D(c)$ to reducible form by a real similarity transformation, so that

$$SD(c)S^{-1} = \begin{pmatrix} a & d \\ 0 & b \end{pmatrix}$$

Taking the trace and determinant of both sides then gives the two equations $a + b = -1$ and $ab = 1$, leading to $a^2 + a + 1 = 0$. But this equation has no real roots ⊗.

3.4 $$D^G(g_1)D^G(g_2) = D^{G/N}(g_1N)D^{G/N}(g_2N) = D^{G/N}((g_1N)(g_2N)) = D^{G/N}(g_1g_2N).$$

3.5 Suppose there are two bases $\{\boldsymbol{e}_1, \boldsymbol{e}_2, \ldots, \boldsymbol{e}_n\}$ and $\{\boldsymbol{f}_1, \boldsymbol{f}_2, \ldots, \boldsymbol{f}_m\}$, with $n < m$. Then consider the set $\{\boldsymbol{f}_1, \boldsymbol{e}_1, \boldsymbol{e}_2, \ldots, \boldsymbol{e}_n\}$. This is linearly dependent, so one of the $\boldsymbol{e}$'s, $\boldsymbol{e}_j$ say, can be expressed in terms of the others, $\{\boldsymbol{f}_1, \boldsymbol{e}_1, \boldsymbol{e}_2, \ldots, \boldsymbol{e}_{j-1}, \boldsymbol{e}_{j+1}, \ldots, \boldsymbol{e}_n\}$, which still form a basis. Now adjoin $\boldsymbol{f}_2$ to this set and eliminate another $\boldsymbol{e}$ in favour of $\boldsymbol{f}_2$. Continue in this way until all the $\boldsymbol{e}$'s have been eliminated. This procedure leaves us with $\{\boldsymbol{f}_1, \boldsymbol{f}_2, \ldots, \boldsymbol{f}_n\}$ as a basis, which means that the set $\{\boldsymbol{f}_1, \boldsymbol{f}_2, \ldots, \boldsymbol{f}_m\}$, with $m > n$, is linearly dependent, contrary to hypothesis.

3.6 Under two successive transformations $x' = Ax + c$, $x'' = A'x' + c'$ we have $x'' = A'(Ax + c) + c' = A'Ax + (A'c + c')$. Thus $A'' = A'A$ and $c'' = A'c + c'$, which is exactly what results from the product

$$\begin{pmatrix} A' & c' \\ 0 & 1 \end{pmatrix}\begin{pmatrix} A & c \\ 0 & 1 \end{pmatrix} = \begin{pmatrix} A'A & A'c + c' \\ 0 & 1 \end{pmatrix}$$

Chapter 4

4.1 The matrices of the [2] representation are

$$D(e) = \begin{pmatrix} 1 & 0 \\ 0 & 1 \end{pmatrix} \qquad D(c) = \begin{pmatrix} -1/2 & -\sqrt{3}/2 \\ \sqrt{3}/2 & -1/2 \end{pmatrix}$$

$$D(c^2) = \begin{pmatrix} -1/2 & \sqrt{3}/2 \\ -\sqrt{3}/2 & -1/2 \end{pmatrix}$$

and, say,

$$D(b) = \begin{pmatrix} 1 & 0 \\ 0 & -1 \end{pmatrix} \qquad D(bc) = \begin{pmatrix} -1/2 & -\sqrt{3}/2 \\ -\sqrt{3}/2 & 1/2 \end{pmatrix}$$

$$D(bc^2) = \begin{pmatrix} -1/2 & \sqrt{3}/2 \\ \sqrt{3}/2 & 1/2 \end{pmatrix}$$

taking b to represent a rotation about the x axis. These provide the following four orthogonal 6-vectors $\{D_{ij}(g)\}$:

$$\{D_{11}(g)\} = (1, -1/2, -1/2, 1, -1/2, -1/2)$$

$$\{D_{12}(g)\} = (0, -\sqrt{3}/2, \sqrt{3}/2, 0, -\sqrt{3}/2, \sqrt{3}/2)$$

$$\{D_{21}(g)\} = (0, \sqrt{3}/2, -\sqrt{3}/2, 0, -\sqrt{3}/2, \sqrt{3}/2)$$

$$\{D_{22}(g)\} = (1, -1/2, -1/2, -1, 1/2, 1/2)$$

The [1] irreps A_1 and A_2 provide the further orthogonal vectors

$$\{\chi^{A_1}(g)\} = (1, 1, 1, 1, 1, 1)$$

$$\{\chi^{A_2}(g)\} = (1, 1, 1, -1, -1, -1)$$

4.2 The defining property of a representation is that $D(g_1)D(g_2) = D(g_1g_2)$. Taking the complex conjugate shows that D^* is also a representation. If $D^*(g) = C^{-1}D(g)C$, then $D(g) = (C^*)^{-1}D^*(g)C^* = (CC^*)^{-1}D(g)(CC^*)$. Hence, by Schur's first lemma, $CC^* = \lambda\mathbb{1}$.

If D is unitary, then taking the Hermitian conjugate of $(D^*)^{-1} = C^{-1}D^{-1}C$ gives $D^* = C^\dagger D(C^\dagger)^{-1}$. Multiplying by $C(\ )C^{-1}$ we obtain $D = (CC^\dagger)D(CC^\dagger)^{-1}$. So again, by Schur's first lemma, $CC^\dagger = \mu\mathbb{1}$.

In this last relation μ must be positive, by considering the determinant. Thus we can always redefine C by dividing by $\sqrt{\mu}$, which makes C unitary. Then $C = \lambda(C^*)^{-1} = \lambda C^T$. Taking the transpose again gives $C = \lambda^2 C$. Hence $\lambda = \pm 1$, so that C is either symmetric or antisymmetric.

4.3 Consider $D^{(\nu)}(h)B_i^\nu(D^{(\nu)}(h))^{-1}$ for any $h \in G$. It is

$$\sum_{g\in K_i} D^{(\nu)}(h)D^{(\nu)}(g)(D^{(\nu)}(h))^{-1} = \sum_{g\in K_i} D^{(\nu)}(hgh^{-1}) = \sum_{g'\in K_i} D^{(\nu)}(g')$$

since K_i is a conjugacy class. Thus B_i^ν commutes with all the elements of an irreducible representation and hence, by Schur's first lemma again, $B_i^\nu = \lambda_i^\nu \mathbb{1}$. Taking the trace, we obtain $\lambda_i^\nu n_\nu = g_i \chi_i^\nu$, where g_i is the number of elements in K_i. Substituting for χ_k^ν in $\Sigma_k g_k |\chi_k^\nu|^2 = [g]$ gives $n_\nu^2 = [g]/(\Sigma_k |\lambda_k^\nu|^2/g_k)$.

4.4 Recall that $\chi_i = \Sigma_\mu a_\mu \chi_i^{(\mu)}$. Thus

$$\sum_i g_i |\chi_i|^2 = \sum_i g_i \sum_\mu a_\mu \chi_i^{(\mu)} \sum_\nu a_\nu^* \chi_i^{(\nu)*} = \sum_{\mu,\nu} a_\mu a_\nu^* \sum_i g_i \chi_i^{(\mu)} \chi_i^{(\nu)*}$$

$$= [g] \sum_\mu |a_\mu|^2, \text{ by orthonormality}$$

D is irreducible iff a single $a_\mu = 1$, all others 0. Hence the result.

4.5 Referring to problem 2.2, the conjugacy classes are (e), $(c, c^3) = 2C_4$, $(c^2) = C_4^2$, $(b, bc^2) = 2C_2$, $(bc, bc^3) = 2C_2'$, five in all. So the dimensionalities of the irreps satisfy $n_2^2 + n_3^2 + n_4^2 + n_5^2 = 7$, whose only solution is $n_2 = n_3 = n_4 = 1$, $n_5 = 2$ up to permutations. Thus there are four [1] representations, in which b and c are represented by numbers, satisfying $b^2 = c^2 = 1$, and we can fill in the table except for the [2] representation. Let the entries for E be $(2, \alpha, \beta, \gamma, \delta)$. By orthogonality it is clear that we must have $\beta = \gamma = \delta = 0$, and hence $\alpha = -2$.

4.6 In spite of the different structure of the two groups, the classes, and hence the characters, are the same.

4.7 The members of the conjugacy classes are $E = (e)$, $2C_6 = (c, c^5)$, $2C_6^2 = (c^2, c^4)$, $C_6^3 = (c^3)$, $3C_2 = (b, bc^2, bc^4)$ and $3C_2' = (bc, bc^3, bc^5)$, while the cosets are $E := \{e, c^3\}$, $C := c^2E = \{c^2, c^5\}$, $C^2 = c^4E = \{c^4, c\}$, $B := bE = \{b, bc^3\}$, $BC = \{bc^2, bc^5\}$ and $BC^2 = \{bc^4, bc\}$, with the structure of D_3. So as far as the character table of D_3 is concerned, the classes there are to be identified with E, $2C_3 = (C, C^2)$ and $3C_2 = (BC, BC^2, BC^4)$. Thus the character of E of D_3 is inherited by E and C_6^3 of D_6, that of $2C_3$ of D_3 by $2C_6$ and $2C_6'$ of D_6, and that of $3C_2$ of D_3 by $3C_2$ and $3C_2'$ of D_6.

4.8 The conjugacy classes are (e), $(b, cbc^2, c^2bc) = 3C_2$, $(c, bc, cb, c^2bc^2) = 4C_3$ and $(c^2, c^2b, bc^2, cbc) = 4C_3^2$, four in all. So the dimensionalities of the irreps satisfy $n_2^2 + n_3^2 + n_4^2 = 11$, whose only solution is $n_2 = n_3 = 1$, $n_4 = 3$ up to permutations. Thus there are three [1] representations, in which b and c are represented by numbers, satisfying $b^2 = c^3 = (bc)^3 = 1$. Hence b is represented by 1, and c by one of the cube roots of unity. For the 3-dimensional representation, let the entries be $(3, \alpha, \beta, \gamma)$. By orthogonality $\beta = \gamma = 0$, leaving $\alpha = -1$.

4.9 The character of $E \otimes E$ is (4, 4, 0, 0, 0). By inspection, or using $a_\sigma = \langle \chi^{(\sigma)}, \chi \rangle$, we see that $E \otimes E = A_1 \oplus A_2 \oplus B_1 \oplus B_2$.

Chapter 5

5.1

(i) The group is $\mathrm{gp}\{c, b \equiv \sigma_v\}$, with $c^3 = b^2 = e$. As permutations of the vertices, $c = (ABC)$ and $b = (BC)$. $\therefore$ $bc = (BC)(ABC) = (AC)$, a reflection in the plane OBD. Hence $(bc)^2 = e$ and $C_{3v} \cong D_3$.

(ii) The representation matrix for c is as before (equation (3.6)). Taking OA as the x axis the matrix for b is

$$D(b) = \begin{pmatrix} 1 & 0 & 0 \\ 0 & -1 & 0 \\ 0 & 0 & 1 \end{pmatrix}$$

(iii) We have to check whether this representation contains the trivial representation. The relevant characters are

C_{3v}	E	$2C_3$	$3\sigma_v$
$D^{(0)} = A_1$	1	1	1
$D^3 = V$	3	0	1

from which $a_0 = (1/6)(1\cdot3\cdot1 + 2\cdot0\cdot1 + 3\cdot1\cdot1) = 1$.

(iv) The character for an axial vector is instead $\chi = (3, 0, -1)$, and now $a_0 = (1/6)(3 - 3) = 0$.

5.2 It is easiest to start with the 2-fold rotations. If b denotes a π rotation about the x axis, then $D(b) = \mathrm{diag}(1, -1, -1)$.

$$\therefore\ D(b)\sigma = \begin{pmatrix} \sigma_{11} & \sigma_{12} & \sigma_{13} \\ -\sigma_{21} & -\sigma_{22} & -\sigma_{23} \\ -\sigma_{31} & -\sigma_{32} & -\sigma_{33} \end{pmatrix} \quad \text{versus}$$

$$\sigma D(b) = \begin{pmatrix} \sigma_{11} & -\sigma_{12} & -\sigma_{13} \\ \sigma_{21} & -\sigma_{22} & -\sigma_{23} \\ \sigma_{31} & -\sigma_{32} & -\sigma_{33} \end{pmatrix}$$

Equating the two gives $\sigma_{12} = \sigma_{21} = \sigma_{13} = \sigma_{31} = 0$.

$$\therefore\ D(c)\sigma = \begin{pmatrix} -1/2 & -\sqrt{3}/2 & 0 \\ \sqrt{3}/2 & -1/2 & 0 \\ 0 & 0 & 1 \end{pmatrix} \begin{pmatrix} \sigma_{11} & 0 & 0 \\ 0 & \sigma_{22} & \sigma_{23} \\ 0 & \sigma_{32} & \sigma_{33} \end{pmatrix}$$

$$= \frac{1}{2} \begin{pmatrix} -\sigma_{11} & -\sqrt{3}\sigma_{22} & -\sqrt{3}\sigma_{23} \\ \sqrt{3}\sigma_{11} & -\sigma_{22} & -\sigma_{23} \\ 0 & 2\sigma_{32} & 2\sigma_{33} \end{pmatrix}$$

to be compared with

$$\sigma D(c) = \begin{pmatrix} \sigma_{11} & 0 & 0 \\ 0 & \sigma_{22} & \sigma_{23} \\ 0 & \sigma_{32} & \sigma_{33} \end{pmatrix} \begin{pmatrix} -1/2 & -\sqrt{3}/2 & 0 \\ \sqrt{3}/2 & -1/2 & 0 \\ 0 & 0 & 1 \end{pmatrix}$$

$$= \frac{1}{2} \begin{pmatrix} -\sigma_{11} & -\sqrt{3}\sigma_{11} & 0 \\ \sqrt{3}\sigma_{22} & -\sigma_{22} & 2\sigma_{23} \\ \sqrt{3}\sigma_{32} & -\sigma_{32} & 2\sigma_{33} \end{pmatrix}$$

Thus $\sigma_{23} = \sigma_{32} = 0$ and $\sigma_{11} = \sigma_{22}$, so that σ is of the form $\mathrm{diag}(\alpha, \alpha, \beta)$.

5.3 Let φ_1, φ_2 be the inclinations of the two sections of the string to the vertical, as shown below.

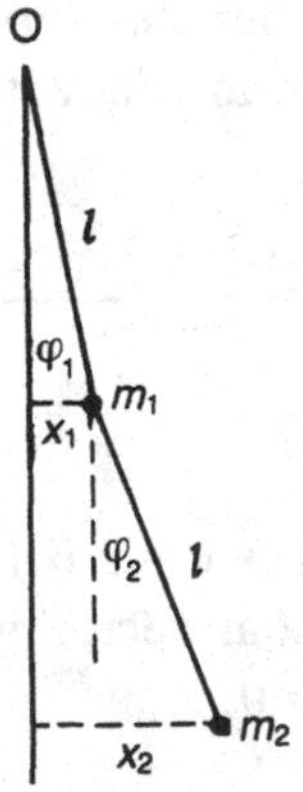

Then to first order in small quantities

$$x_1 = l\varphi_1$$
$$x_2 = l(\varphi_1 + \varphi_2)$$

so that

$$T = \tfrac{1}{2}m_1 l^2\dot{\varphi}_1^2 + \tfrac{1}{2}m_2 l^2(\dot{\varphi}_1 + \dot{\varphi}_2)^2$$
$$= \tfrac{1}{2}l^2(M\dot{\varphi}_1^2 + 2m\dot{\varphi}_1\dot{\varphi}_2 + m\dot{\varphi}_2^2)$$

where $M = m_1 + m_2$, $m = m_2$. The potential energy is

$$V = -m_1 gl\cos\varphi_1 - m_2 gl(\cos\varphi_1 + \cos\varphi_2) = \tfrac{1}{2}gl(M\varphi_1^2 + m\varphi_2^2) + \text{const}$$

Thus the matrices M and K are

$$M = l^2\begin{pmatrix} M & m \\ m & m \end{pmatrix} \qquad K = gl\begin{pmatrix} M & 0 \\ 0 & m \end{pmatrix}$$

and the normal frequencies are given by

$$\det\begin{pmatrix} Mg - Ml\omega^2 & -ml\omega^2 \\ -ml\omega^2 & mg - ml\omega^2 \end{pmatrix} = 0$$

with solutions $\omega_\pm^2 = (Mg/lm_1)[1 \pm (m/M)^{1/2}]$.

Then $\omega_\pm^2 - g/l = (Mg/lm_1)[m/M \pm (m/M)^{1/2}] = \pm(m/M)^{1/2}\omega_\pm^2$, and the normal mode displacements are $(x_1, x_2) = a(\sqrt{M}, -\sqrt{m})$ for ω_+ and $b(\sqrt{M}, \sqrt{m})$ for ω_-. Any arbitrary displacement is a linear combination of these:

$$(x_1(t), x_2(t)) = A(\sqrt{M}, -\sqrt{m})\cos(\omega_+ t + \varphi_+) + B(\sqrt{M}, \sqrt{m})\cos(\omega_- t + \varphi_-)$$

5.4 The character for $l = 1$ is $\chi^{(l=1)} = 2\cos\theta + 1 = (3, -1, 1, -1, -1)$. By inspection $D^{(l=1)} = A_2 \oplus E$. Thus the degeneracy is broken according to $3 \rightarrow 1 + 2$.

5.5

$$\chi^{(l=1)} = 2\cos\theta + 1 = (3, 2, 0, -1, -1, -1) = \chi^{A_2} + \chi^{E_1}$$

while

$$\chi^{(l=2)} = \chi^{(l=1)} + 2\cos 2\theta = (5, 1, -1, 1, 1, 1) = \chi^{A_1} + \chi^{E_1} + \chi^{E_2}$$

So the degeneracies are split according to $3 \rightarrow 1 + 2$, $5 \rightarrow 1 + 2 + 2$ in a crystal environment with D_6 symmetry. When this is reduced to D_3, only the characters containing powers of c^2 are relevant. Then

$$\chi^{(l=1)} = (3, 0, -1) = \chi^{A_2} + \chi^{E}$$
$$\chi^{(l=2)} = (5, -1, 1) = \chi^{A_1} + 2\chi^{E}$$

with no further splitting. However, C_3 is an Abelian group, whose irreps are all 1-dimensional. Thus in general the degeneracy will be completely removed when the symmetry is reduced to C_3.

Chapter 6

6.1 In the defining [2] representation S is represented by

$$\begin{pmatrix} 1 & 0 \\ 0 & -1 \end{pmatrix}$$

i.e. $x \to x$, $y \to -y$. Then

$$SR(\varphi)S^{-1} = \begin{pmatrix} 1 & 0 \\ 0 & -1 \end{pmatrix} \begin{pmatrix} \cos\varphi & -\sin\varphi \\ \sin\varphi & \cos\varphi \end{pmatrix} \begin{pmatrix} 1 & 0 \\ 0 & -1 \end{pmatrix}$$

$$= \begin{pmatrix} \cos\varphi & \sin\varphi \\ -\sin\varphi & \cos\varphi \end{pmatrix} = R(-\varphi)$$

whereas if the group were commutative we would have $SR(\varphi)S^{-1} = R(\varphi)$. Hence conjugation by S of $e^{-im\varphi}$ takes it to $e^{im\varphi}$, and we must therefore combine the representations $D^{(m)}$ and $D^{(-m)}$ of SO(3). This gives a [2] representation with matrices

$$D^{|m|}(\varphi) = \begin{pmatrix} e^{im\varphi} & 0 \\ 0 & e^{-im\varphi} \end{pmatrix} \qquad D^{|m|}(S) = \begin{pmatrix} 0 & 1 \\ 1 & 0 \end{pmatrix}$$

or equivalently

$$D'^{|m|}(\varphi) = \begin{pmatrix} \cos m\varphi & -\sin m\varphi \\ \sin m\varphi & \cos m\varphi \end{pmatrix} \qquad D'^{|m|}(S) = \begin{pmatrix} 1 & 0 \\ 0 & -1 \end{pmatrix}$$

In either case the characters are $\chi^{|m|}(\varphi) = 2\cos m\varphi$, $\chi^{|m|}(S) = 0$. For $m = 0$ the [1] representation remains irreducible and we have the two possibilities $S = +1$ or $S = -1$. The characters are then $\chi^{A_+}(\varphi) = 1$, $\chi^{A_+}(S) = 1$, the trivial representation, and $\chi^{A_-}(\varphi) = 1$, $\chi^{A_-}(S) = -1$. The first is realized by the z component of a true vector, the second by that of an axial vector.

6.2 Recall that $SR_{\boldsymbol{n}}(\alpha)S^{-1}$ is a rotation through α about the rotated axis $S\boldsymbol{n}$. So

$$e^{-iX_3''\psi} = e^{-iX_2'\theta}\, e^{-iX_3\psi}\, e^{iX_2'\theta} \qquad \text{and} \qquad R(\varphi, \theta, \psi) = e^{-iX_2'\theta}\, e^{-iX_3(\varphi+\psi)}$$

But also

$$e^{-iX_2'\theta} = e^{-iX_3\varphi}\, e^{-iX_2\theta}\, e^{iX_3\varphi}$$

Hence

$$R(\varphi, \theta, \psi) = e^{-iX_3\varphi}\, e^{-iX_2\theta}\, e^{-iX_3\psi}$$

This is the standard form for tabulation of the rotation matrices $D^{(j)}$. Thus

$$D^{(j)}_{m'm}(\varphi, \theta, \psi) = \mathrm{e}^{-\mathrm{i}m'\varphi} D^{j}_{m'm}(R_2(\theta))\,\mathrm{e}^{-\mathrm{i}m\psi}$$

in which the only non-trivial part is

$$d^{j}_{m'm}(\theta) = D^{j}_{m'm}(R_2(\theta))$$

6.3 $d^{1/2}(\theta) = \mathrm{e}^{-(1/2)\mathrm{i}\sigma_2\theta} = \mathbb{1}\cos\frac{1}{2}\theta - \mathrm{i}\sigma_2 \sin\frac{1}{2}\theta$, from the exponential series, noting that $(\sigma_2)^2 = \mathbb{1}$. Thus

$$d^{1/2}(\theta) = \begin{pmatrix}\cos\frac{1}{2}\theta & 0\\ 0 & \cos\frac{1}{2}\theta\end{pmatrix} + \begin{pmatrix}0 & -\sin\frac{1}{2}\theta\\ \sin\frac{1}{2}\theta & 0\end{pmatrix}$$

$$= \begin{pmatrix}\cos\frac{1}{2}\theta & -\sin\frac{1}{2}\theta\\ \sin\frac{1}{2}\theta & \cos\frac{1}{2}\theta\end{pmatrix}$$

6.4 We use the general formula

$$D^{j}_{m'm}(R)$$

$$= \sum_{m_1, m_1'} C(j_1 j_2 j;\, m_1', m' - m_1') C(j_1 j_2 j;\, m_1, m - m_1) D^{(j_1)}_{m_1' m_1}(R) D^{(j_2)}_{m_2' m_2}(R)$$

with, in this case,

$$C(\tfrac{1}{2}\,\tfrac{1}{2}\,1;\, \tfrac{1}{2}, \tfrac{1}{2}) = C(\tfrac{1}{2}\,\tfrac{1}{2}\,1;\, -\tfrac{1}{2}, -\tfrac{1}{2}) = 1$$

$$C(\tfrac{1}{2}\,\tfrac{1}{2}\,1;\, \tfrac{1}{2}, -\tfrac{1}{2}) = C(\tfrac{1}{2}\,\tfrac{1}{2}\,1;\, -\tfrac{1}{2}, \tfrac{1}{2}) = 1/\sqrt{2}$$

$$\therefore\ d^{1}_{11}(\theta) = (d^{1/2}_{1/2,1/2}(\theta))^2 = \cos^2\tfrac{1}{2}\theta = \tfrac{1}{2}(1 + \cos\theta)$$

$$d^{1}_{10}(\theta) = 2 d^{1/2}_{1/2,1/2}(\theta)(1/\sqrt{2}) d^{1/2}_{1/2,-1/2}(\theta)$$

$$= \sqrt{2}\cos\tfrac{1}{2}\theta\,(-\sin\tfrac{1}{2}\theta) = -(1/\sqrt{2})\sin\theta$$

$$d^{1}_{1,-1}(\theta) = (d^{1/2}_{1/2,-1/2}(\theta))^2 = \sin^2\tfrac{1}{2}\theta = \tfrac{1}{2}(1 - \cos\theta)$$

etc., giving

$$D^{1}_{m'm}(\theta) = \begin{pmatrix}\frac{1}{2}(1+\cos\theta) & -(1/\sqrt{2})\sin\theta & \frac{1}{2}(1-\cos\theta)\\ (1/\sqrt{2})\sin\theta & \cos\theta & -(1/\sqrt{2})\sin\theta\\ \frac{1}{2}(1-\cos\theta) & (1/\sqrt{2})\sin\theta & \frac{1}{2}(1+\cos\theta)\end{pmatrix}$$

Referred to the usual Cartesian basis, the matrix of the vector representation is

$$R_{ij}(\theta) = \begin{pmatrix}\cos\theta & 0 & \sin\theta\\ 0 & 1 & 0\\ -\sin\theta & 0 & \cos\theta\end{pmatrix}$$

with $x'_i = R_{ij}x_j$. Then

$$\begin{aligned}-(1/\sqrt{2})(x' + iy') &= -(1/\sqrt{2})(R_{1j} + iR_{2j})x_j \\ &= -(1/\sqrt{2})(x\cos\theta + iy + z\sin\theta) \\ &= \tfrac{1}{2}(1 + \cos\theta)[-(x + iy)/\sqrt{2}] \\ &\quad - [(1/\sqrt{2})\sin\theta]z \\ &\quad + \tfrac{1}{2}(1 - \cos\theta)[(x - iy)/\sqrt{2}]\end{aligned}$$

giving the top line of the matrix $d^1_{m'm}(\theta)$. Similarly for the other rows.

6.5 $|j\ j\rangle = \alpha|j\ j\rangle|1\ 0\rangle + \beta|j,\ j-1\rangle|1\ 1\rangle$. This must satisfy $J^+|j\ j\rangle = 0$, which gives $0 = \alpha\sqrt{2}|j\ j\rangle|1\ 1\rangle + \beta\sqrt{(2j)}|j\ j\rangle|1\ 1\rangle$. Thus $\alpha = -\beta\sqrt{j}$ and we can take $|j\ j\rangle = (\sqrt{j}|j\ j\rangle|1\ 0\rangle - |j,\ j-1\rangle)/\sqrt{(j+1)}$. Then with $M = 0$ we have $\hbar j = [j/(j+1)]^{1/2}\langle j\|\boldsymbol{J}\|j\rangle$, so that $\langle j\|\boldsymbol{J}\|j\rangle = \hbar\sqrt{[j(j+1)]}$. Alternatively, with $M = 1$ $(m = j - 1)$ we have $-(1/\sqrt{2})\hbar\sqrt{(2j)} = -[1/\sqrt{(j+1)}]\langle j\|\boldsymbol{J}\|j\rangle$, giving the same result.

6.6 The character table of C_{3v}, with $D^{[3]}$ appended, is

C_{3v}	E	$2C_3$	$3\sigma_v$
A_1	1	1	1
A_2	1	1	−1
E	2	−1	0
$D^{[3]}$	3	0	1

By inspection, $D^{[3]} = A_1 \oplus E$. Clearly z transforms according to the identity representation A_1, leaving E for the subspace (x, y).

(i) We need the CG decomposition of $D^{(v)} \otimes A_1$. But this is trivial: $D^{(v)} \otimes A_1 = D^{(v)}$. Hence for radiation polarized along the z direction the only allowed transitions are $A_i \rightarrow A_i$, $E \rightarrow E$.

(ii) We need the CG decomposition of $D^{(v)} \otimes E$. Well, clearly $A_1 \otimes E = E$. Also, from the character table $A_2 \otimes E = E$. Finally $\chi^{E\otimes E} = (4,\ 1,\ 0)$. $\therefore$ $a_1 = (4+2)/6 = a_2$, $a_E = (8-2)/6 = 1$, so that $E \otimes E = A_1 \oplus A_2 \oplus E$. Hence for radiation polarized along the x or y direction the allowed transitions are $A_i \leftrightarrow E$, $E \rightarrow E$.

6.7 The character of the vector representation is $\chi^{V} = (3, -1, 0, 0)$, so that in fact V = T. Then clearly $V \otimes A = T$, $V \otimes E_i = T$. The character of $V \otimes T$ is $\chi^{V \otimes T} = (9, 1, 0, 0)$, from which it is easy to see that $V \otimes T = A \oplus E_1 \oplus E_2 \oplus 2T$. Hence the selection rules are $A \leftrightarrow T$, $E_i \leftrightarrow T$, $T \rightarrow T$.

Chapter 7

7.1 The $(1s)^2$ and $(2s)^2$ states are completely occupied and do not contribute to the degeneracy. For $(2p)^3$ the degeneracy is the number of ways of choosing three different indices out of six (three for l_z, two for s_z), namely ${}_6C_3 = 20$. The terms are classified by the total spin S, which can be 3/2 or 1/2, and the total angular momentum L, which can have the values 3, 2, 1 or 0. The $S = 3/2$ states are completely symmetric ($\mathscr{S}$) (e.g. $|3/2, 3/2\rangle = |\uparrow\rangle|\uparrow\rangle|\uparrow\rangle$), while the $S = 1/2$ states have mixed symmetry ($\mathscr{M}$). Of the orbital states, $L = 3$ is completely symmetric and $L = 0$ is completely antisymmetric ($\mathscr{A}$), while $L = 2$ and $L = 1$ again have mixed symmetry. To achieve the overall antisymmetry required by Fermi statistics we must combine the spin and orbital angular momentum states in the combinations $\mathscr{S}\mathscr{A}$ or $\mathscr{M}\mathscr{M}$. The first possibility is realized by $S = 3/2$, $L = 0$, i.e. ^{4}S, and the second by $S = 1/2$ together with $L = 2$ (^{2}D) or $L = 1$ (^{2}P). So the term structure is

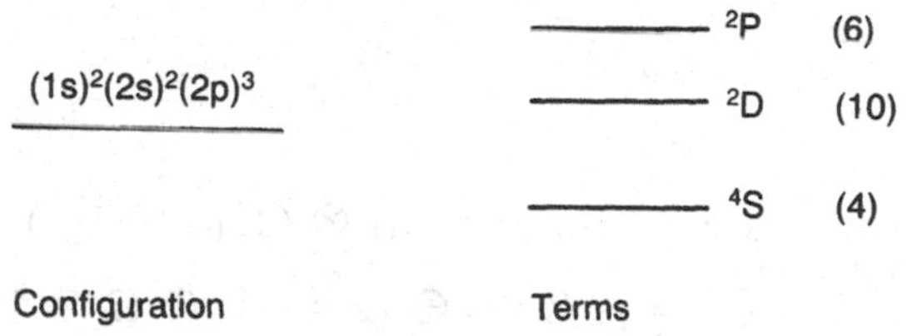

When the $\boldsymbol{L}\cdot\boldsymbol{S}$ interaction is taken into account, the terms split according to $^2P \rightarrow {}^2P_{3/2}, {}^2P_{1/2}$ and $^2D \rightarrow {}^2D_{5/2}, {}^2D_{3/2}$.

7.2

(i) There is only one invariant amplitude, since $I = 1/2$ occurs only once in the CG series $1 + 1/2 \rightarrow 3/2, 1/2$.

(ii) We need the appropriate Clebsch–Gordan coefficients α and β in $|\frac{1}{2}, \frac{1}{2}\rangle = \alpha|1, 1\rangle|\frac{1}{2}, -\frac{1}{2}\rangle + \beta|1, 0\rangle|\frac{1}{2}, \frac{1}{2}\rangle$. They are determined by the condition $I_+|\frac{1}{2}, \frac{1}{2}\rangle = 0$, which gives $0 = (\alpha + \beta\sqrt{2})|1, 1\rangle|\frac{1}{2}, \frac{1}{2}\rangle$, so that $\alpha = -\beta\sqrt{2}$. Thus $T(\mathrm{N}^{*+} \to \pi^+\mathrm{n}) = -\sqrt{2}T(\mathrm{N}^{*+} \to \pi^0\mathrm{p})$.

7.3

(i) The total isospin for πN can be 3/2 or 1/2, while that for $\pi\Delta$ can be 5/2, 3/2 or 1/2. There are two channels in common, namely $I = 3/2, 1/2$. So the process is governed by *two* $SU(2)_I$-invariant amplitudes.

(ii) NN can have $I = 1$ or 0, while NΔ can have $I = 2$ or 1. So this process is governed by just one invariant amplitude.

7.4 In NN scattering there are two invariant amplitudes T_I, namely T_1 and T_0. The eigenstates of total isospin are $|1, 1\rangle = \mathrm{pp}$, $|1, 0\rangle = (\mathrm{pn} + \mathrm{np})/\sqrt{2}$, $|1, -1\rangle = \mathrm{nn}$ and $|0, 0\rangle = (\mathrm{pn} - \mathrm{np})/\sqrt{2}$. So in terms of those eigenstates $\mathrm{pn} = (|1, 0\rangle + |0, 0\rangle)/\sqrt{2}$ and $\mathrm{np} = (|1, 0\rangle - |0, 0\rangle)/\sqrt{2}$. Hence

$$\langle \mathrm{pn}|T|\mathrm{np}\rangle = \tfrac{1}{2}(T_1 - T_0)$$
$$\langle \mathrm{pn}|T|\mathrm{pn}\rangle = \tfrac{1}{2}(T_1 + T_0)$$

so that

$$\langle \mathrm{pn}|T|\mathrm{np}\rangle = T_1 - \langle \mathrm{pn}|T|\mathrm{pn}\rangle = \langle \mathrm{pp}|T|\mathrm{pp}\rangle - \langle \mathrm{pn}|T|\mathrm{pn}\rangle$$

Chapter 8

8.1

$$\begin{aligned} 3 \otimes \bar{3} &= (2_{1/3} \oplus 1_{-2/3}) \otimes (2_{-1/3} \oplus 1_{2/3}) \\ &= (2 \otimes 2)_0 \oplus (2 \otimes 1)_1 \oplus (2 \otimes 1)_{-1} \oplus (1 \otimes 1)_0 \\ &= (3 \oplus 1)_0 \oplus 2_1 \oplus 2_{-1} \oplus 1_0 \end{aligned}$$

Thus

$$8 = 3_0 \oplus 2_1 \oplus 2_{-1} \oplus 1_0 \qquad (\text{e.g. } \pi + \mathrm{K} + \bar{\mathrm{K}} + \eta)$$

$$\begin{aligned} 3 \otimes 3 \otimes 3 &= (2_{1/3} \oplus 1_{-2/3})^3 \\ &= ((3 \oplus 1)_{2/3} \oplus 2 \times 2_{-1/3} \oplus 1_{-4/3}) \otimes (2_{1/3} \oplus 1_{-2/3}) \\ &= ((4 \oplus 2 \times 2)_1 \oplus 2 \times (3 \oplus 1)_0 \oplus 2_{-1}) \\ &\quad \oplus ((3 \oplus 1)_0 \oplus 2 \times 2_{-1} \oplus 1_{-2}) \end{aligned}$$

Thus, after subtracting $2 \times 8 \oplus 1$,

$$10 = 4_1 \oplus 3_0 \oplus 2_{-1} \oplus 1_{-2} \qquad (\text{e.g. } \Delta + \Sigma^* + \Xi^* + \Omega^-)$$

8.2

(*a*)(i) Taking for definiteness the states Δ^{++}, Σ^{*+}, Ξ^{*0} and Ω^-, the various contributions are:

Δ^{++}: $dD^{ab3}D_{ab3} \rightarrow 0$, so that $m(\Delta) = m_0$, the unperturbed mass;
Σ^{*+}: $dD^{113}D_{113} \rightarrow m(\Sigma^*) - m_0 = (1/3)d$;
Ξ^{*0}: $d(D^{133}D_{133} + D^{313}D_{313}) \rightarrow m(\Xi^*) - m_0 = (2/3)d$;
Ω^-: $dD^{333}D_{333} \rightarrow m(\Omega^-) - m_0 = d$.

(*a*)(ii) Considering p, Σ^+, Ξ^0 and Λ for definiteness, the various mass shifts are $m(\text{N}) - m_0 = a$, $m(\Sigma) - m_0 = 0$, $m(\Xi) - m_0 = b$ and $m(\Lambda) - m_0 = (2/3)(a + b)$. Eliminating m_0, a and b gives the required relation.

(*b*)(i) In terms of quark content we would expect $m(\Delta) = 3m_\text{u}$, $m(\Sigma^*) = 2m_\text{u} + m_\text{s}$, $m(\Xi^*) = m_\text{u} + 2m_\text{s}$ and $m(\Omega^-) = 3m_\text{s}$. This gives the same equal spacing rule, with the identification $d = 3(m_\text{s} - m_\text{u})$.

(*b*)(ii) We would expect $m(\text{N}) = 3m_\text{u}$, $m(\Sigma) = m(\Lambda) = 2m_\text{u} + m_\text{s}$, $m(\Xi) = m_\text{u} + 2m_\text{s}$, which again gives equal spacing, with $b = -a = 3(m_\text{s} - m_\text{u})$.

8.3 $8 \otimes 8$ gives a tensor φ^{ik}_{jl}, which is traceless in (ij) and (kl) but has non-zero contractions in (il) and (kj). Thus $\varphi^{ik}_{jl} \rightarrow \hat{\varphi}^{ik}_{jl} + \delta^i_l \hat{\varphi}^k_j + \delta^k_j \hat{\varphi}^i_l + \delta^k_j \delta^i_l \varphi$. Here the second and third terms are octets, and the last term a singlet. The first term still needs to be symmetrized in its upper and lower indices. Symmetrizing in the upper indices gives $\hat{\varphi}^{ik}_{jl} \rightarrow \hat{\varphi}^{(ik)}_{jl} + \varepsilon^{ikn}\varphi_{(jln)}$, in which the last term is a decuplet. Finally, symmetrizing $\hat{\varphi}^{(ik)}_{jl}$ in its lower indices gives $\hat{\varphi}^{(ik)}_{jl} \rightarrow \hat{\varphi}^{(ik)}_{(jl)} + \varepsilon_{jlm}\varphi^{(ikm)}$. The last term is another decuplet $(\overline{10})$, and the first is a 27 (see the following problem).

8.4 We have to fill m boxes with indices 1, 2 or 3, the order being irrelevant. The number of indices equal to 1 can be 0, 1, 2, . . ., m. Let this be $m - r$, leaving r boxes to be filled with indices 2 or 3. There are $r + 1$ ways of doing this, as the number of indices equal to 2 can be 0, 1, 2, . . ., r. So

$$\dim \varphi^{(i_1 i_2 \cdots i_m)} = \sum_{r=0}^{m} (r+1) = \tfrac{1}{2}(m+1)(m+2)$$

Then

$$\begin{aligned}
\dim \hat{\varphi}^{(i_1 i_2 \cdots i_m)}_{(j_1 j_2 \cdots j_n)} &= \dim \varphi^{(i_1 i_2 \cdots i_m)}_{(j_1 j_2 \cdots j_n)} - \dim \varphi^{(i_1 i_2 \cdots i_{m-1})}_{(j_1 j_2 \cdots j_{n-1})} \\
&= \tfrac{1}{4}(m+1)(m+2)(n+1)(n+2) \\
&\quad - \tfrac{1}{4}m(m+1)n(n+1) \\
&= \tfrac{1}{4}(m+1)(n+1)[(m+2)(n+2) - mn] \\
&= \tfrac{1}{2}(m+1)(n+1)(m+n+2)
\end{aligned}$$

8.5 Using the dimensionality rules of §8.4 we get

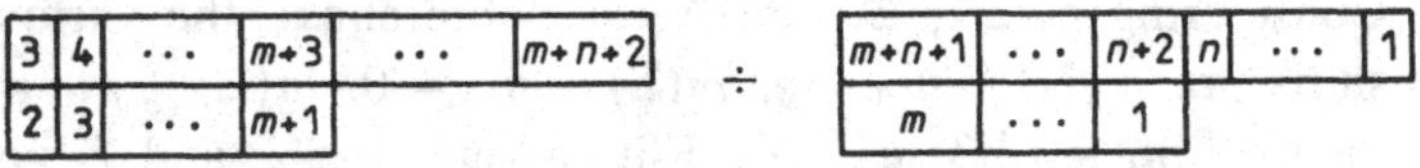

$$= \tfrac{1}{2}(m+n+2)(n+1)(m+1)$$

where the first three factors come from the ratio of the top rows, and the last factor comes from the ratio of the bottom rows.

8.6

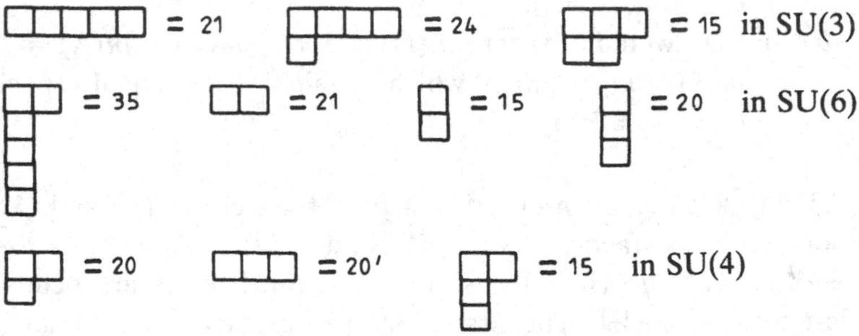
= 21
= 24
= 15 in SU(3)
= 35
= 21
= 15
= 20 in SU(6)
= 20
= 20′
= 15 in SU(4)

8.7 In SU(3):

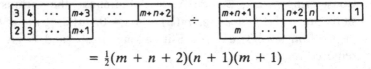

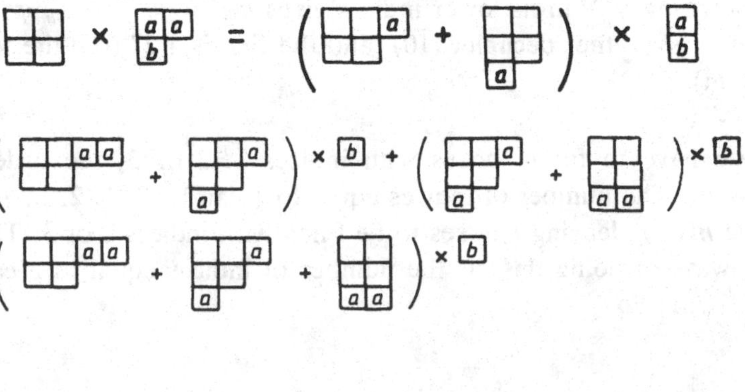

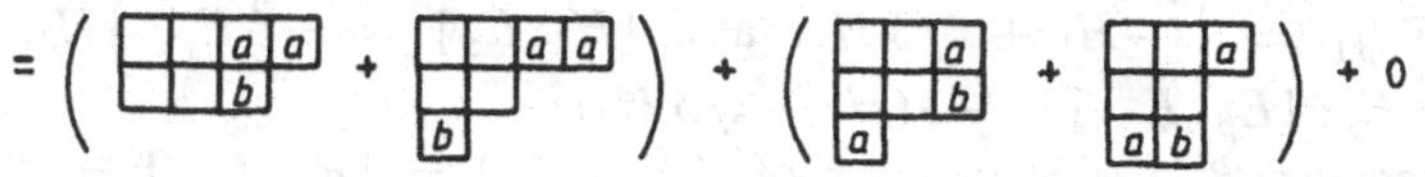

i.e. $\overline{6} \otimes 8 = 24 \oplus \overline{15} \oplus \overline{6} \oplus 3$;

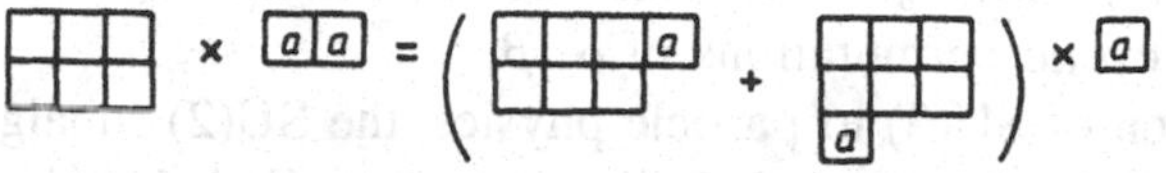

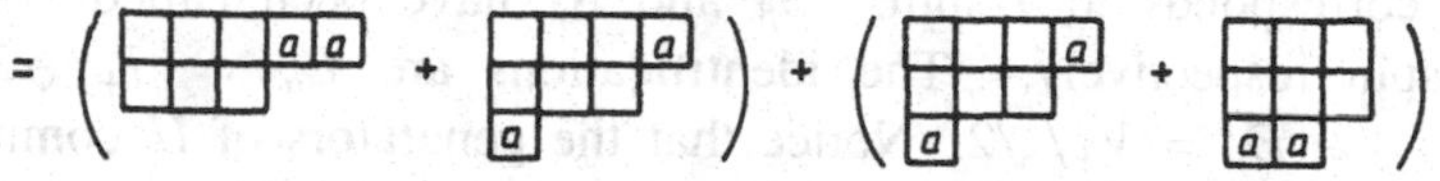

i.e. $\overline{10} \otimes 6 = 42 \oplus 15 \oplus \overline{3}$.

In SU(6): $\overline{6} \otimes 6 = 35 \oplus 1$ and $6 \otimes 6 \otimes 6 = 56 \oplus 70 \oplus 70 \oplus 20$.

8.8 $8 \otimes 8 = 27 \oplus 10 \oplus \overline{10} \oplus 8 \oplus 8 \oplus 1$ and $10 \otimes 8 = 35 \oplus 27 \oplus 10 \oplus 8$, with the irreps 27, 10 and 8 in common. Thus the process is described by *three* SU(3)-invariant amplitudes.

8.9 Without showing the working (!), $35 \otimes 56 = 1134 \oplus 700 \oplus 70 \oplus 56$ and $35 \otimes 70 = 1134 \oplus 560 \oplus 540 \oplus 70 \oplus 70 \oplus 56 \oplus 20$. So *three* SU(6)- invariant amplitudes are required, corresponding to the irreps 1134, 70 and 56.

Chapter 9

9.1 From equation (9.14) the matrices representing the J_i are $D_A(J_k)_i^j = i\varepsilon_{ijk}$. These are in fact the negatives of the matrices X_k given in (6.15), (6.17) and (6.19). Then, for example, $(X_1^2)_i^j = -\varepsilon_{1ik}\varepsilon_{1kj} = \delta_{i2}\delta_{j2} + \delta_{i3}\delta_{j3}$, so that $g_{11} = \mathrm{Tr}(X_1^2) = 2$. Similarly for g_{22} and g_{33}. For the typical off-diagonal element g_{12} we have $(X_1X_2)_i^j = -\varepsilon_{1ik}\varepsilon_{2kj} = -\delta_{i2}\delta_{j1}$, so that $g_{12} = \mathrm{Tr}(X_1X_2) = 0$. Thus, $g_{ij} = 2\delta_{ij}$.

9.2 The components of α, β and γ are $\alpha = (1, 0)$, $\beta = \frac{1}{2}(-1, -\sqrt{3})$ and $\gamma = -\frac{1}{2}(1, \sqrt{3})$. Therefore $[H_1, E_\alpha] = E_\alpha$, $[H_2, E_\alpha] = 0$, $[E_\alpha, E_{-\alpha}] = H_1$. Similarly $[H_1, E_\beta] = -\frac{1}{2}E_\beta$, $[H_2, E_\beta] = \frac{1}{2}\sqrt{3}\, E_\beta$,

$[E_\beta, E_{-\beta}] = \frac{1}{2}(-H_1 + \sqrt{3}\,H_2)$ and $[H_1, E_\gamma] = -\frac{1}{2}E_\gamma$, $[H_2, E_\gamma] = -\frac{1}{2}\sqrt{3}\,E_\gamma$, $[E_\beta, E_{-\beta}] = -\frac{1}{2}(H_1 + \sqrt{3}\,H_2)$.

Because $\beta - \alpha$ is not a root, $[E_{-\alpha}, E_\beta] = [E_\alpha, E_{-\beta}] = 0$. Then taking $\mu = -\beta$ in equation (9.68), in which $N_{\alpha,-\beta} = 0$, we have $|N_{\alpha,\gamma}|^2 = -\alpha \cdot \beta = \frac{1}{2}$. We can choose the phase so that $N_{\gamma,\alpha} = 1/\sqrt{2}$. Thus, $[E_\gamma, E_\alpha] = (1/\sqrt{2})E_{-\beta}$. The other non-vanishing commutators can be obtained by cyclic permutations of α, β, γ.

In the application of SU(3) to particle physics, the SU(2) subalgebra S_α corresponds to I-spin. S_β and S_γ have been called U-spin and V-spin respectively. The identifications are $E_\alpha = I_+/\sqrt{2}$, $E_\beta = U_+/\sqrt{2}$, $E_\gamma = V_+/\sqrt{2}$. Notice that the generators of U commute with $H_1 + H_2/\sqrt{3} = I_3 + \frac{1}{2}Y = Q$, the charge operator.

9.3 The Weyl reflection σ_α is a reflection in the plane perpendicular to α, and similarly for σ_β, σ_γ. These can be identified with the reflections b_1, b_2, b_3 of the dihedral group D_3 (see section 1.3(ii)), which is isomorphic to S_3. The element $\sigma_\alpha\sigma_\beta$ is a $2\pi/3$ rotation, which can be identified with c. Inversion in the origin, which interchanges $\{\alpha\}$ with $\{-\alpha\}$, is not contained within the Weyl group. This is ultimately the reason why the quark and antiquark representations of SU(3) are distinct (whereas they are equivalent in SU(2)).

9.4 α_1 and α_2 are simple roots, so $\alpha_2 - \alpha_1$ is not a root. Then by equation (9.44), with $q = 0$, the length of the α_1-string through α_2 is $1 - 2\alpha_1 \cdot \alpha_2/\alpha_1^2 = 4$. This string contains α_2, $\alpha_2 + \alpha_1 = \alpha_3$, say, $\alpha_2 + 2\alpha_1 = \alpha_4$, say, and $\alpha_2 + 3\alpha_1 = \alpha_5$, say.

Now let us look for α_2-strings throught these newly generated points. The string through α_3 merely takes us back to α_1, and there is no string through α_4 because $\alpha_2 \cdot \alpha_4 = 0$. In the case of α_5, the difference $\alpha_5 - \alpha_2 = 3\alpha_1$ is not a root, so by equation (9.44) the α_2-string through α_5 has length $1 - 2\alpha_2 \cdot \alpha_5/\alpha_2^2 = 2$. The root thus generated is $2\alpha_2 + 3\alpha_1 = \alpha_6$, say, the long vertical root in the diagram. The remaining roots are generated by the Weyl reflection σ_{α_6}. It is easy to see that any further strings will merely reproduce roots already found.

9.5 Anticipating the result, write $\alpha = (e_a - e_b)/\sqrt{2} = (1, -1, 0)/\sqrt{2}$ in the original coordinate system. Similarly $\beta = (0, 1, -1)/\sqrt{2}$ and $\gamma = (-1, 0, 1)/\sqrt{2}$. The rotation $x \to (x - y)/\sqrt{2}$, $y \to (x + y)/\sqrt{2}$ transforms α into $\alpha' = (1, 0, 0)$, as required, while the transforms of β and γ are $\beta' = \frac{1}{2}(-1, 1, -\sqrt{2})$ and $\gamma' = \frac{1}{2}(-1, -1, \sqrt{2})$. The further rotation $y \to (y - \sqrt{2}z)/\sqrt{3}$, $z \to (\sqrt{2}y + z)/\sqrt{3}$ does not affect α', but transforms β' and γ' into $\beta'' = \frac{1}{2}(-1, \sqrt{3}, 0)$ and $\gamma'' = \frac{1}{2}(-1, -\sqrt{3}, 0)$.

For a general vector $v = (v_1, v_2, v_3)$ the combined effect of the two rotations is $v_1'' = (v_1 - v_2)/\sqrt{2}$, $v_2'' = (v_1 + v_2 - 2v_3)/\sqrt{6}$ and $v_3'' = (v_1 + v_2 + v_3)/\sqrt{3}$. Applied to H_i this gives $H_1'' = \mathrm{diag}(1, -1, 0)/\sqrt{2} = \sqrt{2}\, I_3$, $H_2'' = \mathrm{diag}(1, 1, -2)/\sqrt{6} = \sqrt{(3/2)}\, Y$ and $H_3'' = 0$.

9.6 Recall that a single link corresponds to an angle of 120°, a double link to 135° and a triple link to 150°, while two roots which are not connected are at right angles. So in the SU(4) diagram the sum of the angles is $2 \times 120 + 90 = 330$, which is allowed. The SO(7) diagram has $120+135+90 = 345$, which is again allowed. If in the latter diagram the single link were changed to a double link, the sum of the angles would become $2 \times 135 + 90 = 360°$. The roots would then have to be coplanar, i.e. linearly dependent, which is not allowed. If instead the double link were changed to a triple link the sum would become $120+150+90 = 360$, again not allowed. Any augmentation of links beyond these two cases would give a total angle in excess of 360°. The other possibility might be a triangular diagram, with each vertex connected by a single link, but this would be a closed loop of single links, which we ruled out in section 9.5. In this case, an alternative proof is that the sum of the angles would be $3 \times 120 = 360$. Any triangular configuration containing double or triple links would have a total angle greater than 360°.

9.7 Since α and β are connected by a single line, they are of equal length and $2\alpha \cdot \beta/\alpha^2 = -1$. Thus $(\alpha + \beta)^2 = \alpha^2(1 + 1 - 1) = \alpha^2 = \beta^2$. Any other root γ in the original diagram can be connected to α or β, but not both (otherwise we would have a triangular sub-diagram, which was ruled out in the previous problem). If γ is connected with α then $\gamma \cdot (\alpha + \beta) = \gamma \cdot \alpha$. Similarly, if it is connected with β then $\gamma \cdot (\alpha + \beta) = \gamma \cdot \beta$, while if it is connected to neither $\gamma \cdot (\alpha + \beta) = 0$ as well. Thus, if the scalar products in the original diagram are all legal, so are those in the reduced diagram in which the element α ○—○ β is replaced by $(\alpha + \beta)$ ○.

9.8 In the weight lattice we can set up the equivalence relation $\mu \sim \mu'$ iff $\mu - \mu' \in R$. Moreover, every weight, μ, can be expressed as $\mu = m_1\lambda_1 + m_2\lambda_2$, where $\lambda_1 + \lambda_2 = \alpha_1 + \alpha_2 \in R$. So, if $m_1 = m_2$, $\mu \sim 0$. Similarly, if $m_1 = m_2 + 1$, $\mu \sim \lambda_1$. But since $3\lambda_1 = 2\alpha_1 + \alpha_2 \in R$ this is also true if $m_1 = m_2 + 1 \pmod 3$. Again, since $3\lambda_2 = \alpha_1 + 2\alpha_2 \in R$, $m_1 = m_2 - 1 \pmod 3 \Leftrightarrow \mu \sim \lambda_2$.

The 6 representation has $m_1 = 2$, $m_2 = 0$ and hence has $t = 2$, or -1. This is also clear from the decomposition $3 \times 3 = 6 + \overline{3}$. The octet representation $(1, 1)$, occurring in $3 \times \overline{3}$, has $t = 0$. The irrep 10 is $(3, 0)$, and hence has $t = 0$, as does $\overline{10}$ $((0, 3))$. Finally 27, or $(2, 2)$, which occurs in the decomposition of 8×8, also has $t = 0$.

9.9 Take $\alpha_1 = (-1, 1)$ and $\alpha_2 = (1, 0)$. The fundamental weights λ_i, determined by $2\lambda_1 \cdot \alpha_j/\alpha_j^2 = \delta_{ij}$, are then $\lambda_1 = (0, 1)$ and $\lambda_2 = \frac{1}{2}(1, 1)$. Consider first the irrep with heightest weight λ_1. This has $m_1 = 1$, $m_2 = 0$ and cannot be raised by E_{α_1} or E_{α_2}. So the length of the α_1-string is $m_1 + 1 = 2$, and that of the α_2-string $m_2 + 1 = 1$. The α_1-string produces the additional weight α_2, with $m_1 = -1$, $m_2 = 2$. This again cannot be raised by E_{α_2}, so the α_2 string has length 3 and produces the additional weights $0, -\alpha_2$. The latter has $m_1 = 1$, $m_2 = -2$. It cannot be raised by E_{α_1} for the same reason that α_2 cannot be lowered by $E_{-\alpha_1}$. So the α_1-string is a doublet, with an additional weight $a_2 - \alpha_1 = -\lambda_1$. At this point the process ends (by symmetry). So the representation is 5-dimensional, the defining representation of SO(5). The weights form a diamond shape, with an additional weight in the centre. All the weights except the latter could have been obtained by Weyl refelections.

Starting from λ_2 instead, a doublet α_2-string produces $\lambda_2 - \alpha_2 = \lambda_1 - \lambda_2$, with $m_1 = 1$, $m_2 = -1$. A doublet α_1-string takes this to $-\lambda_1 + \lambda_2$, with $m_1 = -1$, $m_2 = 1$. Finally, a doublet α_2-string takes this to $-\lambda_2$. The weights form a square, and in this case they could all have been obtained by Weyl reflections. This four-fold representation is a spinor representation of the algebra of SO(5), or the defining representation of the algebra of Sp(4), which is isomorphic to it. The relationship between the groups is similar to that between SO(3) and SU(2), namely $SO(5) \cong Sp(4)/Z_2$.

Chapter 10

10.1
$$v'_\mu = \tfrac{1}{2}\mathrm{Tr}(\tilde{\sigma}_\mu V') = \tfrac{1}{2}\mathrm{Tr}(\tilde{\sigma}_\mu A V A^\dagger) = \tfrac{1}{2}\mathrm{Tr}(\tilde{\sigma}_\mu A \sigma_\nu A^\dagger) v^\nu$$

$$\therefore \Lambda_{\mu\nu}(A) = \tfrac{1}{2}\mathrm{Tr}(\tilde{\sigma}_\mu A \sigma_\nu A^\dagger)$$

$$V'' = BV'B^\dagger = BAVA^\dagger B^\dagger = (BA)V(BA)^\dagger$$

$$\therefore \Lambda(BA) = \Lambda(B)\Lambda(A)$$

The kernel of the homomorphism is defined by $V' = V$, i.e. $AVA^\dagger = V$ $\forall V$. Take $V = \mathbb{1}$. Then $A^\dagger = A^{-1} \Rightarrow A \in SU(2)$, and the condition becomes $AV = VA$. Now take $V = \boldsymbol{v}\cdot\boldsymbol{\sigma}$. Then $[A, \sigma_i] = 0 \Rightarrow A = a_0\mathbb{1}$, with $a_0^2 = 1$.

10.2 $V' = e^{-(1/2)i\sigma_3\theta}(v_0 - \boldsymbol{v}\cdot\boldsymbol{\sigma})e^{(1/2)i\sigma_3\theta} = v_0 - \boldsymbol{v}\cdot\boldsymbol{\sigma}'$, where $\boldsymbol{\sigma}' = (\sigma_1\cos\theta + \sigma_2\sin\theta, \sigma_2\cos\theta - \sigma_1\sin\theta, \sigma_3)$. Writing $V' = v_0' - \boldsymbol{v}'\cdot\boldsymbol{\sigma}$, we have $v_0' = v_0$, $v_3' = v_3$, $v_1' = v_1\cos\theta - v_2\sin\theta$, $v_2' = v_2\cos\theta + v_1\sin\theta$.

10.3 We require $\sigma\cdot p = AmA^\dagger$. Let $A = (\sigma\cdot p/m)^{1/2}U$. Then $\mathbb{1} = UU^\dagger$, so that U is a unitary matrix.

10.4 $D(B) = \exp(-\frac{1}{2}\boldsymbol{\sigma}\cdot\boldsymbol{\zeta}) = \cosh\frac{1}{2}\zeta - \boldsymbol{\sigma}\cdot\hat{\boldsymbol{\zeta}}\sinh\frac{1}{2}\zeta$. If $\sinh\zeta = |\boldsymbol{p}|/m$ then $\cosh\zeta = E/m$ and $\cosh^2\frac{1}{2}\zeta = \frac{1}{2}(E/m + 1)$, $\sinh^2\frac{1}{2}\zeta = \frac{1}{2}(E/m - 1)$. So

$$\begin{aligned} D(B) &= [(E+m)^{1/2} - (E-m)^{1/2}\boldsymbol{\sigma}\cdot\hat{\boldsymbol{p}}]/(2m)^{1/2} \\ &= (E + m - \boldsymbol{\sigma}\cdot\boldsymbol{p})/[2m(E+m)]^{1/2} \\ &= (\sigma\cdot p + m)/[2m(E+m)]^{1/2} \end{aligned}$$

Then $D(B^2) = [(E^2 - 2E\boldsymbol{\sigma}\cdot\boldsymbol{p} + \boldsymbol{p}^2) + 2m\sigma\cdot p + m^2]/2m(E+m)$, in which $\boldsymbol{p}^2 = E^2 - m^2$

$$\therefore\ D(B^2) = [2(E+m)\sigma\cdot p]/2m(E+m) = \sigma\cdot p/m$$

as required.

10.5

$$\begin{aligned} [\tfrac{1}{2}\sigma_{\mu\nu}, \tfrac{1}{2}\sigma_{\lambda\kappa}] &= -\tfrac{1}{4}[\gamma_\mu\gamma_\nu - \eta_{\mu\nu}, \gamma_\lambda\gamma_\kappa - \eta_{\lambda\kappa}] = -\tfrac{1}{4}[\gamma_\mu\gamma_\nu, \gamma_\lambda\gamma_\kappa] \\ &= -\tfrac{1}{4}[\gamma_\mu\gamma_\nu\gamma_\lambda\gamma_\kappa - \gamma_\lambda\gamma_\kappa\gamma_\mu\gamma_\nu] \end{aligned}$$

First write $\gamma_\nu\gamma_\lambda = 2\eta_{\nu\lambda} - \gamma_\lambda\gamma_\nu$ in the first term and $\gamma_\kappa\gamma_\mu = \eta_{\kappa\mu} - \gamma_\mu\gamma_\kappa$ in the second, to give $-\frac{1}{4}[2(\eta_{\nu\lambda}\gamma_\mu\gamma_\kappa - \eta_{\kappa\mu}\gamma_\lambda\gamma_\nu) - \gamma_\mu\gamma_\lambda\gamma_\nu\gamma_\kappa + \gamma_\lambda\gamma_\mu\gamma_\kappa\gamma_\nu]$. Then interchange $\gamma_\mu\gamma_\lambda$ and $\gamma_\kappa\gamma_\nu$ using the anticommutation relations, and collect terms.

10.6

$$J_3 = \tfrac{1}{2}i\gamma_1\gamma_2 = \tfrac{1}{2}i\begin{pmatrix}0 & -\sigma_1\\ \sigma_1 & 0\end{pmatrix}\begin{pmatrix}0 & -\sigma_2\\ \sigma_2 & 0\end{pmatrix} = \tfrac{1}{2}i\begin{pmatrix}-i\sigma_3 & 0\\ 0 & -i\sigma_3\end{pmatrix}$$

$$= \begin{pmatrix}\tfrac{1}{2}\sigma_3 & 0\\ 0 & \tfrac{1}{2}\sigma_3\end{pmatrix}, \text{ etc.}$$

$$K_i = \tfrac{1}{2}i\gamma_0\gamma_i = \tfrac{1}{2}i\begin{pmatrix}0 & \mathbb{1}\\ \mathbb{1} & 0\end{pmatrix}\begin{pmatrix}0 & -\sigma_i\\ \sigma_i & 0\end{pmatrix} = \tfrac{1}{2}i\begin{pmatrix}\sigma_i & 0\\ 0 & -\sigma_i\end{pmatrix}$$

$$= \begin{pmatrix}\tfrac{1}{2}i\sigma_i & 0\\ 0 & -\tfrac{1}{2}i\sigma_i\end{pmatrix}$$

10.7

$$D(B) = \exp(-i\boldsymbol{K}\cdot\boldsymbol{\zeta})$$
$$= \begin{pmatrix} E + m + \boldsymbol{\sigma}\cdot\boldsymbol{p} & 0 \\ 0 & E + m - \boldsymbol{\sigma}\cdot\boldsymbol{p} \end{pmatrix} \Big/ [2m(E + m)]^{1/2}$$

Now

$$-\boldsymbol{p}\cdot\boldsymbol{\gamma}\gamma_0 = \begin{pmatrix} 0 & \boldsymbol{\sigma}\cdot\boldsymbol{p} \\ -\boldsymbol{\sigma}\cdot\boldsymbol{p} & 0 \end{pmatrix} \begin{pmatrix} 0 & \mathbb{1} \\ \mathbb{1} & 0 \end{pmatrix} = \begin{pmatrix} \boldsymbol{\sigma}\cdot\boldsymbol{p} & 0 \\ 0 & -\boldsymbol{\sigma}\cdot\boldsymbol{p} \end{pmatrix}$$

So

$$D(B) = (E + m - \boldsymbol{p}\cdot\boldsymbol{\gamma}\gamma_0)/[2m(E + m)]^{1/2}$$
$$= (p\cdot\gamma\gamma_0 + m)/[2m(E + m)]^{1/2}$$

Then

$$D(B^2) = [p\cdot\gamma\gamma_0(2E - \gamma_0 p\cdot\gamma) + 2mp\cdot\gamma\gamma_0 + m^2]/2m(E + m)$$
$$= p\cdot\gamma\gamma_0/m, \text{ using } (p\cdot\gamma)^2 = m^2$$

10.8 We have to verify the anticommutation relations $\{\gamma_\mu, \gamma_\nu\} = 2\eta_{\mu\nu}$. Well, $\gamma_0^2 = -\gamma_i^2 = 1$, while

$$\gamma_0\gamma_i = \begin{pmatrix} 0 & \sigma_i \\ \sigma_i & 0 \end{pmatrix} = -\gamma_i\gamma_0$$

and $\gamma_i\gamma_j = -\gamma_j\gamma_i$ by virtue of $\sigma_i\sigma_j = -\sigma_j\sigma_i$.

10.9

$$T_{fi} = \langle \hat{\boldsymbol{q}}; \lambda_3\lambda_4 | T(m) | \boldsymbol{p} = 0; s_3 = \lambda \rangle$$
$$= \sum_{J,M,J',M'} \langle \hat{\boldsymbol{q}}; \lambda_3\lambda_4 | qJM; \lambda_3\lambda_4 \rangle \langle qJM; \lambda_3\lambda_4 | T(m) | J'M'; \lambda \rangle$$
$$\times \langle J'M'; \lambda | \boldsymbol{p} = 0; s_3 = \lambda \rangle$$

on inserting two complete sets of angular momentum states. Here $\langle qJM; \lambda_3\lambda_4 | T(m) | J'M'; \lambda \rangle = \delta_{JJ'}\delta_{MM'}\langle \lambda_3\lambda_4 | T^J(m) | \lambda \rangle$, while

$$\langle J'M'; \lambda | \boldsymbol{p} = 0; s_3 = \lambda \rangle$$

is proportional to $\delta_{J's}\delta_{M'\lambda}$.

$$\therefore\ T_{fi} \propto (D^s_{\lambda\mu}(R))^* = d^s_{\lambda\mu}(\theta)\exp[i(\lambda - \mu)\varphi]$$

Chapter 11

11.1

$$
\begin{aligned}
(\nabla \times \boldsymbol{A}(\boldsymbol{x}))_i &= -\varepsilon_{ijk}\partial_j\varepsilon_{klm}\int_0^1 \mathrm{d}t\, x_l t B_m(\boldsymbol{x}') \\
&= -(\delta_{il}\delta_{jm} - \delta_{im}\delta_{jl})\partial_j x_l \int_0^1 \mathrm{d}t\, t B_m(\boldsymbol{x}') \\
&= 2\int_0^1 \mathrm{d}t\, t B_i(\boldsymbol{x}') - (\delta_{il}\delta_{jm} - \delta_{im}\delta_{jl}) x_l \int_0^1 \mathrm{d}t\, t^2\, \partial B_m(\boldsymbol{x}')/\partial x'_j \\
&= 2\int_0^1 \mathrm{d}t\, t B_i(\boldsymbol{x}') + \int_0^1 \mathrm{d}t\, t^2\, \mathrm{d}B_i(\boldsymbol{x}')/\mathrm{d}t \\
&= B_i(\boldsymbol{x}) \text{ on integration by parts}
\end{aligned}
$$

11.2 The phase difference is $\delta = (e/\hbar)(\int_{P_2} - \int_{P_1})\boldsymbol{A}(\boldsymbol{x}')\cdot\mathrm{d}\boldsymbol{x}' = (e/\hbar)\oint \boldsymbol{A}(\boldsymbol{x}')\cdot\mathrm{d}\boldsymbol{x}' = (e/\hbar)\int \boldsymbol{B}\cdot\mathrm{d}\boldsymbol{S}$, by Stokes' theorem

11.3 Recall that $F_{i0} = E_i/c$ and $F_{ij} = -\varepsilon_{ijk}B_k$. First consider the inhomogeneous equation $\partial^\mu F_{\mu\nu} = \mu j_\nu$:

$$\nu = 0 \rightarrow \mu c\rho = \nabla_i F_{i0} = \nabla\cdot\boldsymbol{E}/c \qquad \text{i.e. } \nabla\cdot\boldsymbol{E} = \mu c^2\rho = \rho/\varepsilon$$

$$\nu = i \rightarrow (1/c)\partial F_{0i}/\partial t + \nabla_j F_{ij} = -(1/c^2)\partial E_i/\partial t + (\nabla\times\boldsymbol{B})_i$$

i.e. $\nabla\times\boldsymbol{B} = \mu\boldsymbol{j} + \varepsilon\mu\partial\boldsymbol{E}/\partial t$

Now for the homogeneous equation $\varepsilon_{\mu\nu\rho\sigma}\partial^\nu F^{\rho\sigma} = 0$:

$$\mu = 0 \rightarrow \varepsilon_{ijk}\nabla_i\varepsilon_{jkl}B_l = 0 \qquad \text{i.e. } \nabla\cdot\boldsymbol{B} = 0$$

$$\mu = i \Rightarrow \nu = 0 \quad \text{or } j$$

So

$$0 = -(1/c)\partial(\varepsilon_{ijk}F_{jk})/\partial t + 2\varepsilon_{ijk0}\nabla_j F_{k0}$$

$$\text{i.e. } 0 = -(2/c)\partial B_i/\partial t - (2/c)(\nabla\times\boldsymbol{E})_i \qquad \text{or } \nabla\times\boldsymbol{E} = -\partial\boldsymbol{B}/\partial t$$

11.4

$$L_1 = J_1 + K_2 = \tfrac{1}{2}(\sigma_1 - \mathrm{i}\sigma_2) = \sigma_-$$

$$L_2 = J_2 - K_1 = \tfrac{1}{2}(\sigma_2 + \mathrm{i}\sigma_1) = \mathrm{i}\sigma_-$$

So in each case $L_1\varphi + L_2\varphi = 0$ if

$$\varphi \propto \begin{pmatrix} 0 \\ 1 \end{pmatrix}$$

11.5

$$[\gamma_\mu, \Sigma_{\rho\sigma}] = \tfrac{1}{2}i[\gamma_\mu, \gamma_\rho\gamma_\sigma - \eta_{\rho\sigma}] = \tfrac{1}{2}i(\gamma_\mu\gamma_\rho\gamma_\sigma - \gamma_\rho\gamma_\sigma\gamma_\mu)$$
$$= i(\eta_{\mu\rho}\gamma_\sigma - \eta_{\mu\sigma}\gamma_\rho)$$

to be compared with equation (10.60). This commutator is exactly what is involved when $\psi \to \exp(ia_{\rho\sigma}\Sigma^{\rho\sigma})\psi(\Lambda x)$ and correspondingly $\bar{\psi} \to \bar{\psi}(\Lambda x)\exp(-ia_{\rho\sigma}\Sigma^{\rho\sigma})$. Hence the result.

11.6

$$\gamma_0\gamma_5\gamma_0 = \gamma_0\gamma_0\gamma_1\gamma_2\gamma_3\gamma_0 = -\gamma_5$$
$$\therefore\ \bar{\psi}(x)\gamma_5\psi(x) \to \bar{\psi}(\Pi x)\gamma_0\gamma_5\gamma_0\psi(\Pi x) = -\bar{\psi}(\Pi x)\gamma_5\psi(\Pi x)$$

11.7 When gauged, $\partial_\mu \to D_\mu = \partial_\mu - ieA_\mu$. So $\mathscr{L} = (D_\mu\varphi)(D^\mu\varphi^\dagger) - m^2\varphi\varphi^\dagger - \frac{1}{4}F_{\mu\nu}F^{\mu\nu} - (\partial\cdot A)^2/2\xi$. This contains both 3-point and 4-point interactions with A_μ, represented graphically by

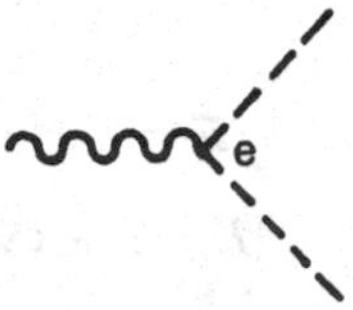

11.8

$$U^{-1}D'_\mu U = U^{-1}[\partial_\mu + U(D_\mu U^{-1})]U$$
$$= U^{-1}[\partial_\mu + U(\partial_\mu U^{-1}) - igUW_\mu U^{-1}]U$$
$$= U^{-1}(\partial_\mu U) + \partial_\mu + (\partial_\mu U^{-1})U - igW_\mu$$
$$= \partial_\mu - igW_\mu = D_\mu,\ \text{since}\ \partial_\mu(U^{-1}U) = 0$$

Index

9 780750 30504

Printed in the United States
by Baker & Taylor Publisher Services